MAKING CLIMATE CHANGE WOR

European Perspectives on Adaptation and Mitig⸱⸱⸱⸱⸱⸱⸱⸱

Making Climate Change Work for Us is an introduction to the main challenges and opportunities of developing local, regional and global strategies for addressing climate change, and explains many of the dilemmas faced when converting strategies into policies.

The book provides a synthesis of the findings of the three-year ADAM (Adaptation and Mitigation Strategies: Supporting European Climate Policy) research project. Written from a European perspective by many of the continent's leading inter-disciplinary climate change research teams, European strategies for tackling climate change are placed within a global context. The volume addresses questions such as 'How is European climate policy made?', 'How feasible are very low emissions scenarios?', 'What is the role of policy in adaptation?', 'How can the goals of climate change and development policy be brought into alignment?' and 'What options are there for an international climate agreement after 2012?' The book explains and illustrates the differences between adaptation and mitigation, offers regional and global case studies of how adaptation and mitigation are inter-linked, and suggests five different metaphors for thinking about the strategic options we have for making climate change work for us, rather than against us.

The book is intended for readers interested in finding practical solutions to climate change – both adaptation and mitigation – within the policy contexts in which these solutions have to be implemented. It is valuable reading for researchers in environmental studies, environmental economics, political science, geography, international relations, integrated assessment, and risk analysis, as well policy-makers in government, industry and NGOs.

Three other books arise from the ADAM project, all published by Cambridge University Press and, together with this volume, derive from research funded by DG-RTD as part of the Sixth Framework Programme of the European Commission.

Global Climate Governance Beyond 2012: Architecture, Agency and Adaptation
  Edited by Frank Biermann, Philipp Pattberg and Fariborz Zelli

Climate Change Policy in the European Union: Confronting the Dilemmas of Adaptation and Mitigation?
  Edited by Andrew Jordan, Dave Huitema, Harro van Asselt, Tim Rayner and Frans Berkhout

Mainstreaming Climate Change in Development Cooperation: Theory, Practice and Implications for the European Union
Edited by Joyeeta Gupta and Nicolien van der Grijp

MIKE HULME is Professor of climate change in the School of Environmental Sciences at the University of East Anglia and was the Founding Director of the Tyndall Centre for Climate Change Research from 2000 to 2007. His research interests include representations of climate change in history, society and the media, the design and uptake of climate scenarios, and the interaction between climate change science and policy. His previous book – *Why We Disagree About Climate Change* – was published by Cambridge University Press in 2009. He has prepared climate scenarios and reports for the UK Government (including the UKCIP98 and UKCIP02 scenarios), the European Commission, the IPCC, UNEP, UNDP and WWF-International. He has published over 120 peer-reviewed journal papers and over 35 book chapters on these and other topics, together with over 230 reports and popular articles about climate change. He is editor-in-chief of the newly launched *Wiley's Interdisciplinary Reviews – Climate Change*. He delivered the prestigious Queen's Lecture in Berlin in 2005 and won the Hugh Robert Mill Prize in 1995 from the Royal Meteorological Society.

HENRY NEUFELDT is Head of the climate change program of the World Agroforestry Centre (ICRAF) in Nairobi, Kenya. Between 2006 and 2009 he was based in the School of Environmental Sciences at University of East Anglia, and was a Senior Research Co-ordinator in the Tyndall Centre for Climate Change Research, where he managed the ADAM Project. His research interest is in global climate change, vulnerability and sustainable development; in particular, mitigation and adaptation in land management in the context of science and policy. He has worked primarily in Germany, Brazil and Paraguay. He has published over 30 peer-reviewed journal papers and book chapters as well as numerous reports on sustainable land use in the tropics and climate change mitigation and adaptation policies.

THE ADAM BOOK SERIES FROM CAMBRIDGE UNIVERSITY PRESS

Making Climate Change Work for Us: European Perspectives on Adaptation and Mitigation Strategies
Edited by Hulme, M. and Neufeldt, H.

Climate Change Policy in the European Union: Confronting the Dilemmas of Mitigation and Adaptation?
Edited by Jordan, A., Huitema, D., van Asselt, H., Rayner, T. and Berkhout, F.

Global Climate Governance Beyond 2012: Architecture, Agency and Adaptation
Edited by Biermann, F., Pattberg, P. and Zelli, F.

Mainstreaming Climate Change in Development Cooperation: Theory, Practice and Implications for the European Union
Edited by Gupta, J. and van der Grijp, N.

MAKING CLIMATE CHANGE WORK FOR US

European Perspectives on Adaptation and Mitigation Strategies

Edited by

MIKE HULME

and

HENRY NEUFELDT

CAMBRIDGE
UNIVERSITY PRESS

CAMBRIDGE UNIVERSITY PRESS
Cambridge, New York, Melbourne, Madrid, Cape Town,
Singapore, São Paulo, Delhi, Mexico City

Cambridge University Press
The Edinburgh Building, Cambridge CB2 8RU, UK

Published in the United States of America by Cambridge University Press, New York

www.cambridge.org
Information on this title: www.cambridge.org/9780521119412

First published 2010
Paperback edition published 2012

Printed and Bound in Great Britain by the MPG Books Group

A catalogue record for this publication is available from the British Library

ISBN 978-0-521-11941-2 Hardback
ISBN 978-1-107-67138-6 Paperback

Additional resources for this publication at www.cambridge.org/9780521119412

Cambridge University Press has no responsibility for the persistence or
accuracy of URLs for external or third-party internet websites referred to
in this publication, and does not guarantee that any content on such
websites is, or will remain, accurate or appropriate.

Every effort has been made in preparing this book to provide accurate and up-to-date information
which is in accord with accepted standards and practice at the time of publication. Although case
histories are drawn from actual cases, every effort has been made to disguise the identities of the
individuals involved. Nevertheless, the authors, editors and publishers can make no warranties that the
information contained herein is totally free from error, not least because clinical standards are
constantly changing through research and regulation. The authors, editors and publishers therefore
disclaim all liability for direct or consequential damages resulting from the use of material contained in
this book. Readers are strongly advised to pay careful attention to information provided by the
manufacturer of any drugs or equipment that they plan to use.

Contents

List of contributing authors	*page* ix
Foreword: from EU Director-General José Manuel	
Silva Rodríguez	xvii
Preface: The ADAM project	xix
Acknowledgements	xxvi
List of abbreviations	xxvii
Part I Concepts and scenarios	1
1 Climate policy and inter-linkages between adaptation and mitigation	3
Henry Neufeldt et al.	
2 Climate change appraisal in the EU: current trends and future challenges	31
Duncan Russel et al.	
3 Scenarios as the basis for assessment of mitigation and adaptation	54
Detlef P. van Vuuren et al.	
4 National responsibilities for adaptation strategies: lessons from four modelling frameworks	87
Asbjørn Aaheim et al.	
5 Learning to adapt: re-framing climate change adaptation	113
Jochen Hinkel et al.	
Part II Strategies within Europe	135
6 How do climate policies work? Dilemmas in European climate governance	137
Frans Berkhout et al.	
7 Transforming the European energy system	165
Gunnar S. Eskeland et al.	
8 A risk management approach for assessing adaptation to changing flood and drought risks in Europe	200
Reinhard Mechler et al.	

9   Mainstreaming adaptation in regional land use and water
    management   230
    Saskia E. Werners et al.
    Part III   Strategies beyond Europe   261
10 Global climate governance beyond 2012: architecture, agency and
    adaptation   263
    Frank Biermann et al.
11 The economics of low stabilisation: implications for technological
    change and policy   291
    Brigitte Knopf et al.
12 Mainstreaming climate change in development co-operation
    policy: conditions for success   319
    Joyeeta Gupta et al.
13 Insurance as part of a climate adaptation strategy   340
    Joanne Linnerooth-Bayer et al.
    Part IV   Synthesis   367
14 What can social science tell us about meeting the challenge of
    climate change? Five insights from five years that might make a
    difference   369
    Anthony Patt et al.
    Appendix: Description of models   389
    Index   408

Colour plates are to be found between pp. 220 and 221.

Contributors

Co-ordinating lead authors

Asbjørn Aaheim is an economist, educated at University of Oslo, Norway. He is now Research Director of the Unit of Impacts, Adaptation and Vulnerability at CICERO. He has published papers on 'green accounting', including treatment of national wealth and income from the extraction of natural resources. He was a Lead Author of the IPCC Second Assessment report on the applicability of cost–benefit analysis. His current activities are mainly related to integrated assessment modelling.

Frans Berkhout is Director of IVM at VU University, Amsterdam and has extensive research and research management experience. His recent work has been concerned with technology, policy and sustainability, with special emphasis on the links between technological innovation and environmental performance in firms, the measurement of sustainability performance, futures scenario studies, business adaptation to environmental change and policy frameworks for innovation and the environment.

Frank Biermann is a Professor of political science and Professor of environmental policy sciences at VU University, Amsterdam. He specialises in global environmental governance, with emphasis on climate negotiations, UN reform, public-private governance mechanisms, North–South relations, and trade and environment conflicts. He holds a number of research management positions, including being Head of the Department of Environmental Policy Analysis at IVM of VU University Amsterdam, and Director-General of the Netherlands Research School for the Socio-economic and Natural Sciences of the Environment (SENSE), a national research network of nine institutes with 150 scientists and 350 doctoral students. Frank Biermann is also the Founding Chair of the Berlin Conferences on the Human Dimensions of Global Environmental Change; Founding Director of the Global Governance Project; and Chair of the Earth System Governance Project, a new ten-year core research activity under the International Human Dimensions Programme on Global Environmental Change (IHDP).

Gunnar S. Eskeland is Professor of economics at the Norwegian School of Economics and Business Administration, and Research Director for Energy and Climate. He has formerly held senior research and director positions at the World Bank and CICERO, respectively. His research interests are in theoretical and applied welfare economics, with most of his applications in environmental policy. His applied publications include areas such as health effects and valuation of environmental change, management of environment in the transportation sector, adaptation responses to climate change in the energy sector, climate policy and technological change in the energy sector, and the effects of climate and environmental policies on trade and investment. His theoretical interests include optimal taxation, contract theory, institutional economics, decentralization and co-operation. He is leading several multi-party research projects, and has been task manager for the electricity sector case study in the ADAM project.

Joyeeta Gupta is Professor of climate change law and policy at the VU University, Amsterdam and of water law and policy at the UNESCO-IHE Institute for Water Education, Delft, the Netherlands. She is editor-in-chief of *International Environmental Agreements: Politics, Law and Economics* and is on the editorial board of journals such as *Carbon and Law Review, International Journal on Sustainable Development, Environmental Science and Policy*, and *International Community Law Review*. She was lead author in the Intergovernmental Panel on Climate Change and of the Millennium Ecosystem Assessment which won the Zaved Second Prize. She has published extensively on climate change. She is on the scientific steering committees of many different international programmes including the Global Water Systems Project and the Project on Earth System Governance of the International Human Dimensions Programme.

Jochen Hinkel is a senior researcher at the Potsdam Institute for Climate Impact Research (PIK), Germany where he leads a group on climate change vulnerability and adaptation. He holds a Ph.D. in environmental sciences (Wageningen University, the Netherlands) and a Masters in geo-ecology (Karlsruhe University, Germany). His research interests include transdisciplinary knowledge integration, coastal vulnerability, mathematical formalisation, and meta-analysis of impact, vulnerability and adaptation case studies. Jochen Hinkel coordinates the development of the DIVA model, an integrated model for assessing coastal vulnerability and adaptation. Prior to his academic engagement, he was working as a development practitioner, software developer and information technology consultant.

Brigitte Knopf is a senior researcher at the Potsdam Institute for Climate Impact Research (PIK), Germany. Her scientific work focuses on low concentration pathways of CO_2 emissions for mitigating climate change. Her main interest is the transformation towards a low carbon economy and the economic consequences and technological requirements for mitigation. She co-ordinated the work package M2 within the ADAM project and is leader of the PIK activity LOWC on low stabilisation scenarios. She holds a Ph.D. in physics and has a strong background in climate modelling, especially the Indian Monsoon, and in the assessment of uncertainties. She is involved in the ongoing project on climate change and global poverty, which links the issues of climate change and justice.

Joanne Linnerooth-Bayer is based at the International Institute for Applied Systems Analysis (IIASA), Austria, where she leads a programme on Risk and Vulnerability. She is an economist by training and holds degrees from Carnegie-Mellon University, USA and from the University of Maryland, USA. Her current interest is improving the financial management of catastrophe risks on the part of households, farmers and governments in transition and developing countries. She has recently led research projects on this topic in the Tisza river region, Hungary, and the Dongting Lake region, China, and she has consulted widely with organisations such as the World Bank, DFID and Oxfam America. Joanne Linnerooth-Bayer is Associate Editor of the *Journal for Risk Research* and on the editorial board of *Risk Analysis and Risk Abstracts*, holds positions at Beijing Normal University and is also a member of the Science Committee of the Chinese Academy of Disaster Reduction and Emergency Management.

Reinhard Mechler is an economist at the International Institute for Applied Systems Analysis (IIASA), Austria, where he leads the research group on Disasters and Development in the Risk and Vulnerability Programme. Specific interests of his include catastrophe risk modelling, the impacts of extreme events and climate change on development, the use of novel risk financing mechanisms for globally sharing disaster risks as well as the interaction of climate mitigation and adaptation policy. He has published one book and various journal articles and has acted as a reviewer of the IPCC Fourth Assessment Report. Reinhard Mechler has been leading and contributing to projects for many international organisations and teaches at the University of Karlsruhe, Germany and University of Vienna, Austria. He studied economics, mathematics, and English and holds a diploma in economics (University of Heidelberg) and a Ph.D. in economics (University of Karlsruhe).

Henry Neufeldt is now leading the climate change programme of the World Agroforestry Centre (ICRAF) in Nairobi, Kenya. From 2006 to 2009 he was based in the School of Environmental Sciences at University of East Anglia UK, and was a Senior Research Co-ordinator in the Tyndall Centre for Climate Change Research where he was manager of the ADAM Project. His general research interest is global climate change, vulnerability and sustainable development,' in particular, mitigation and adaptation in land management in the context of science and policy. He has worked primarily in Germany, Brazil and Paraguay. He has published over 30 peer-reviewed journal papers and book chapters as well as numerous reports on sustainable land use in the tropics and climate change mitigation and adaptation policies.

Anthony Patt received a doctorate degree in law from Duke University, USA and a Ph.D. in public policy from Harvard University, USA. In addition to being a member of the Risk and Vulnerability Programme at IIASA, Austria, he is Assistant Research Professor at Boston University, USA. Anthony Patt studies decision making under uncertainty, especially with respect to climate change adaptation and mitigation and he has published extensively.

Duncan Russel is lecturer in public policy, climate change and sustainability at the University of Exeter, UK. He was formerly a senior researcher in the Centre of Social and Economic Research on the Global Environment (CSERGE) and the Tyndall Centre for Climate Research, both based at UEA, UK. He has researched and published in fields of the politics of policy appraisal, environmental policy integration and the politicisation of knowledge in policy processes. His work in the ADAM project entailed researching the current practice and future practice of the appraisal of climate polices in the European Union.

Detlef P. van Vuuren works as senior researcher at the Netherlands Environmental Assessment Agency (PBL). His work concentrates on integrated assessment of global environmental change and more specifically on long-term projection of climate change. He was involved as Co-ordinating Lead Author and Lead Author in several international assessments including the Millennium Ecosystem Assessment and the IPCC Fourth Assessment Report. He has published over 50 articles in peer reviewed journals. He is also involved in activities of the Stanford University-based Energy Modelling Forum.

Saskia E. Werners is at the Centre for Water and Climate, Wageningen University, the Netherlands. Her main research interest is adaptation to climate change in water management. Her research is firmly rooted in the global change community branching out into institutional as well as biophysical aspects. In her work, Saskia Werners seeks to identify robust land and water management strategies and opportunities to implement these strategies at the regional scale. In particular, she studies diversification of water and land use as a strategy to reduce climate-related risks, and the role of individuals in realising new policy strategies. Her graduate studies on Environmental Sciences, Experimental Physics and Water Management and

Engineering are complemented by her practical experience, working in the national government and the private sector.

Contributing authors

Nigel Arnell, University of Reading, UK

Christoph Bals, Germanwatch, Bonn, Germany

Ilona Banaszak, Polish Academy of Sciences, Poland

Terry Barker, University of Cambridge, UK

Nico Bauer, Potsdam Institute for Climate Impact Research (PIK), Germany

Lavinia Baumstark, Potsdam Institute for Climate Impact Research (PIK), Germany

Marco Bindi, University of Florence, Italy

Sandy Bisaro, Potsdam Institute for Climate Impact Research (PIK), Germany

Ingrid Boas, VU University Amsterdam, the Netherlands

Giacomo Catenazzi, CEPEETH Zurich, Switzerland

Bertrand Château, ENERDATA, France

Adam Choryński, Polish Academy of Sciences, Poland

Francesc Cots, Autonomous University of Barcelona (UAB), Spain

Patrick Criqui, CNRS-University of Grenoble, France

Xingang Dai, Institute of Atmospheric Physics, Chinese Academy of Sciences, PR China

Therese Dokken, Norwegian University of Life Sciences, Norway

Thomas E. Downing, Stockholm Environment Institute, UK

Ottmar Edenhofer, Potsdam Institute for Climate Impact Research (PIK), Germany

Wolfgang Eichhammer, Fraunhofer-Institute for Systems and Innovation Research (ISI), Germany

Zsuzsanna Flachner, Hungarian Academy of Sciences, Hungary

Christian Flachsland, Potsdam Institute of Climate Impact Research (PIK), Germany

Elisabetta Genovese, Université Laval, Quebec, Canada

Nitu Goel, The Energy and Resources Institute (TERI), India

Constanze Haug, VU University Amsterdam, the Netherlands

Alex Haxeltine, University of East Anglia, UK

Anne Held, Fraunhofer-Institute for Systems and Innovation Research (ISI), Germany

Henk Hilderink, Netherlands Environmental Assessment Agency (PBL – Planbureau voor de Leefomgeving), the Netherlands

Roger Hildingsson, Lund University, Sweden

Stefan Hochrainer, International Institute for Applied Systems Analysis (IIASA), Austria

Andries Hof, Netherlands Environmental Assessment Agency (PBL – Planbureau voor de Leefomgeving), the Netherlands

Mareen E. Hofmann, Potsdam Institute for Climate Impact Research (PIK), Germany

Dave Huitema, VU University Amsterdam, the Netherlands

Morna Isaac, Netherlands Environmental Assessment Agency (PBL – Planbureau voor de Leefomgeving), the Netherlands

Martin Jakob, CEPE, ETH Zurich, Switzerland

Anne Jerneck, Lund University, Sweden

Eberhard Jochem, Fraunhofer-Institute for Systems and Innovation Research (ISI), Germany

Andrew Jordan, University of East Anglia, UK

Harvir Kalirai, International Institute for Applied Systems Analysis (IIASA), Austria

Alban Kitous, ENERDATA, France

Richard J. T. Klein, Stockholm Environment Institute (SEI), Sweden

Tom Kram, Netherlands Environmental Assessment Agency (PBL – Planbureau voor de Leefomgeving), the Netherlands

Zbigniew W. Kundzewicz, Polish Academy of Sciences, Poland

Socrates Kypreos, Paul Scherrer Institute, Switzerland

Carlo Lavalle, European Commission – Joint Research Centre (JRC–IES), Italy

Marian Leimbach, Potsdam Institute for Climate Impact Research (PIK), Germany

Kristin Linnerud, Center for International Climate and Environmental Research – Oslo (CICERO), Norway

Kate Lonsdale, UK Climate Impacts Programme (UKCIP), UK

Nicola Lugeri, European Commission – Joint Research Centre (JRC–IES), Italy

Bertrand Magné, International Energy Agency (IEA), France

Eric Massey, VU University Amsterdam, the Netherlands

Piotr Matczak, Polish Academy of Sciences, Poland

Darryn McEvoy, International Centre for Integrated Assessment and Sustainable Development (ICIS), University of Maastricht and Global Cities Research Institute, RMIT University, Australia

Torben K. Mideksa, Center for International Climate and Environmental Research – Oslo (CICERO), Norway

Silvana Mima, CNRS-University of Grenoble, France

Suvi Monni, European Commission – Joint Research Centre (JRC–IES), Italy

Marco Moriondo, University of Florence, Italy

Gert-Jan Nabuurs, Alterra Wageningen University and Research Centre, the Netherlands

Måns Nilsson, Stockholm Environment Institute (SEI), Sweden

Lennart Olsson, Lund University, Sweden

Philipp Pattberg, VU University Amsterdam, the Netherlands

Åsa Persson, Stockholm Environment Institute (SEI), Sweden

Maciej Radziejewski, Adam Mickiewicz University, Poznan, Poland and Polish Academy of Sciences, Poland

Tim Rayner, University of East Anglia, UK

Diana Reckien, Potsdam Institute for Climate Impact Research (PIK), Germany

Ulrich Reiter, Paul Scherrer Institute, Switzerland

Nathan Rive, Center for International Climate and Environmental Research – Oslo (CICERO), Norway

Dirk Rübbelke, Center for International Climate and Environmental Research – Oslo (CICERO), Norway

Håkon Sælen, Center for International Climate and Environmental Research – Oslo (CICERO), Norway

Wolfgang Schade, Fraunhofer-Institute for Systems and Innovation Research (ISI), Germany

Mart-Jan Schelhaas, Alterra Wageningen University and Research Centre, the Netherlands

Serban Scrieciu, University of Cambridge, UK

Johannes Stripple, Lund University, Sweden

Malgorzata Szwed, Polish Academy of Sciences, Poland

J. David Tàbara, Autonomous University of Barcelona (UAB), Spain

Michael Thomspon, International Institute for Applied Systems Analysis, Austria

Thure Traber, German Institute for Economic Research (DIW), Germany

Giacomo Trombi, University of Florence, Italy

Hal Turton, Paul Scherrer Institute, Switzerland

Harro van Asselt, VU University Amsterdam, the Netherlands

Nicolien van der Grijp, VU University Amsterdam, the Netherlands

Paul Watkiss, Stockholm Environment Institute (SEI), Oxford, UK

Taoyuan Wei, Center for International Climate and Environmental Research – Oslo (CICERO), Norway

Jennifer West, Center for International Climate and Environmental Research – Oslo (CICERO), Norway

Anita Wreford, Scottish Agricultural College (SAC), UK

Markus Wrobel, Potsdam Institute for Climate Impact Research (PIK), Germany

Fariborz Zelli, German Development Institute, Germany

Foreword

Climate change has become one of the essential political, social and economic challenges of our times. This was a challenge that the European Union was quick to recognise in the late 1980s and one that we have continued to place close to the heart of our strategic thinking and policy-making, at the same time as the EU has enlarged and strengthened as a political entity. During these 20 years or more, the European Commission has funded a significant number of research projects exploring the scientific, economic, social and political dimensions of the problem. Our contribution to the international body of knowledge about climate change has been impressive. Within the Sixth RTD Framework Programme of the European Community (2002–2006), new opportunities were created for large-scale Integrated Projects to be implemented, which brought together significant European research capacity to address strategic questions of high scientific and political significance. The ADAM project – Adaptation and Mitigation Strategies: Supporting European Climate Policy – was one such project. I am very pleased to see the results of this project now appear in this edited volume at such a timely moment in the evolution of our thinking and decision-making about climate change. It is published during COP 15 in Copenhagen, where the signatories to the UNFCCC will attempt to forge a forward-looking deal that will break the policy deadlock and provide the necessary instruments to tackle climate change more effectively.

I sincerely hope that this book – and the three others in the ADAM book series – fulfils its goal of bringing the insights of European integrated climate change researchers into the wide arena of international climate change deliberation, debate and decision making.

José Manuel Silva Rodríguez
Director-General of the
Directorate-General for Research
European Commission
Brussels, September 2009

Preface

The ADAM Project

www.adamproject.eu

Changes in climate induced by human emissions of greenhouse gases, and other climate changing agents, into the atmosphere have introduced a new political and cultural dynamic at the beginning of the twenty-first century. Debates about public policy, the development of business strategies and the deliberations of new social and environmental movements and organisations are now conducted with considerations about climate change very much in evidence. Anthropogenic climate change not only changes the nature – frequency and intensity – of climate risks to which societies have long been exposed, but introduces the possibility at some indeterminate point in the future of prospective changes to climate which lie well outside the experience of human history. These prospects and possibilities introduce new challenges for all levels of governance – for public authorities from local and regional/city scales, through to national to international scales; for small businesses and multinational corporations; and for elected and non-elected sovereign governments.

Making Climate Change Work for Us: European Perspectives on Adaptation and Mitigation Strategies offers a synthesis of recently completed research which addresses these challenges. The research upon which this book is based was completed in the project 'Adaptation and Mitigation Strategies: Supporting European Climate Policy' (ADAM), a project funded by the European Commission under the Sixth Framework Research Programme of the European Union (EU). The ADAM project involved 24 of the continent's leading inter-disciplinary climate change research institutions, plus two partner institutions from China and India. The research described in this edited volume was completed during the period 2006 to 2009 and involved some 150 researchers from across Europe and beyond.

The significance of *Making Climate Change Work for Us* is twofold. Firstly, the book offers an inter-disciplinary perspective – drawing upon environmental economics, policy sciences, geography, technology analysis, integrated assessment and other social and natural science disciplines – on the ideas and dilemmas surrounding the development and deployment of adaptation and mitigation strategies for addressing climate change, and on the methods and tools used to investigate them. Secondly, it offers this unique perspective from a cohort of Europe's leading integrated climate change research experts who have developed their analytical and intellectual skills over many years as close observers and participants in vibrant EU and international science and policy debates about climate change.

The research described here is contextualised by current EU and international developments, dilemmas and debates about climate change and about the relationship between climate science and policy. Our point of departure is the EU's policy goal of restricting anthropogenic global warming to no more than 2 °C above pre-industrial temperature. Yet the analyses in this book examine a wider range of questions and concerns. They are set in the context of a contested and slowly evolving global climate regime, against a back-drop of growing interest in adapting societies around the world to be more resilient to climate risks, and are fully aware of the changing international climate diplomacy in search of a new global framework agreement for the post-2012 period. The chapters navigate through various combinations of these scientific, political, economic and ethical uncertainties, exploring them at different scales and reporting new ideas, new findings and new possibilities from an integrated research perspective and from within European culture.

The title of this volume – *Making Climate Change Work for Us* – is intended to reflect a positive stance in relation to climate change. The editors firmly believe that the risks and challenges of climate change must be viewed as opportunities to improve quality of life for all peoples, both now and in the future, i.e. as a means of moving towards greater sustainability, rather than portrayed as the first signs of an inevitable global catastrophe. It is important that the unique characteristics of anthropogenic climate change – the global drivers and consequences of change and the demand for a multi-decadal if not multi-generational perspective – are used powerfully to re-think and re-shape the ways in which local, national, regional and international strategic planning and policy making are conducted in the early twenty-first century. While not being directly addressed in this volume, the current financial and economic crisis provides just one such opportunity. By now investing heavily in transformations of energy systems worldwide, new possibilities arise for avoiding high-end climate change scenarios.

In the context of other books

The number and diversity of books about climate change has increased almost exponentially over the last few years. Each of the book publishing categories of textbooks, popular science, polemical, journalistic, coffee-table and academic research are now well populated with climate change offerings. *Making Climate Change Work for Us* falls clearly into the category of academic research yet by focusing specifically on adaptation *and* mitigation strategies, and by being rooted in a large inter-disciplinary research project (ADAM), *Making Climate Change Work for Us* makes a unique contribution to the literature.

This volume should be viewed as a logical supplement to the earlier Cambridge University Press books edited by John Schellnhuber and colleagues (Schellnhuber *et al.*, 2006) arising from the February 2005 Exeter Conference on dangerous climate change, and by Neil Adger and colleagues (Adger *et al.*, 2009) arising from the February 2008 Tyndall Centre Conference on limits to adaptation. The former focused on the dangers of climate change, the latter on limits and barriers to adapting to these dangers, while *Making Climate Change Work for Us* examines the range of adaptation and mitigation strategies, at different scales, that can be pursued to avoid, defuse or otherwise manage such dangers. Collectively, these three research-based and edited volumes make a valuable triumvirate contribution to our understanding of climate change, global ecology and human society.

Making Climate Change Work for Us is itself supplemented by three further books emerging from the ADAM research project and also published by Cambridge University Press: *Climate Change Policy in the European Union* (edited by Andrew Jordan and colleagues), *Global Climate Governance Beyond 2012* (edited by Frank Biermann and colleagues) and *Mainstreaming Climate Change in Development Cooperation* (edited by Joyeeta Gupta and Nicolein van der Grijp). These three volumes provide more in-depth analyses of the policy dimensions of climate change as examined within Europe (Jordan *et al.*, 2010), from an international perspective (Biermann *et al.*, 2010) and from a development perspective (Gupta *et al.*, 2010). Taken together, these four books from the ADAM project constitute a substantial advance in our understanding of the policy implications of climate change as viewed from the end of the first decade of the twenty-first century. The research completed in the ADAM project, and which informs this book series, is also reported in two journal special issues: in *The Energy Journal* ('The economics of low stabilisation' edited by Ottmar Edenhofer and colleagues) and in *Mitigation and Adaptation Strategies for Global Change* ('Assessing adaptation to extreme weather events in Europe' edited by Zbigniew W. Kundzewicz and Reinhard Mechler).

Structure and contents

Making Climate Change Work for Us is built around 14 substantive and original chapters. The first five of these introduce some of the concepts and scenarios used in the ADAM project. Four chapters in Part II of the book then explore strategies for responding to climate change within Europe, followed by four chapters in Part III, which extend this exploration of strategic options beyond the boundaries of the European Union. The volume is completed by an integrating synthesis chapter.

In Part I of the book, five chapters introduce some of the concepts and scenarios used in the ADAM project: concepts used as the basis for identifying and analysing mitigation and adaptation strategies, and scenarios used as the basis for framing possible future states of Europe and the world so as to be amenable for strategic and policy investigations. Together, these chapters build the conceptual and methodological framework for later analyses of climate change strategies. These opening perspectives go beyond current state-of-the-art: they benefit from new insights emerging from recent climate policy analysis and integrated assessment research and they are oriented to illuminate climate change decision making and policy deliberations.

Chapter 1 (co-ordinated by Henry Neufeldt from the Tyndall Centre and School of Environmental Sciences at the University of Anglia in the UK) offers a conceptual basis for discussing adaptation and mitigation by looking at the different kinds of challenges that need to be addressed when dealing with both adaptation and mitigation climate policies: synergies, conflicts and trade-offs as played out in different sectors and over different scales. Chapter 2 (co-ordinated by Duncan Russel also from the School of Environmental Sciences at the University of East Anglia) provides an analysis of current trends and future challenges for climate change appraisal processes in the EU, drawing upon empirical evidence of recent climate policy appraisals conducted in Europe at different scales and contexts. Chapter 3 (co-ordinated by Detlef van Vuuren from the Netherlands Environmental Assessment Agency) introduces the global society–energy–climate–environment scenarios used in the ADAM project and which frame the analysis consistently throughout the project. This chapter outlines the recent development of recursive scenarios that take into account the impacts of climate change and a certain level of future adaptation. Such scenarios of adaptation are further investigated in Chapter 4 (co-ordinated by Asbjørn Aaheim from the Center for International Climate and Environmental Research in Norway) using different top-down and bottom-up modelling approaches to explore climate impacts and adaptation in Europe. It is suggested that the common perception of adaptation taking place at local levels will lead to significant underestimation of the actual costs of adaptation because of the existing market imperfection: for example locality and extreme weather events or limits to moving stranded assets. National and international adaptation strategies may instead be needed. In contrast, Chapter 5 (co-ordinated by Jochen Hinkel from the Potsdam Institute for Climate Impact Research in Germany) takes a bottom-up approach to examining adaptive capacity and the barriers to adaptation practice. Illustrated through four different decision-making contexts, the chapter focuses on the social and institutional processes of adaptation learning. These illustrations are drawn from the ADAM project's case studies, as well as from a meta-analysis of existing literature.

The four chapters in Part II of the book explore strategies to deal with a number of challenges related to European climate change policy at different scales and for varying contexts. Yet these are representative of similar challenges facing other regions of the world: climate governance, the energy system, weather risks and extremes and regional land use and water management. Chapter 6 (co-ordinated by Frans Berkhout from the Institute of Environmental Studies in the VU University Amsterdam) introduces the concept of governance dilemmas (i.e. making choices between equally favourable or equally disagreeable alternatives) as applied to EU climate mitigation policies. Chapter 7 (co-ordinated by Gunnar Eskeland from the Center for International Climate and Environmental Research in Norway) discusses how Europe can devise strategies that enable a transition towards a low-carbon energy system while still operating effectively within a global context. The chapter explores questions of energy efficiency, low-carbon technology, land use changes and the direct impacts on electricity supply and demand of the changing climate. Chapter 8 (co-ordinated by Reinhard Mechler from the International Institute for Applied Systems Analysis in Vienna, Austria) examines the changing nature of weather risk in Europe using the theory and practice of disaster risk analysis and management. It focuses on current and future risks emerging from floods, drought and heat waves and illustrates the economic impacts of such events and how structural funds may be used as a form of adaptation. The final chapter in this section of the book – Chapter 9 co-ordinated by Saskia Werners from Wageningen University in the Netherlands – investigates two central issues of regional and spatial planning in the face of climate change and variability: land use change and water distribution. For three regions studied in the ADAM project – the Tisza basin in Hungary, the Guadiana basin in the Iberian Peninsula and the Alxa region in Inner Mongolia, China – the chapter synthesises lessons for adaptation derived from understanding the differing environmental, social and political settings of each region.

Part III of the book comprises four chapters which extend analysis beyond the borders of the EU and provide insights into, respectively, governance, economic/technological, development and financial aspects of climate change at the global level. These chapters investigate a number of adaptation and mitigation strategies that will have to be considered carefully if climate change is to be retained at levels approximating to the EU's policy target of 2 °C. Chapter 10 (co-ordinated by Frank Biermann from the Institute of Environmental Studies at the VU University Amsterdam) establishes a number of avenues to explore regarding global climate governance after 2012. These perspectives include how to involve non-state actors in such a regime and how to strengthen the goals of adaptation in such an international system of governance. Their investigations rely on qualitative policy assessment, formal modelling and participatory methods. Chapter 11 (co-ordinated by Brigitte Knopf from the Potsdam Institute for Climate Impact Research) uses an ensemble of energy-economy models to reveal the technological challenges and political and economic consequences of reaching the 2 °C goal with more than a 50% chance of success. This goal implies negative global emissions at some point this century. Special attention is therefore given to the emissions reduction potentials of bio-energy, non-carbon dioxide gases and carbon capture and storage, and the consequences of these technologies for different global regions and for Europe. Chapter 12 (co-ordinated by Joyeeta Gupta from the Institute of Environmental Studies at the VU University Amsterdam) explores the relationship between climate change and European development assistance. It examines the possibilities and barriers to mainstream considerations of climate change and variability into development policies and how best to improve EU development cooperation in the future. One specific option for mainstreaming – risk-sharing through insurance mechanisms – is investigated in Chapter 13 (co-ordinated by Joanne Linnerooth-Bayer from the International Institute for Applied Systems Analysis). Such mechanisms require global public–private partnerships to be effective at different scales and the chapter describes examples of such insurance-based adaptation at local, national and regional scales that manage climate-related risks for developing countries. The analysis also explores the limits of such insurance-based instruments for reaching the poorest of the poor.

The final chapter of the book – Chapter 14 co-ordinated by Anthony Patt from the International Institute for Applied Systems Analysis – draws on many of the arguments, analyses and insights from the ADAM project to offer five guideposts for thinking about successful climate strategies. These guideposts are elaborated using a different metaphor for each case: describing priorities between mitigation and adaptation policies rather than optimal trade-offs; describing mitigation as the need to invest in strategies that go far beyond picking low-hanging fruit; describing climate policies as trial-and-error approaches out of which may emerge robust solutions; describing the technological changes necessitated by climate change as an opportunity to secure future sustainable development while eliminating many convenient, but inadequate, 'crutches'; and, finally, describing climate change policies as a game of winners and losers where the losers will have to be compensated to continue to play the game. These strategic guideposts offer a vision of how we can – deploying collective wisdom, political will and human ingenuity – 'make climate change work for us'.

How the book was produced

Each chapter in *Making Climate Change Work for Us* was led by a co-ordinating lead author who had overall responsibility for the chapter. With the exceptions of the opening and closing chapters – which frame (Chapter 1) and synthesise (Chapter 14) the entire project – each chapter is rooted in one of the primary areas of work conducted within the ADAM project. The full writing teams for each chapter were drawn, however, from across the ADAM consortium and reflect the inter-disciplinary and institutionally collaborative character of the ADAM project. Each chapter was peer reviewed twice: an initial internal review in which researchers in the ADAM project were required formally to review the work of colleagues in different domains of the project, followed by a second, external, review in which two independent reviewers selected from institutions in Europe and North America not involved in the ADAM project were asked to conduct a full evaluation of the merits and deficiencies of the draft chapters. The editors of the book required authors to respond formally in writing to each cycle of review comments and they ensured that corrections and improvements to each chapter were subsequently implemented.

The 101 authors of this volume are drawn from some of Europe's leading inter-disciplinary climate change research institutions, many of whom have had prominent roles in either the Third, Fourth or Fifth Assessment Reports of the Intergovernmental Panel on Climate Change. Their affiliations are included above.

Mike Hulme
Henry Neufeldt
Norwich, April 2009

References

Adger, W. N., O'Brien, K. and Lorenzoni, I. (eds.) (2009) *Adapting to Climate Change Thresholds, Values, Governance*. Cambridge, UK: Cambridge University Press.

Biermann, F., Pattberg, P. and Zelli, F. (eds.) (2010) *Global Climate Governance Beyond 2012: Architecture, Agency and Adaptation*. Cambridge, UK: Cambridge University Press.

Gupta, J. and van der Grijp, N. (eds.) (2010) *Mainstreaming Climate Change in Development Cooperation: Theory, Practice and Implications for the European Union*. Cambridge, UK: Cambridge University Press.

Jordan, A. J., Huitema, D., van Asselt, H., Rayner, T. and Berkhout, F. (eds.) (2010) *Climate Change Policy in the European Union: Confronting the Dilemmas of Mitigation and Adaptation?* Cambridge, UK: Cambridge University Press.

Schellnhuber, H. J., Cramer, W., Nakicenovic, N., Wigley, T. M. L. and Yohe, G. (eds) (2006) *Avoiding Dangerous Climate Change*. Cambridge, UK: Cambridge University Press.

Acknowledgements

The ADAM project was funded by DG-RTD under the EU's Sixth Framework Programme, Contract Number 018476 (GOCE). The project officers were Ger Klassen and Wolfram Schrimpf and we thank them for ensuring efficient liaison was maintained with the Commission during the project lifetime.

The editors and authors of the book are immensely grateful to Helen Colyer at the University of East Anglia for the many hours of work spent checking, indexing and proof-reading for the book. Her patience and diligence were exemplary. Angela Ritchie contributed to some of the final stages of the manuscript preparation and she also played a huge role in keeping the ADAM project in good administrative shape during its latter years, while Emanuela Elia played a similar crucial role during ADAM's earlier stages.

Twenty-six external reviewers invested time and effort in undertaking reviews of the drafts of these book chapters and we thank each of them for their insightful and constructive comments. Listed in alphabetical order they are: Roberto Acosta, Steinar Andresen, Barry Barnett, Olivia Bina, Ian Burton, Stéphane Hallegatte, Donald A. Hanson, Julia Hertin, Monique Hoogwijk, Einar Hope, Klaus Jacob, Andre Jol, Norichika Kanie, Bo Lim, Andreas Löschel, Brian O'Neill, Hans Opschoor, Jon Padgham, Claudia Pahl-Wostl, Keywan Riahi, Roberto Roson, Peter Russ, Joachim Schleich, Roger Street, Rob Swart and Anegret Thieken.

At Cambridge University Press (CUP) we are grateful to Matt Lloyd for his efforts in enabling the book – and the ADAM book series – to appear with CUP and for keeping it on track through the production cycle. We also acknowledge the role played by Laura Clark, Abigail Jones and Mary Sanders at CUP in managing the production process.

Abbreviations

A2	IPCC SRES scenario
AAD	Annual average damages
ACEA	European Automobile Manufacturers Association
ADAM	Adaptation and mitigation strategies: supporting European climate policy (EU FP6 research project)
ADB	Asian Development Bank
AD-RICE	Adaptation in regional dynamic integrated model of climate change and the economy (version of DICE)
AD-DICE	Adaptation in dynamic integrated model of climate change and the economy (see model appendix)
ALTENER	an EU programme aimed at promoting the use of renewable energy sources
AOSIS	Alliance of small island states
AR4	IPCC Fourth Assessment Report
ART	Alternative risk transfer
ASTRA	A strategic integrated assessment model (see model appendix)
B2	IPCC SRES scenario
BSAEU	Burden sharing agreement
C&D	Climate and development
CATSIM	Catastrophe simulation model (see model appendix)
CBA	Cost–benefit analysis
CCA	Climate change agreement
CCPMs	Common and co-ordinated policies and measures
CCRIF	Caribbean catastrophe risk insurance facility
CCS	Carbon capture and storage
CDAC	Commission for the Convention Development and Application

CDM	Clean development mechanism
CEC	Commission of the European Communities
CGE	Computable general equilibrium model
CI	Carbon intensity
CIDA	Canadian International Development Agency
CIP	Climate insurance pool
CO_2	Carbon dioxide
CO_2e	Carbon dioxide equivalent
COP	UNFCCC Conference of the Parties
Cropsyst	A multi-year, multi-crop, daily time-step crop-growth simulation-model (see model appendix)
DAC	Development Assistance Committee
Defra	UK Department for Environment, Food and Rural Affairs
DG	Directorate General (of the EU)
DICE	Dynamic integrated model of climate change and the economy
DIVA	Dynamic and interactive vulnerability assessment model (see model appendix)
DPSIR	Driver–pressure–state-impact-response
E3ME	Energy–environment–economy model of Europe (see model appendix)
E3MG	Energy–environment–economy modelling at the global level (see model appendix)
EAC	Environmental Audit Committee
EC	European Commission
ECAs	Energy conservation agreement schemes
ECCP	European Climate Change Programme
EDI	Ethiopia Drought Index
EEA	European Environment Agency
EFISCEN	European forest information scenario model
EI	Energy intensity
EMELIE	model assessing the European electricity market (see model appendix)
EMF	Stanford Energy Modelling Forum
ETS	Emissions trading scheme
EU	European Union
EU-15	Austria, Belgium, Denmark, Finland, France, Germany, Greece, Ireland, Italy, Luxembourg, the Netherlands, Portugal, Spain, Sweden, United Kingdom

EU-27	EU-15 countries + Bulgaria, Czech Republic, Cyprus, Estonia, Hungary, Latvia, Lithuania, Malta, Poland, Romania, Slovakia, Slovenia
EU-27+2	EU-27 countries + Norway and Switzerland
EuroMM	European Multi-regional MARKAL energy-conversion model (see model appendix)
EUSF	European Union Solidarity Fund
EV	Equivalent variation
FAIR	Climate policy model (see model appendix)
FES	Future energy solutions
FIT	Feed in tariff
FoEE	Friends of the Earth Europe
FPPP	Full polluter pays principle
FUND	An integrated assessment model of the climate and the economy
G77	Seventy-seven developing country signatories of the 'Joint Declaration of the Seventy-Seven Countries' on 15 June 1964
GDP	Gross domestic product
GEF	Global environment facility
GHG	Greenhouse gas
GIRF	Global index reinsurance facility
GIS	Geographical information system
GNI	Gross national income
GP	EU Adaptation Green Paper
GRACE	Global responses to anthropogenic change in the environment (see model appendix)
GRACE-EL	model based on GRACE, developed for the ADAM project (see model appendix)
GTAP	Global trade analysis project
GTZ	German Technical Co-operation Agency
HadCM3	Hadley Centre coupled climate model, version 3 – coupled atmosphere-ocean general circulation model
HIRHAM	Regional atmospheric climate model, with a pan-Arctic domain
IAM	Integrated assessment models
ICFD	International conference for financing in development
IEA	International energy agency
IFI	International financial institutions

IMAGE	Integrated model to assess the global environment (see model appendix)
IPCC	Intergovernmental Panel on Climate Change
IRI	International Research Institute for Climate and Society (Columbia University, New York)
IS	Industry energy system model simulating distinct conservation options and industrial processes (see model appendix)
ITC	Induced technological change
JAMA	Japanese Automobile Manufacturers Association
JI	Joint implementation
KAMA	Korea Automobile Manufacturers Association
MARA/ARMA	Malaria suitability model (see model appendix)
MATEFF	A model simulating potentials of material efficiency of energy-intensive materials (see model appendix)
MCII	Munich climate insurance initiative
MERGE	Model for evaluating regional and global effects (see model appendix)
MERGE-ETL	A modified version of MERGE5 (see model appendix)
MESSAGE	A model that embeds the world energy system within a macroeconomic framework
MMARM	Ministerio de Medio Ambiente Rural y Marino, Madrid
NAPA	National adaptation plan of action
NDRC	National Development and Reform Commission
NGO	Non-governmental organisation
NHS	National Health Service
NUTS	Nomenclature of territorial units for statistics
ODA	Official development assistance
OECD	Organisation for Economic Co-operation and Development
ORASECOM	Orange-Senqu River Commission
PAGE	Policy analysis of the greenhouse effect model
PAMs	EU climate change policies and measures
PESETA	Project – Projection of economic impacts of climate change in sectors of the European Union based on bottom-up analysis
POLES	A global sectoral model of the world energy system (see model appendix)
PowerACE	ResInvest, an agent-based sector model (see model appendix)
ppm	parts per million

PRECIS	Providing regional climates for impact studies – regional climate model from the Hadley Centre
PRSP	Poverty reduction strategy papers
PSI	Policy Studies Institute
R&D	Research and development
REMIND	also known as REMIND-R being an inter-temporal optimising energy–economy–environment model (see model appendix)
RESAPPLIANCE	A model simulating the appliances in the residential sector (see model appendix)
RESIDENT	A model simulating the non-appliances demand in the residential sector (see model appendix)
RET	Renewable energy technologies
RICE	Regional dynamic integrated model of climate change and the economy (version of DICE)
RMCP	Regional modelling comparison project
RWI	Rhineland-Westfalen Institute for Economic Research
SAVE	Specific actions for vigorous energy efficiency programme
SERVE	A model calculating the energy demand and investment in the commercial sector (see model appendix)
SRES	IPCC Special Report on Emissions Scenarios
SRU	Expert Commission on the Environment
SWAps	Sector-wide approaches
SWCC	Second World Climate Conference
TCPA	Town and Country Planning Association UK
TGC	Tradable green certificates
TIMER	Global energy model, part of IMAGE model (see model appendix)
UK	United Kingdom
UKCIP	UK climate impacts programme
UN	United Nations
UNDAF	United Nations Development Assistance Framework
UNDG	United Nations Development Group
UNDP	United Nations Development Programme
UNFCCC	United Nations Framework Convention on Climate Change
UNGA	United Nations General Assembly
UNISDR	United Nations International Strategy for Disaster Reduction
USA	United States of America

USAID	United States Agency for International Development
VA	Voluntary Agreements
VAHAVA	Valtozas-Hatas-Valaszadas: change–impact–response project
WETO-H_2	World energy technology outlook scenario
WFP	World food programme
WHO	World Health Organization
WSSD	World summit on sustainable development
WTO	World Trade Organization

Part I

Concepts and scenarios

1

Climate policy and inter-linkages between adaptation and mitigation

Lead authors:

HENRY NEUFELDT[1], EBERHARD JOCHEM, JOCHEN HINKEL,
DAVE HUITEMA, ERIC MASSEY, PAUL WATKISS,
DARRYN MCEVOY, TIM RAYNER, ANDRIES HOF,
KATE LONSDALE

[1]Co-ordinating lead author

Contributing authors:

TERRY BARKER, ANNE HELD, MIKE HULME, ULRICH REITER,
HAL TURTON, DETLEF P. VAN VUUREN, SASKIA E. WERNERS

Summary

The objective of this chapter is to provide a better understanding of the inter-linkages, trade-offs and synergies between adaptation and mitigation, building on the work of the Intergovernmental Panel on Climate Change (IPCC) Fourth Assessment Report. The chapter elaborates on three different perspectives for assessing the two domains and illustrates these with examples from the ADAM research: the analysis of mitigation and adaptation in the European energy system based on integrated assessment models; mitigation and adaptation opportunities and barriers in the context of urban planning to reflect on social learning and capacity building; and an analysis of present and future climate governance challenges in the EU as an example of institutional and policy analysis. In an explorative section, the chapter then provides a meta-analysis of European climate policies of the past ten years. The analysis shows that the inter-relationships between the two domains of adaptation and mitigation are complex and may involve different temporal, spatial and organisational scales. This leads us to conclude that: (i) mitigation efforts today may lead to climate vulnerabilities in the future if the life cycles specific to each sector are not adequately taken into account; (ii) development of response capacity in one domain does not lead to capacity in the other because adaptation and mitigation involve mostly different sectors, actors and institutions; (iii) climate impacts may lead to growing welfare inequalities, which can be balanced through co-ordinated policies at higher levels, but which need to overcome existing institutional barriers; and (iv) synergies between adaptation and mitigation are most easily found where mitigation efforts are reinforced by behavioural changes, which lead to

Making Climate Change Work for Us: European Perspectives on Adaptation and Mitigation Strategies, ed. Mike Hulme and Henry Neufeldt. Published by Cambridge University Press © Cambridge University Press 2010.

an overall increase in resilience by implementing broad concepts of sustainability. Further consideration of these adaptation–mitigation linkages is a research priority. Policy innovation will be needed to capture synergistic benefits and avoid the introduction of new climate vulnerabilities or accelerated emissions of greenhouse gases.

1.1 Introduction

Climate change adaptation and mitigation strategies in the context of existing policies and policy development stand at the centre of this volume. A critical discussion of the concepts and perspectives used for assessing adaptation and mitigation is therefore essential to place the contributions of the following chapters into perspective. In particular, there is a need for a better understanding of the inter-relationships between adaptation and mitigation as policies in both domains become more developed. The objective of the chapter is therefore to provide some evidence of the inter-linkages, synergies and trade-offs between adaptation and mitigation. This will lead to more effective decision making in the European policy arena.

Linkages between mitigation and adaptation have been explored in the *IPCC Fourth Assessment report*, in particular in Chapter 18 of Working Group II 'Inter-relationships between adaptation and mitigation' (Klein *et al.*, 2007) and chapter 3 of Working Group III 'Issues related to mitigation in the long-term context' (Fisher *et al.*, 2007). One important limitation of these earlier works, however, is that the linkages have not been sufficiently explored in the context of the very distinct research approaches that are applied for assessing mitigation and adaptation policies. This chapter therefore analyses these linkages in the context of three main approaches: integrated assessment modelling, social learning and institutional and policy analysis. These three approaches are complementary and have been used equally throughout the ADAM project in order to provide a multi-faceted picture of European climate policy and its challenges and trade-offs.

Integrated assessment models frame mitigation and adaptation either from a top-down or bottom-up decision analytical perspective (see Chapters 3 van Vuuren *et al.*, 4 Aaheim *et al.*, 7 Eskeland *et al.*, 8 Mechler *et al.* and 11 Knopf *et al.*) and provide information on the direct and indirect economic effects of climate policies, together with the technological settings needed to achieve them. However, there is increasing recognition of the value of the social-learning approach as complex problems generally involve many actors at different temporal and spatial scales (Tàbara and Pahl-Wostl, 2007). This complexity makes it difficult and often impossible to predict outcomes of actions and suggests that the decision-analytical framework should be embedded in a social-learning framework that can better take account of the complexity of nested decision situations (see Chapters 2 Russel *et al.* and 5 Hinkel *et al.*). In such cases, case studies can often help understand the different temporal, spatial, and institutional scales involved by focusing on how actors interact and learn and on the cross-cutting themes that transcend the analysed cases (see Chapter 9 Werners *et al.*). Finally, institutional and policy analysis can identify important barriers to, and opportunities for, climate policies by analysing the norms, institutions, policies and measures of climate governance (see Chapters 6 Berkhout *et al.*, 10 Biermann *et al.*, 12 Gupta *et al.* and 13 Linnerooth-Bayer *et al.*, as well as the companion ADAM volume edited by Jordan *et al.*, 2010).

For each of the three perspectives, we explore the inter-linkages and trade-offs between adaptation and mitigation by focusing on one representative case study from the ADAM research. These are:

(i)   The analysis of mitigation and adaptation in the European energy system based on bottom-up technology integrated assessment models (IAMs). The chapter focuses on IAMs because serious attempts have only recently been made to integrate adaptation into mitigation frameworks, whilst most modelling frameworks still focus solely on mitigation (Fisher *et al.*, 2007);

(ii) Mitigation and adaptation opportunities and barriers in the context of urban infrastructure. This examines the need for social learning and capacity building to achieve synergies and avoid technological lock-ins whilst dealing with the different perspectives of multiple stakeholders; and

(iii) An analysis of present and future climate governance challenges in the EU. This describes how climate policies emerged in the context of enabling and constraining factors in the EU and its Member States and how mitigation and adaptation could be better integrated.

The chapter is organised as follows. Section 1.2 first summarises the IPCC understanding of the inter-linkages between adaptation and mitigation and then briefly introduces the three perspectives: modelling, social learning and policy analysis. Section 1.3 then presents the three cases, illustrating these perspectives. Section 1.4 analyses the status of European mitigation and adaptation policy based on a meta-analysis of European and EU Member State climate policies to shed some light on the way the two domains influence each other at the policy level. Section 1.5 discusses and synthesises the findings focusing on the synergies and trade-offs between adaptation and mitigation at different scales and, finally, Section 1.6 draws some overarching conclusions.

1.2 Perspectives on the assessment of adaptation and mitigation

1.2.1 The IPCC perspective

Even the most ambitious mitigation policies will lead to some climate change that requires a certain level of adaptation (see also Chapter 3 van Vuuren *et al.*, and van

---

Box 1.1. IPCC definitions of adaptation and mitigation

Working Group II of the *Fourth Assessment Report of the Intergovernmental Panel on Climate Change* (IPCC, 2007a) defines adaptation as 'adjustment in natural or human systems in response to actual or expected climatic stimuli or their effects, which moderates harm or exploits beneficial opportunities' whereas mitigation is defined as 'an anthropogenic intervention to reduce the anthropogenic forcing of the climate system; it includes strategies to reduce greenhouse gas sources and emissions and enhance greenhouse gas sinks'. The definitions of Working Group III (IPCC, 2007b) do not differ significantly from the ones cited above.

Vuuren *et al.*, 2008). The impacts of a changing climate (i.e. higher temperatures and altered patterns and amounts of precipitation, extreme weather events and sea-level rise) will have negative effects on regions and sectors that are sensitive to climate alterations, and will be felt most by those who are most vulnerable and lack the means to protect themselves (IPCC, 2007a). Adaptation is therefore most likely to be implemented at scales where direct impacts will occur. This will help to bring immediate and long-term benefits to those who are affected, either by raising their resilience or by reducing their exposure to climate impacts (Adger, 2006).

Mitigation, on the other hand, requires co-ordinated action, predominantly at international and national levels, and is mainly applied to sectors with high greenhouse gas emissions. The benefits of these actions accrue at the global scale and only after considerable lag times. The exceptions are, however, the immediate welfare benefits of investing in mitigation technologies and measures, such as improved air quality or reduced fuel poverty. Hence, adaptation and mitigation have different problem structures and pose considerably different challenges to policies and management. The formal IPCC definitions of adaptation and mitigation are given in Box 1.1.

Nevertheless, there are important linkages between adaptation and mitigation, which have an impact on two key factors: the implementation of measures and the fostering of 'response capacity' (Tompkins and Adger, 2005).

In regards to implementation of adaptation and mitigation measures, the IPCC (Klein *et al.*, 2007) distinguishes four relationships: (i) adaptation actions that have consequences for mitigation; for instance, higher temperatures may induce more air conditioning which in turn will increase energy demand and thus the need to mitigate emissions; (ii) mitigation actions that have consequences for adaptation; for example, increased use of biofuels will likely have effects on food supply and prices, particularly for poor countries, lowering their ability to adapt; (iii) decisions that include trade-offs or synergies between adaptation and mitigation, such as large-scale mitigation with effects on impacts and adaptation; and (iv) processes that have consequences for both adaptation and mitigation; for instance, afforestation initiatives can act as a carbon sink, protect against flash flooding, and provide income for local communities (hence increasing local adaptive capacity).

The achievement of adaptation and mitigation measures also requires sufficient capacity for measures to take hold. Tompkins and Adger, 2005 refer to this as 'response capacity', which includes both mitigative capacity as 'the ability to reduce anthropogenic greenhouse gases or enhance natural sinks' (Winkler *et al.*, 2007) and adaptive capacity, as 'the ability or potential of a system to respond successfully to climate variability and change' (Sathaye *et al.*, 2007). It is based on the observation that despite some differences in spatial scale and sectoral focus, both mitigative and adaptive capacities are driven by similar sets of factors. In particular, it is recognised that the response capacity is determinate upon a broad range of socio-technical, economic and institutional resources. These, in turn, can be influenced by socio-cultural dimensions, as well as by other important factors such as infrastructure, risk perception, and political will, which allow a given group (community, society, etc.) to respond adequately to a threat. Therefore, Klein *et al.* (2007) suggest that 'the influence of each determinant of capacity is highly location-specific and path-dependent'.

1.2.2 *The integrated assessment modelling perspective*

Integrated assessment models (IAMs) have become a common tool for assessing the costs and benefits of climate change policy over long time horizons. They traditionally focus on questions related to defining mitigation targets (e.g. stabilisation levels), assessing the costs and benefits of reaching different targets (e.g. social costs of carbon, costs of residual damages) and the type of measures needed to achieve certain targets.

Mitigation is represented differently in different types of IAMs. Process-oriented IAMs represent mitigation strategies on the basis of the emission reduction potential and costs of a wide range of specific mitigation measures. Examples of such models are MESSAGE (Riahi *et al.*, 2007) and IMAGE (Bouwman *et al.*, 2006). Other models focus more on economic consistency and represent mitigation costs in terms of production functions. Examples include FUND (Tol, 2006) and DICE (Nordhaus, 2007).

Adaptation in IAMs is represented more crudely. Most IAMs that include a description of damages, in fact, implicitly assume optimal adaptation. In these IAMs adaptation is therefore not a policy variable like mitigation, but a direct function of the mitigation target. These models do not describe the impact of different adaptation strategies nor do they provide a description of the measures and costs needed to inform adaptation policies.

Only a few IAMs include adaptation as a policy variable, namely PAGE (Hope, 2006) and AD-DICE (de Bruin *et al.*, 2009). These models use a much aggregated

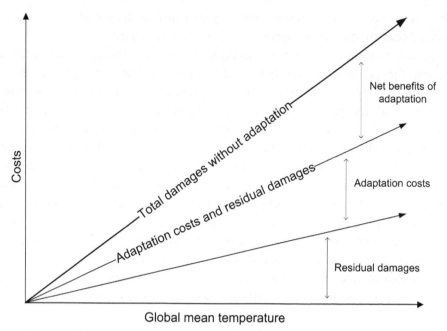

Figure 1.1. Schematic representation of adaptation (adopted from Stern (2006)). The total costs with adaptation are the sum of adaptation costs and residual damages. The net benefit of adaptation is the difference between these costs and what these costs would have been in the absence of adaptation.

approach to adaptation. Adaptation in PAGE affects the rate and level of temperature change at which an onset of impacts begins, and can reduce the severity of these impacts (Warren *et al.*, 2006). AD-DICE disaggregates the damage function of DICE into adaptation costs and residual damages (see Figure 1.1). The model selects the preferred combination of mitigation and adaptation in response to climate impacts. However, there is an important difference: mitigation reduces climate change damages only in the long run, while adaptation can also reduce damages in the short run.

Given this crude treatment of adaptation, IAMs continue to provide only limited insights into adaptation. They indicate how much to adapt, but not on how to adapt[1]. Based on the assumption that damage functions include some form of optimal adaptation, Patt *et al.* (2009) recently suggested that IAMs are likely to overestimate the amount of adaptation that will occur and therefore also overestimate the benefits obtained from adaptation. The authors also argue that global IAMs cannot identify the costs and benefits of adaptation measures accurately because to do so would require local or regional level detail. Adding a better

---

[1] Chapter 4 (Aaheim *et al.*) provides a discussion of opportunities and challenges of representing adaptation in IAMs.

description of adaptation to IAMs would therefore improve the overall cost–benefit analysis of mitigation strategies, but would not help develop adaptation strategies. In order to do this, IAMs would need to include more data on the amount and distribution of adaptation costs.

1.2.3 The social learning perspective

IAMs formally represent action–outcome linkages that allow prediction of the outcome of chosen actions. However, the point of departure for the social-learning approach is that socio-ecological systems are too complex to be fully understood or formally represented in action–outcome linkages. The unpredictability of action – outcome linkages requires learning (i.e. act, observe outcome and learn, then act again), and the interdependence between the nested decision situations requires learning to be a social (institutional) process amongst different actors (Tàbara and Pahl-Wostl, 2007).

Approaches that model action–outcome linkages, such as those of IAMs described above may be appropriate when the problems are 'tame', i.e. when issues can be easily deconstructed into cause and effect, and there is a clear understanding of how to fix the problem. These approaches however, are not sufficient to deal with some of the highly uncertain and complex situations associated with adaptation and mitigation to climate change (e.g. see Funtowicz and Ravetz, 1991, Gallopín, 1999, Darwin *et al.*, 2002). Many of these situations can be considered 'unbounded' (or 'wicked' as opposed to 'tame'; see Rittel and Weber, 1973). Chapman (2002) has described them as problems where there is no clear argument about what exactly the problem is; where there is uncertainty and ambiguity as to how improvements might be made; and where the problem has no limits in terms of the time and resources it could absorb.

Unbounded problems require a different approach to planning and implementing solutions that recognise (rather than ignore) disagreement and uncertainty between different groups affected. This requires a process of dialogue in which the actors involved can listen to and understand the perspectives of others (Senge, 1990). Figure 1.2 illustrates means of stakeholder involvement along axes of increasing impact potential and uncertainty. As the stakes are raised and the complexity and uncertainty of the information to be dealt with also increase, more stakeholders and institutions are brought in to participate in the social learning process. The role of the moderator of the process then becomes increasingly important as someone who can, from a neutral position (or accepted non-neutral position), encourage and support processes of dialogue and engagement (Snowden, 2005).

Moreover, by seeing adaptation and mitigation as processes of 'social learning', we are enquiring into the ways in which individual actors make sense of the situation in

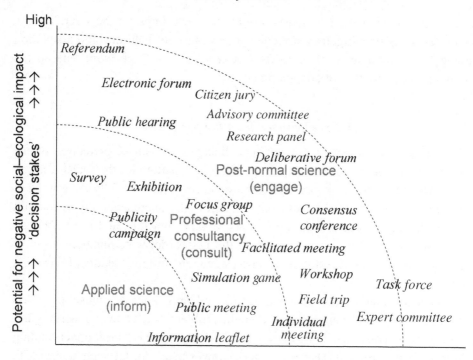

Figure 1.2. A typology of methods. After Forrester *et al*. (2008).

which they find themselves, and how they are supported or constrained in taking action. Therefore, for processes of social learning to be effective, opportunities for effective dialogue between the relevant groups must exist. This is not easy to achieve in practice and Cuppen *et al.* (2006) and Jochem *et al.* (2000) have identified a range of barriers to the process. These include power relationships, lack of knowledge and understanding, traditional routines and attitudes.

Finally, social learning is often about values and other 'higher-order' concepts such as norms, responsibilities, goals, and the framing of issues in terms of causes and effects (Kemp and Weehuiszen, 2005). When dealing with unbounded problems, the process of problem framing is therefore also important as this can pre-configure what are seen to be the available solutions (Rittel and Weber, 1973; see Hulme (2009) for an application of this thinking to climate change). The process of problem framing can thus become an exercise in power (Slovic, 2000).

1.2.4 *The institutional and policy analysis perspective*

The preceding section has noted the distinctive, 'wicked' features of the climate 'problem structure'. To the complexity of the problems themselves must also be added the complexity of the institutional arrangements, or governance frameworks, through which they must be addressed. The dilemmas facing policy makers when handling such complexities are examined in an ADAM companion volume (Jordan *et al.*, 2010).

Within the domain of climate governance, there is a general recognition of two elements needed for successful policy implementation: first, the building of capacity, i.e. creating the information and conditions that are needed to support subsequent actions; and second, delivering these actions or measures. The concepts of capacity building have been applied to explain the success of mitigation policy for some time (see e.g. Weidner, 2002; Fisher *et al.*, 2007), and the concept is now well developed in the adaptation literature, (Klein *et al.*, 2007).

Following Jänicke (1997), the capacity for environmental policies arises from two factors: the strength, competence and configuration of organized governmental and non-governmental proponents of environmental protection; and the cognitive–informational, political–institutional and economic–technological framework conditions. It means that governance capacity increases as the proponents in an area become better organised and competent, as monitoring and information gathering leads to a better understanding of the problem, as political attention to the problem increases, and as economic and technological capabilities become more easily available. However, the capacity for environmental governance does not always translate directly into action. It also depends on the strategy, will and skill of proponents and on their situative opportunities (Jänicke, 1997). Action is also dependent on the perceived urgency of the problem, the resources available, and the availability of options and their costs. In the case of climate change, additional factors to be considered are its global characteristics and extremely long time frames.

By providing frameworks that institutionalise policy development, policy makers attempt to hedge the uncertainties surrounding climate change and the difficulties in providing policies that deliver effective and cost-efficient measures. The adaptation policy cycle (Horrocks *et al.*, 2006) presented in Figure 1.3 is an example of how the UK Government attempts to bridge the gap between building capacity, which includes elements of learning, and delivering adaptation policies that will eventually lead to planned adaptation measures. The process starts with the setting of the overall adaptation goal[2]. This is set, for example, in the second European Climate Change Programme (ECCP II) 'to identify good, cost-effective practice in the development of adaptation policy and to foster learning' (European Commission, 2005). To accommodate the challenges posed by adaptation, the process allows for

---

[2] Formulating policy goals requires many normative choices, such as deciding on the value of risk aversion, aversion to inequity, etc.

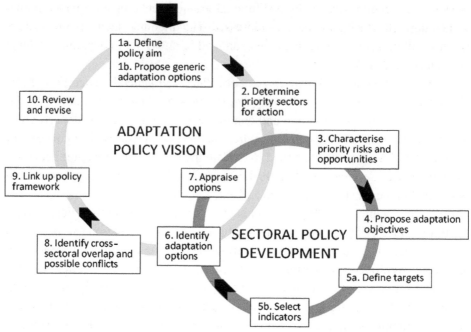

Figure 1.3. Objective setting for climate change adaptation policy. (Source: Horrocks *et al.*, 2006.)

circular and iterative learning. It also requires input from individual sectors and engagement with a range of stakeholders at various stages in its application.

For a number of reasons, mitigation policies are also likely to benefit from such an engagement of stakeholders. Firstly, engagement of the public might help to reduce local resistance to policy or technological innovations such as carbon-related taxes or wind turbines[3]. Secondly, it might help the implementation of no-regret mitigation measures as found, for example, in the building sector. Finally, mitigation in some cases requires a considerable amount of institutional learning or transformation of the existing economic–technological framework. A framework that effectively connects the policy with the sectoral and public domains will therefore probably lead to more integrated mitigation policies.

1.3 Examples of adaptation and mitigation inter-linkages in Europe

This section uses the European energy sector, infrastructure planning in urban areas, and an analysis of European climate policies as respective examples of the three

---

[3] The film 'Age of Stupid' (released March 2009) refers dramatically to such a case; www.ageofstupid.net.

approaches described above (i.e. modelling, learning and policy analysis) to high-light the differences and similarities between adaptation and mitigation policies and measures. These three cases analyse the existence, potential and challenges of adaptation–mitigation linkages in different settings and explore opportunities for integrating adaptation and mitigation into planning and decision making (i.e. mainstreaming).

1.3.1 The European energy sector

This case study examines adaptation in the European energy sector and how this relates to mitigation[4]. It is also an example of how modelling, in particular IAMs, can inform climate policy making.

The energy sector[5] contributes about two thirds to total greenhouse gas emissions at both European and global levels and is therefore a key focus for mitigation. However, climate change is also expected to have direct effects on both the supply and demand for energy and this is leading to an increasing interest in adaptation in the sector. The level of necessary adaptation will differ between EU Member States, depending on their geographical location, socio-economic development and future climates. In the short to medium term, i.e. 2020 to 2030, the predicted impacts from climate change to the energy sector will be modest. Depending on the rate of climate change, however, these impacts will significantly increase thereafter because of the relatively long lifetime of the energy supply infrastructure.

On the energy supply side, rising temperatures and changed precipitation patterns due to climate change will lead to significant impacts (Jochem *et al.*, 2009). Higher summer temperatures and decreased precipitation are expected to reduce the effi-ciencies of power plants where rivers provide the cooling waters (i.e. the majority of European power generation). This is particularly the case for southern Europe. Where future water temperatures consistently exceed regulations for cooling water thresholds, a policy choice may have to be made, changing current regulations to meet energy demands at the expense of environmental protection of the affected ecosystems. The expected increasing number and intensity of extreme weather events, such as floods or heatwaves (see also Chapter 8, Mechler *et al.*), are likely to increase the risks of electricity supply disruptions. This will require planned adaptation, for example decentralisation of power generation and investments to modernise the transmission and distribution grids to cope with the growing risks.

---

[4] Quantitative estimates of adaptation costs are only feasible in the case of increasing temperatures, whereas changes in extreme events can so far only be discussed in qualitative terms because their predictability is still too limited (also see Chapter 8, Mechler *et al.*).

[5] The energy sectors as defined in Jochem *et al.*, 2009 (also see Chapter 7, Eskeland *et al.*) include: residential; services; transport; industrial; and energy provision.

The increasing number of droughts, particularly in southern Europe, would also have detrimental effects on crop yields and forest productivity and thereby reduce the biomass energy potentials.

On the other hand, rising precipitation in countries north of the Alps and melting alpine glaciers are likely to increase water discharge and thereby increase the potential for hydropower generation in these regions. At the same time, the rising temperatures will increase biomass growth in northern Europe and add to the biomass energy potential. Finally, increasing average wind velocities are expected to lead to minor improvements in wind converter outputs thereby raising the wind energy potential.

On the energy demand side, rising mean annual temperatures will lead to increased space cooling and decreased heating (Aebischer *et al.*, 2007; see also Chapter 7, Eskeland *et al.*). For much of Europe increases in electricity demand for cooling will be balanced by reductions in the need for heating energy. The net result in ADAM's 4 °C Scenario (van Vuuren *et al.*, 2007; see also Chapter 3, van Vuuren *et al.*) is − 3.3 per cent final energy demand by 2050 (Jochem *et al.*, 2009). However, since the electricity used in space cooling is presently far more carbon-intensive than the energy used for heating, the net carbon dioxide emissions in some Member States could rise slightly despite the decrease in energy demand.

In a 4 °C Scenario, total electricity demand is expected to decrease slightly (0.5 per cent) in Nordic and Baltic countries by 2050 (Jochem *et al.*, 2009). However, an additional electricity demand of 7 per cent by 2050 is expected for Mediterranean countries. This may lead to a greater need to balance summer electricity flows via the trans-European electricity grid, particularly during extreme heat waves. It also means that equity issues driven by the distributional imbalance of climate impacts between northern and southern European countries are likely to have an influence on Europe's adaptation and mitigation policies.

Adaptation to heating and cooling will largely be autonomous, driven by individual behaviour, and by companies addressing demands. Without public policy-driven incentives to develop innovative solutions such as passive ventilation or integrated spatial planning for 'cool cities', short-term autonomous adaptation measures such as air conditioning may lead to 'lock in' situations with long-term consequences on energy demand and mitigation needs (Jochem *et al.*, 2009; Hallegatte *et al.*, 2007). Given the long lifetimes of the building stock, thermal power plants, or transport infrastructure, this is clearly a priority area for early policy considerations (also see Chapters 4: Aaheim *et al.* and 7: Eskeland *et al.*). These considerations are likewise valid for other cross-sectoral linkages between adaptation and mitigation, which have the potential to increase energy demand. For instance, continued high water use in southern Europe will raise energy demand for pumping, desalinisation, recycling and water transfer as precipitation is predicted to decrease in the future (see Chapter 9, Werners *et al.*).

The impacts of climate change on the energy system and the linkages to mitigation require integrated adaptation–mitigation policy responses that take account of the various stakeholders involved and address the cross-sectoral nature of the problem (Tàbara, 2009). There may also be unintended co-effects that increase or decrease vulnerability to climate change, for example the intermittency of renewables and peak summer day electricity demand (Jochem *et al.*, 2009). Another issue may be the risk of enhanced weather extremes on long-term energy infrastructure investments (e.g. power plants, high voltage transmission lines) and associated adaptation responses, though the lack of knowledge and the uncertainty in prediction of such events presently challenges in-depth exploration.

How much adaptation will be implemented in the European energy system will also depend on the present and forthcoming policy efforts in mitigation. The longer governments of the industrialised and emerging countries postpone mitigation policies to curb global greenhouse gas emissions, the more European policy makers and businesses will tend to invest in adaptation. There is a possibility that adaptation strategies will gain dominance over mitigation actions as they can be more easily implemented at the national level, particularly in industrialised countries such as in Europe (Eskeland *et al.*, 2008).

1.3.2 Adaptation and mitigation 'fit' in our towns and cities

This section examines adaptation and mitigation activities in towns and cities, providing a valuable area of investigation for three main reasons. Firstly, urban systems are a major source of greenhouse gas emissions, particularly carbon dioxide. Secondly, they are the places where most people live and work, hence adaptation of the form and functioning of cities for future conditions is vitally important (GLA, 2008; McEvoy, 2007). Thirdly, it allows analysis to move beyond individual sectoral 'silos' and to consider some of the more complex cross-sectoral implications of responses which seek to address the causes and potential impacts of climate change.

Decarbonisation of cities typically revolves around two predominant activities: reducing the carbon intensity of energy and, once supplied, improving the efficiency of its use. Reducing carbon intensity involves switching fuels to less carbon intensive options, with the most obvious examples being renewable technologies such as solar panels, wind turbines and bio-fuels. Improvements to efficiency tend to be tackled according to the type of end-user group, typically broken down as domestic, commercial, industrial and transport. Much mitigation action can therefore be seen to target either technological or behavioural solutions, involving a limited number of easily identifiable key actors. The resulting emissions are also influenced by historical planning and design. The desire for less carbon-intensive cities in the future is expressed through the advocacy of high density, mixed used development (also known as the 'compact' city) and its subsequent translation into land use policy across Europe (Williams, 2000).

However, whilst densification is central to the urban renaissance agenda, concern has been voiced that consolidation can also directly contradict the adaptation agenda as well as jeopardise wider sustainable development objectives (McEvoy *et al.*, 2006). At the conurbation scale, any increase in the density of urban areas not only intensifies the urban heat island with implications for human health and well-being (as evidenced by the 2003 heatwave in western Europe), but it can also result in problems for urban drainage by reducing infiltration capacity (Gill *et al.*, 2007). Adaptation needs, however, require greater consideration of ecological principles at the city scale in order to adequately plan for climate-related impacts of flooding, drought and heat (in the language of policy-makers 'climate-proofing our towns and cities'). These concerns have also begun to have greater prominence in the public domain. Increasing media attention highlights key issues such as the loss of gardens to urban development pressures, the consequences of this in the face of a changing climate, and the emergence of 'bottom-up' initiatives that are attempting to address this emerging agenda[6].

The relationship between mitigation and adaptation seems to be predominantly one of conflict and trade-offs at the level of strategic city-scale planning. This ultimately requires balance between promoting mitigation attributes and the need to retain ecological functioning that actively moderates the adverse impacts of climate change. There would appear to be more opportunities for synergy between the two, however, at the neighbourhood and individual building scales. Indeed, the design of localities is particularly important for attempting to address the resilience of the built environment to a changing climate, whilst ensuring greater energy efficiency of the building stock[7]. Key factors to be considered with respect to adaptation include the height, location and layout of buildings, the material and albedo used, cooling and ventilation, landscape architecture and provision of out-door spaces (Three Regions Climate Change Group, 2005). All these options can potentially influence local energy demand and it is therefore critical that the synergies that exist are fully exploited, and that maladaptation (including conflict with mitigation objectives) is avoided. An important element of well-informed urban design will be to ensure that the latest technical knowledge is accessible to all the actors involved, in a format suitable for the particular end-user (e.g. best

---

[6] For example, the 'Garden for a Living London' initiative by the London Wildlife Trust involves seven pledges which are specifically designed to address climate change impacts in the urban environment (profiled by the BBC in July 2008).
[7] New urban development opportunities need to be fully exploited; though for many large cities in Europe retro-fitting activity will be a further critical component of the adaptation agenda.

practice guidelines). Where possible, information overload and conflicting advice from different sources should be avoided.

Whilst it is possible to identify synergistic measures at the local scale, stakeholder interviews carried out for the ADAM project provided evidence that, in many cases, significant barriers remain to exploiting this potential (McEvoy *et al.*, 2008; also see Chapter 5, Hinkel *et al.*). These barriers are both technical and institutional in nature. It was highlighted that gaps remain between theory and practice, and that greater 'real world' monitoring is needed to assess how building design and technologies actually perform over the longer term. In other words there is a perceived need for a better understanding of performance before wide-scale imple- mentation. Buildings also continue to be built to standards that reflect historic weather data, with little or no consideration of future climate change. This is further reinforced by the institutional context, including a current political emphasis on mitigation. For example, building regulations make limited reference to the adapta- tion agenda. This suggests that new housing may be unsuitable for future condi- tions. For instance, there is current emphasis on insulation and keeping buildings warmer in winter with little provision for cooling in summer. There is also overuse of glass in new public buildings such as schools. The promotion of mitigation measures in many countries across Europe may therefore actually have a negative impact on adaptation objectives.

To conclude, evidence suggests that, whilst mitigation and adaptation measures are often in conflict at the scale of the conurbation, there are more opportunities for technical and design complementarity at the neighbourhood and building scales. Whilst the mitigation agenda is further advanced and more sharply defined than adaptation, there is a new impetus to develop adaptation strategies and interven- tions, particularly in promoting it as a learning process. Adaptation processes may often be more complex than mitigation processes, with actors coming from a much wider variety of sectors and spatial scales that are sensitive to the impacts of climate change. It is therefore crucial that emerging opportunities for knowledge transfer, training and peer-to-peer learning are not only applied to adaptation but also to mitigation, if we are to move towards truly sustainable cities.

1.3.3 Present and future EU climate governance challenges

The need to co-ordinate mitigation policy efforts at European level is relatively well established. As the Kyoto compliance deadline approached, the Member States of the EU agreed to a range of 'common and co-ordinated policies and measures' (European Commission, 1999). Through this, their joint commitments could be reached in the most cost-effective way, causing least distortion of the European single market. The most significant instrument to achieve these goals has been the

EU emissions trading scheme. With warnings of catastrophic climatic changes growing ever louder, in 2008 the EU adopted a package of climate and energy measures aimed at reducing emissions by 20 per cent from 1990 levels by 2020, boosting the use of renewables, limiting emissions from new cars and encouraging new carbon capture and storage technology (European Commission, 2008). Such steps have been possible, despite the EU lacking competence in key policy areas such as energy and fiscal policy, because Member States have recognised the imperative of common and co-ordinated action and accepted elaborate systems of burden sharing (see Chapter 6, Berkhout *et al.*).

The situation in the adaptation domain, however, is somewhat different. Here, no international treaty deadlines have provided an equivalent impetus to policy development. In the face of considerable uncertainties surrounding the spatial and temporal incidence of future climate impacts, as well as the diversity of regions and national institutional arrangements, a 'one size fits all' approach is regarded as neither necessary nor desirable. The most appropriate level for decision making is therefore widely held to be local (Klein *et al.* 2007; see also Chapters 4 Aaheim *et al.* and 5 Hinkel *et al.*). However, local adaptation decisions are significantly enabled and constrained by institutional arrangements at higher levels, be they national or supranational. Moreover, key policies in a range of sectors significantly at risk from climate change are made at EU level (see also Aaheim *et al.*, 2008). This implies that EU-level institutions may be able to contribute to a framework conducive to adaptation in a range of policy areas. The extent to which an EU adaptation policy role is justified and feasible, however, is complicated by ongoing, well-known tensions between the Commission and Member States in relation to 'task allocation' and 'subsidiarity' issues (Rayner and Jordan, 2010).

Policy making for the EU climate change adaptation 'domain' therefore remains at a very early stage. The Commission only seriously began developing a specific EU adaptation policy response in 2005, with the founding of a dedicated working group of the ECCP II, leading to a Green Paper (European Commission, 2007) and, eventually, a White Paper (European Commission, 2009). This somewhat late response (when compared to mitigation) may be explained by a number of factors (Jordan *et al.*, 2010). Firstly, there is concern that action on adaptation could 'dilute the message' that early and forceful mitigation is necessary. Secondly, there has been a degree of nervousness that adapting to climate change might involve an unpicking of hard-won gains in such areas as biodiversity conservation. Thirdly, opposition to 'competence-creep' into areas traditionally the preserve of national decision making, such as land-use planning, has also acted as a constraint, limiting activity to the so-called open method of co-ordination, under which EU institutions provide a loose framework rather than dictate policy to Member States. This reflects the fact that when it comes to adaptation, Member States are much less reliant on one another to achieve policy results than they are in relation to mitigation. Fourthly, even in policy areas where the EU has a well-established role, and Member States have backed calls to 'mainstream' adaptation concerns, inertia on the part of the better-established Commission services such as the Directorate General for Agriculture may have kept the degree of policy ambition in check (Rankin, 2009).

Among the enabling factors bringing adaptation onto the policy agenda have been increasingly firm scientific evidence of climate change, and a number of high profile catalysing or 'focusing events' such as the 2002 floods and the 2003 heatwave. As a number of pioneering Member States pressed ahead with national adaptation policies, the Commission sensed that an EU-wide response was called for. Although future impacts are likely to be felt differentially (and hence trigger different national and regional responses), adaptation has important cross-national implications that imply a role for some kind of supranational body. With the EU having competence in so many areas that stand to be affected by climate change (including, for example, agriculture and water), and investing in regional development that might be vulnerable to future climate change, it was logical that the Commission should at least consider some role.

It is striking how policy development in the adaptation and mitigation domains have remained largely separate[8]. In future, the EU will be faced with the task of consolidating a more integrated portfolio of climate policies, able to exploit synergies between adaptation and mitigation action (or at least minimise conflicts), and reconcile them with traditional concerns over subsidiarity and global competitiveness. What form this integration takes, and to what extent solidarity between the most and least negatively affected will be possible given prevailing financial constraints, remains to be seen. How far action can be co-ordinated by the EU depends not simply on the nature of the problems arising, but the institutional capacity available to the EU. In the coming decades, this will be determined to a large degree on how far Member States are willing to cede competence to act in areas affected by climate impacts or which influence vulnerability. In a more integrated EU, it is conceivable that the Commission could take a greater degree of responsibility for preserving critical European capital, be that natural/environmental, economic, cultural, or human, against climate change impacts. For example, where habitats are judged significant at a pan-European scale, Member States would be obliged to act to protect them.

Among the possible institutional innovations envisaged, a Commissioner could be given responsibility for climate policy, increasing the likelihood that the Commission integrates climate concerns into key areas of competence, including agriculture, water and regional funding (Acclimatise, 2007). New forms of policy

---

[8] For example, see http://ec.europa.eu/environment/climat/adaptation/index_en.htm.

appraisal and evaluation (e.g. see Chapter 2, Russel *et al.*) to investigate whether particular projects and programmes financed by the Commission deliver greater resilience to climate impacts, or increase vulnerability, could emerge.

The EU is likely to take on a greater role in adaptation policy, even in the context of its existing competencies. The traditional rationales for EU action could be invoked to justify policy development in a range of new areas, as more of the functioning of the internal market comes to be threatened by climate impacts and perhaps by autonomous (mal)adaptation that occurs. Furthermore, there is a sense in which policies of the EU already impinge on key land-use planning issues, and steps are needed to co-ordinate a currently fragmented set of instruments, which often pull in conflicting directions (EEA, 2007).

1.4 Exploring the linkages of European adaptation and mitigation policies

A large number of factors could influence the inter-relationship between adaptation and mitigation. As the consideration of adaptation–mitigation linkages is a new policy area however, there is, as yet, little evidence to provide a basis for these relationships. Nevertheless, it is possible to raise a number of questions that could be explored to investigate these linkages. These questions are likely to vary with the policy level, geographical scale, climatic zone, sector and institutional framework, but at least include the following:

(i) How do measures in the adaptation domain influence measures in the mitigation domain (and vice versa)? In what areas and under what conditions do they interfere with each other, exclude each other or strengthen each other? How strong are these influences?

(ii) How do measures taken in either the adaptation or mitigation domain feed back into capacity for the other domain? Do mitigation measures positively or negatively affect the capacity for adaptation (and vice versa)?

(iii) Could an increase in mitigative capacity also lead to an increase in adaptive capacity (and vice versa), or are the two capacities independent or even mutually exclusive?

As a first step towards addressing these questions, we turn to a meta-analysis of self-reported climate policy activities that was compiled to study past and current EU climate policies in six Member States: Finland, Germany, Italy, Poland, Portugal and the United Kingdom (Haug *et al.*, 2007; see also Chapter 6, Berkhout *et al.*). It covers a broad representation of the socio-economic, political and geographical diversity of the EU. The data collected included UN Framework Convention on Climate Change National Communications as well as national government reports. The meta-analysis structures the policy activities in terms of four major criteria:

domain of activity (adaptation or mitigation), year of introduction, targeted sector (e.g. energy, waste) and, where possible, type of policy instrument (e.g. regulatory, tax). The aggregated data for the six countries is by no means sufficient to answer the above three questions, and we should certainly not confound the rate at which new policies are introduced with their substantive content. The analysis does, however, provide insights as to how the two domains relate to each other, and whether there are synergies or trade-offs between adaptation and mitigation.

1.4.1 The profile of adaptation and mitigation activities over time

Figure 1.4 suggests that mitigation policies have followed a saturation curve with a large increase in number until about 2000, after which the number stabilises. Conversely, adaptation measures and activities rise exponentially. In some sense this shows the changing relative position of mitigation and adaptation as policy priorities: while the early focus was on mitigation, in recent years adaptation has emerged as a major policy priority. After the 2001 IPCC Third Assessment Report (IPCC, 2001) acknowledged that no amount of mitigation could entirely prevent climate change, adaptation entered the international policy agenda in its own right. This has led to a rapid increase in the number of adaptation measures, particularly over the period 2004–2006.

Since the start and end point of the dataset are random, and drivers of policy development may change, it is not currently possible to draw a conclusion to the

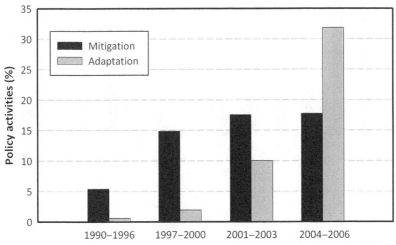

Figure 1.4. Time series analysis of the proportion of adaptation and mitigation activities introduced between 1990 and 2006. 100% represents the sum of adaptation and mitigation activities over the entire period of time. (Source: based on Haug *et al.*, 2007.)

question of how measures in one domain affect those in the other. However, the distribution does allow us to reject the possibility of a zero-sum relationship in which the rise in one domain detrimentally affects the development of another. The question of whether there is an influence of competing priorities also cannot be answered with this dataset as the possible development of mitigation policies over time in the absence of adaptation is unknown. Finally, the question of whether the recent interest in adaptation policies will be sustained at a high level, and thus reflect the public domain's growing interest in climate change, remains to be seen. The future will also tell how the number of newly introduced mitigation measures will develop over time. While a decline could be interpreted as an admission of failure in the mitigation domain, a constant level or even an increase of new measures could reflect the understanding that mitigation policies are perceived as effective.

1.4.2 The relative share of adaptation and mitigation activities by sector

Figure 1.5 shows that the variation of policy activity has a strong sectoral bias. Unsurprisingly, for the sectors that are major emitters, policies have been mitigation focused. Sectors such as energy, transport, waste and manufacturing have had, as of 2006, very little adaptation focus. Conversely, sectors such as land use/forestry/water

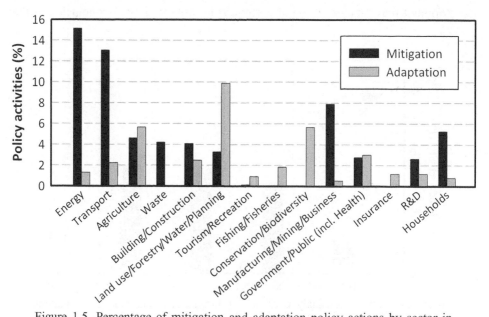

Figure 1.5. Percentage of mitigation and adaptation policy actions by sector in selected European countries in 2006. 100% represents the sum of adaptation and mitigation activities in all sectors. (Source: based on Haug *et al.*, 2007.)

and conservation/biodiversity, tourism and insurance have had proportionately more attention in the field of adaptation. This is again not surprising, as these sectors tend to be more susceptible to climate impacts. Agriculture and building/construction have an emphasis on both mitigation and adaptation, because both sectors are very climate sensitive, but are also potential contributors to emission reductions. The balance between mitigation and adaptation policies in the public sector is probably due to its very diverse range of activities. Interestingly, research and development is still predominantly focusing on mitigation, although the absolute levels of funding would be more reliable indicators for preference.

Again, there is no evidence to sustain the possibility that within a given sector, activities of one domain might interfere with the other. However, the distribution could reflect a labelling of measures according to their primary goal. This might mean that mitigation measures that have adaptation co-benefits are not yet fully categorised in both areas, and that currently, policies that aim for synergies between the two domains are not being highly prioritised. This would be particularly important in those sectors that have both direct mitigation opportunities, and are likely to be affected by climate change (e.g. energy, agriculture, households and, in the future, potentially water).

1.4.3 The instruments used in adaptation and mitigation activities

Figure 1.6 shows the distribution of mitigation and adaptation activities by type of instrument, indicating that mitigation policies follow a broad spectrum of measures. Conversely, for adaptation, most activities are mere recommendations. It will be

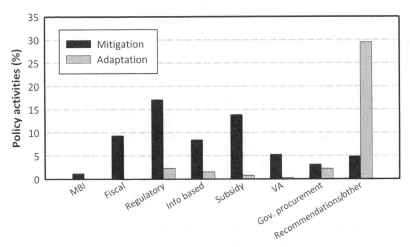

Figure 1.6. Distribution of mitigation and adaptation activities by instruments. MBI = Market-based instruments; VA = Voluntary agreements. 100% represents the sum of adaptation and mitigation activities. (Source: based on Haug *et al.*, 2007.)

interesting to identify in the future whether these recommendations have any effect in the respective sectors to which they apply. It will likewise be worthy to study whether future adaptation measures become as diversified and formally regulated as are current mitigation measures.

The distribution of measures across sectors indicates that adaptation activities are currently at a much lower institutional and organisational level of development than mitigation activities. This suggests that adaptation is at an earlier stage in the policy cycle and that the existence of mitigative capacity does not necessarily lead to adaptive capacity. It appears that adaptation policies may well need to be developed independently. In the sectors where adaptation plays an important role, it may be that the policies will follow a similar temporal and institutional development to that of mitigation policies a decade ago.

1.5 Discussion and policy recommendations

The previous sections summarise the IPCC perspective on the inter-relationship between adaptation and mitigation, and current thinking from three different perspectives of assessing adaptation and mitigation: modelling, social learning and policy analysis. Through three examples from ADAM research, we offered insights into the ways in which climate change is likely to introduce new challenges and opportunities. How climate policies have responded so far provides evidence of opportunities for improved resource and economic management, as well as synergies and trade-offs between adaptation and mitigation. Finally, the meta-analysis of EU climate policy measures demonstrates the inter-relationship between adaptation and mitigation by policy domain, year of introduction, sector and policy instrument. In this section we now reflect on that body of evidence in the context of synergies and trade-offs of adaptation and mitigation policies and possible governance dilemmas.

While the three case studies and the meta-analysis provide valuable lessons individually, they also provide policy messages when considered together. The energy and building cases focus on the potential impact/adaptation[9] of energy and cooling demand, but they come from two different sectors and therefore approach the issue differently. The focus from the energy sector (supply side) is largely decarbonisation through technical solutions, whereas the buildings sector (demand side) advances energy efficiency and a mix of technical and behavioural approaches. In practice, both need to be considered together to advance synergistic adaptation–mitigation policy. The integration of such cross-sectoral policy is challenging, but this is an important policy message: specific policy will be needed to

---

[9] Note that the increased cooling demand can be defined either as an impact or an adaptation.

advance adaptation–mitigation linkages, and this involves overcoming technologi-cal, institutional and political barriers to such holistic thinking.

The cases also highlight an important relationship between autonomous and planned adaptation. The energy case study assesses a largely autonomous response to increased exposure to cooling demand, met through air conditioning. This is unlikely to be the optimal societal response, increasing greenhouse gas emissions and conflicting with mitigation policy. Therefore, a better co-ordinated EU policy could provide frameworks for reducing maladaptation by exploiting the links between mitigation and adaptation. As the urban planning case study reports, there are alternatives to air conditioning which are complementary to mitigation objectives, for example passive ventilation, building design, insulation and plan-ning, but these require planned intervention. They also involve greater barriers to implementation, such as earlier pro-active policy, higher up front capital costs, institutional barriers and behavioural change. This example suggests that planned adaptation in the public realm is key to synergistic adaptation–mitigation policy, and that autonomous adaptation by private actors alone is less likely to exploit such linkages. Further work to explore whether this also applies to other sectors is a priority.

The barriers against the implementation of adaptation and mitigation measures can be very persistent when they are deeply rooted in society, for example beliefs, values or cultural identities (Hulme, 2009). The aversion against wind turbines 'polluting' the landscape is an example for which there is no simple solution and the answers in favour or against will depend on a number of cultural factors (Lorenzoni *et al.*, 2007). Shifting beliefs, values and perceptions will take time, even under favourable political conditions (Beddoe *et al.*, 2009). Whether overcoming these kinds of barriers can be accelerated in a top-down way is an important question that requires further exploration.

Next to the potential synergies reported above, there are significant potential trade-offs between adaptation and mitigation. As far as the presented evidence is concerned, these trade-offs are often not direct (in the sense that increased air conditioning raises energy demand). Instead, they involve different temporal and spatial scales. The energy case study, for instance, illustrates a potential temporal trade-off for adaptation–mitigation actions. Mitigation policy is already in place, with 2020 targets. However, the study finds no significant climate vulnerability for the energy system within this time frame, only occurring in later time periods. This raises a challenge for adaptation–mitigation linkages and policy, as it is difficult to build adaptation into short-term mitigation policy when adaptation benefits only arise in the future, towards the end of the economic lifetime of current plants. However, failure to address this temporal mis-match will lead to lock-in and could increase longer-term vulnerability.

Another indirect potential trade-off between adaptation and mitigation can best be illustrated by the differential impacts that climate change will have on Nordic and Mediterranean countries. In the North, climate change will be beneficial for renewable energy generation through increased runoff and higher biomass potentials and countries will experience lower heating demand. However, the contrary is the case for the South, where higher temperatures and greater drought will increase the need for space cooling and reduce the hydropower potential. These changes will lead to shifts in welfare distribution between the societies in the North and the South if no policy measures are put in place to provide a balance. This will require co-operation at the European level and solidarity between Member States. Given mounting evidence of climate change, it is likely that greater integration between mitigation and adaptation will be sought to exploit synergies and possibly also to address issues of subsidiarity, regional development and intra-European solidarity. However, there are a number of constraints that may impede EU policy development, most importantly the tensions between EU and Member State competencies, the principle of subsidiarity and the explicit role for co-ordinated EU action (see Chapter 6, Berkhout *et al.*).

From the meta-analysis, we have seen that the rise in adaptation measures and activities over the past years is not reflected by an equivalent diversity of instruments as in the mitigation domain. This indicates that the mitigative capacity that has built up over the past years cannot be directly transferred to the adaptation domain largely because different sectors, actors and institutions are involved. Where future policies evolve around sectors with important adaptation–mitigation linkages it will therefore be important to develop more integrative policies that take account of both domains to minimise or manage the kind of temporal and spatial trade-offs described above.

1.6 Conclusions

The above sections allow the following conclusions to be drawn:

(i) Mitigation efforts today may lead to climate vulnerabilities in the future if the life cycles specific to each sector are not adequately taken into account;

(ii) Development of response capacity in one domain does not necessarily lead to capacity in the other because adaptation and mitigation often involve different sectors, actors and institutions;

(iii) Climate impacts may lead to growing welfare inequalities, which can be balanced through co-ordinated policies at higher levels. The policies will, however, need to overcome existing institutional barriers;

(iv) Synergies between adaptation and mitigation are most easily found where mitigation efforts are reinforced by behavioural changes. This can lead to an overall increase in resilience by implementing broad concepts of sustainability;

(v) Finally, further consideration of these adaptation – mitigation linkages is a research priority. Policy innovation will be needed to capture synergistic benefits and avoid the introduction of new climate vulnerabilities or accelerate emissions of greenhouse gases.

This chapter has introduced some of the work conducted within the ADAM project and which is reported in greater detail in the following chapters. It has examined aspects of the nature of adaptation and mitigation in the context of climate change in the EU, in particular examining inter-linkages, trade-offs and synergies. It is complementary to the final chapter in this book (Chapter 14, Patt *et al.*) which also examines the relationships between adaptation and mitigation. Chapter 14 builds on some of the relationships between adaptation and mitigation introduced in this chapter, and further exploits some of the work summarised in the other chapters, to offer five different perspectives on how we may structure and envisage the climate change problem at a global strategic level.

References

Acclimatise (2007) *What Policies Present Barriers to Adaptation in the UK and the Netherlands?* ESPACE Project Final report. Southwell: Acclimatise.

Adger, W. N. (2006) Vulnerability. Special Issue on Vulnerability, Resilience and Adaptation, *Global Environmental Change,* **16**(3), 268–81.

Aebischer, B., Jakob, M., Catenazzi, G., and Henderson, G. (2007) *Impact of Climate Change on Thermal Comfort, Heating and Cooling Energy Demand in Europe.* Colle sur Loup, France: Proceedings ECEEE Summer Study, June 2007.

Aaheim, A., Berkhout, F., McEvoy, D., *et al.* (2008) *Adaptation to Climate Change: Why is it Needed and How can it be Implemented?* CEPS Policy Brief No 161, Brussels: Centre for European Policy Studies.

Beddoe, R., Costanza, R., Farley, J. *et al.* (2009) Overcoming systemic roadblocks to sustainability: The evolutionary redesign of worldviews, institutions, and technologies. *Proceedings of the National Academy of Sciences,* **106**, 2483–9.

Bouwman, L., Kram, T. and Klein-Goldewijk, K. (2006) *Integrated Modelling of Global Environmental Change. An Overview of IMAGE 2.4.* Bilthoven: Netherlands Environmental Assessment Agency.

Chapman, J. (2002) *System Failure: Why Governments Must Learn to Think Differently.* London, UK: Demos.

Cuppen, E., Hisschemoller, M., Dunn, B., Midden, C. and de Kerkof, M. (2006) *Evaluating the Quality of Methods to Facilitate Participatory Assessment.* Edinburgh, UK: Participatory Approaches in Science and Technology Conference, June 2006.

Darwin, J., Johnson, P. and McAuley, J. (2001) *Developing Strategies for Change.* FT Prentice Hall.

de Bruin, K. C., Dellink, R. B. and Tol, R. S. J. (2007) *AD-DICE: An Implementation of Adaptation in the DICE Model.* FEEM Nota di Lavoro Series 51.2007.

EEA (2007) *Land-use Scenarios for Europe: qualitative and quantitative analysis on a European scale.* EEA Technical Report No 9/2007. Copenhagen: European Environment Agency.

Eskeland, G., Jochem, E., Neufeldt, H., *et al.* (2008) *The Future of European Electricity: Choices Before 2020.* CEPS Policy Brief No 164, Brussels: Centre for European Policy Studies.

European Commission (1999) *Preparing for the Implementation of the Kyoto Protocol.*
    COM (99) 230 final. Brussels: Commission of the European Communities.
European Commission (2005) *Winning the Battle Against Global Climate Change.* COM
    (2005) 35. Brussels: Commission of the European Communities.
European Commission (2007) *Green Paper: Adapting to Climate Change in Europe –
    Options for EU Action.* COM (2007) 354 final. Brussels: Commission of the European
    Communities.
European Commission (2008) *20 20 by 2020 – Europe's Climate Change Opportunity.*
    COM (2008) 30 final. Communication from the Commission to the European
    Parliament, the Council, the European Economic and Social Committee and the
    Committee of the Regions.
European Commission (2009) *White Paper. Adapting to climate change – Towards a
    European framework for action.* COM (2009)147/4. Brussels: Commission of the
    European Communities.
Fisher, B. S., Nakicenovic, N., Alfsen, K. *et al.* (2007) Issues related to mitigation in the long
    term context. In *Climate Change 2007: Mitigation. Contribution of Working Group III
    to the Fourth Assessment Report of the Inter-governmental Panel on Climate Change,*
    ed. B. Metz, O. R. Davidson, P. R. Bosch, R. Dave, L. A. Meyer. Cambridge, UK:
    Cambridge University Press, pp. 169–250.
Forrester, J., Gerger Swartling, Å. and Lonsdale, K. (2008) *Stakeholder Engagement and
    the Work of SEI: An Empirical Study.* Stockholm.
Funtowicz, S. O. and Ravetz, J. R. (1991) A new scientific methodology for global
    environmental issues. In *Ecological Economics,* ed. R. Constanza. New York:
    Columbia University Press, pp. 137–52.
Gallopín, G. (1999), Generating, sharing and using science to improve and integrate policy.
    International Journal on Sustainable Development, **2**(3), 397–410.
Gill, S., Handley, J. F., Ennos, A. R. and Pauleit, S. (2007) Adapting cities for climate
    change: the role of the green infrastructure. *Built Environment,* **33**, No 1.
GLA (2008) *The London Climate Change Adaptation Strategy*: Available at: http://www.
    london.gov.uk/mayor/publications/2008/docs/climate-change-adapt-strat.pdf
    (accessed: 6th January, 2009).
Hallegatte, S., Hourcade, J. C. and Ambrosi, P. (2007) Using climate analogues for
    assessing climate change economic impacts in urban areas. *Climatic Change,*
    82, 47–60.
Haug, C., Rayner, T., Huitema, D. *et al.* (2007) *How Effective are European Climate
    Policies? A Meta-analysis of Recent Policy Evaluations.* ADAM project report D-P2.3,
    Amsterdam.
Hope, C. (2006) The marginal impact of CO_2 from PAGE2002: An integrated assessment
    model incorporating the IPCC's five reasons for concern. *Integrated Assessment,*
    6, 19–56.
Horrocks, L., Mayhew, J., Watkiss, P., Hunt, A. and Downing, T. (2006) *Objective Setting
    for Climate Change Adaptation Policy.* London, UK: DEFRA.
Hulme, M. (2009) *Why We Disagree about Climate Change: Understanding Controversy,
    Inaction and Opportunity.* Cambridge, UK: Cambridge University Press.
IPCC (2001) *Climate Change 2001: Synthesis Report. Contribution of Working Groups I,
    II and III to the Third Assessment Report of the International Panel on Climate
    Change.* Geneva.
IPCC (2007a) *Climate Change 2007: Impacts, Adaptation and Vulnerability. Contribution
    of Working Group II to the Fourth Assessment Report of the Intergovernmental Panel*

on Climate Change, ed. M. L. Parry, O. F. Canziani, J. P. Palutikof, P. J. van der Linden and C. E. Hanson. Cambridge, UK: Cambridge University Press.

IPCC (2007b) *Climate Change 2007: Mitigation. Contribution of Working Group III to the Fourth Assessment Report of the Intergovernmental Panel on Climate Change.* ed. B. Metz, O. R. Davidson, P. R. Bosch, R. Dave and L. A. Meyer. Cambridge, UK: Cambridge University Press.

Jänicke, M. (1997) The political system's capacity for environmental policy. In *National Environmental Policies. A Comparative Study of Capacity Building.* Berlin: Springer, pp. 1–24.

Jochem, E., Sathaye, J., Bouille D., ed. (2000): *Society, Behaviour, and Climate Change Mitigation.* Dordrecht, The Netherlands: Kluwer Academic Publishers.

Jochem E., Barker T., Scrieciu S. *et al.* (2009). Report of the Reference (4 °C) Scenario for Europe. ADAM report D-M1.2 Karlsruhe: Fraunhofer ISI.

Jordan, A., D. Huitema, H. van Asselt, F. Berkhout and T. Rayner, eds. (2010) *Climate Change Policy in the European Union: Confronting the Dilemmas of Mitigation and Adaptation.* Cambridge, UK: Cambridge University Press.

Kemp, R. and Weehuiszen, R. (2005) *Policy Learning: what does it mean and how can we study it?* Report on project innovations in the public sector. Oslo: NIFU STEP.

Klein, R. J. T., Huq, S., Denton, F. *et al.* (2007) Inter-relationships between adaptation and mitigation. *Climate Change 2007: Impacts, Adaptation and Vulnerability. Contribution of Working Group II to the Fourth Assessment Report of the Intergovernmental Panel on Climate Change,* ed. M. L. Parry, O. F. Canziani, J. P. Palutikof, P. J. van der Linden and C. E. Hanson. Cambridge, UK: Cambridge University Press, pp. 745–77.

Lorenzoni, I. Nicholsen-Cole, S. and Whitmarsh, L. (2007) Barriers perceived to engaging with climate change among the UK public and their policy implications. *Global Environmental Change,* **17,** 445–59.

McEvoy, D. (2007) Climate change and cities. *Built Environment,* **33**(1), 5–9.

McEvoy, D., Lindley, S. and Handley, J. (2006) Adaptation and mitigation in urban areas: synergies and conflicts. *Municipal Engineer,* **159**(4), 185–191.

McEvoy D., Lonsdale, K. and Matczak, P. (2008) *Adaptation and Mainstreaming of EU Climate Change Policy: An Actor-Based Perspective.* CEPS Policy Brief No 149, Brussels: Centre for European Policy Studies.

Nordhaus, W. D. (2007) *The Challenge of Global Warming: Economic Models and Environmental Policy.* New Haven, Connecticut: Yale University.

Patt, A., van Vuuren, D. P., Berkhout, F. *et al.* (2009), Adaptation in integrated assessment modeling: where do we stand? *Climatic Change* (in press).

Rankin, J. (2009) 'Climate plans provoke backlash in Commission', *European Voice* 5th March.

Rayner, T. and Jordan, A. (2010) Adapting to a Changing Climate: An Emerging European Union Policy?, In *Climate Change Policy in the European Union: Confronting the Dilemmas of Mitigation and Adaptation,* ed. A. Jordan, D. Huitema, H. van Asselt, T. Rayner and F. Berkhout. Cambridge, UK: Cambridge University Press.

Riahi, K., Gruebler, A. and Nakicenovic, N. (2007) Scenarios of long-term socio-economic and environmental development under climate stabilization. *Technological Forecasting and Social Change,* **74,** 887–935.

Rittel, H. W. and Weber, M. M. (1973) Dilemmas in a general theory of planning, *Policy Sciences,* **4**: 155–69.

Sathaye, J., Najam, A., Cocklin, C. (2007) Sustainable development and mitigation. *Climate Change 2007: Mitigation. Contribution of Working Group III to the Fourth Assessment Report of the Intergovernmental Panel on Climate Change*, ed. B. Metz, O. Davidson, P. Bosch, R. Dave and L. Meyer. Cambridge, UK: Cambridge University Press. pp. 691–743.

Senge, P. (1990), *The Fifth Discipline: The Art and Practice of the Learning Organisation*. London, UK: Century.

Slovic, P. (2000), *The Perception of Risk*, London, UK: Earthscan.

Snowden, D. J. (2005), Multi-ontology sense making: a new simplicity in decision making. *Informatics in Primary Care*, **13**, 45–54.

Stern, N. (2006) *The Economics of Climate Change*. Cambridge, UK: Cambridge University Press.

Tàbara, J. D. (2009) Integrated Climate Governance (ICG) and sustainable development. In *Sustainable Development. A Challenge for European Research*. Brussels: European Commission.

Tàbara, J. D. and Pahl-Wostl, C. (2007) Sustainability learning in natural resource use and management. *Ecology and Society* **12**, 3.

Three Regions Climate Change Group (2005) *Adapting to Climate Change: A Checklist for Development*. London, UK: Greater London Authority.

Tol, R. S. J. (2006) Multi-gas emission reduction for climate change policy: an application of FUND. *Energy Journal Special issue* **3**, 235–250.

Tompkins, E. and Adger, N. (2005) Defining a response capacity for climate change. *Environmental Science Policy*, **8**, 562–71.

van Vuuren, D. P., Meinshausen, M., Plattner, G.-K. *et al.* (2008) Temperature increase of 21st century mitigation scenarios. *Proceedings of the National Academy of Sciences*, **105**, 15258–62.

Warren, R., Hope, C., Mastrandrea, M., *et al.* (2006) *Spotlighting impacts functions in integrated assessment*. Tyndall Centre for Climate Change Research Working Paper 91.

Weidner, H. (2002) Capacity building for ecological modernization: lessons from cross-national research, *American Behavioral Scientist*, **45**, 1340–1368.

Williams K. (2000), Does intensifying cities make them more sustainable? In *Achieving Sustainable Urban Form*. ed. K. Williams, E. Burton and M. Jenks. London: E & FN Spon.

Winkler, H., Baumert, K., Blanchard, O., Burch, S. and Robinson, J. (2007) What factors influence mitigative capacity? *Energy Policy*, **35**, 15–28.

2

Climate change appraisal in the EU: current trends and future challenges

Lead authors:
DUNCAN RUSSEL[1], ALEX HAXELTINE, DAVE HUITEMA,
MÅNS NILSSON

[1]Coordinating lead author

Contributing authors:
JOCHEN HINKEL, TIM RAYNER

Summary

Producing effective climate change policy is hampered by scientific uncertainty and by constantly shifting goals as economic, social, political and scientific circumstances change. Effective policy therefore necessitates more reflexive forms of policy appraisal, which assess the suitability of current climate change goals to avoid dangerous climate change. This chapter develops the notion of reflexive appraisal for climate policy and looks for elements of such appraisal in three European jurisdictions, namely: the European Commission, Sweden and the United Kingdom (UK). In addition, an appraisal conducted by the UK Sustainable Development Commission is examined because previous ADAM work suggested it might be reflexive. Overall, appraisal of climate policy in the three political jurisdictions was not very reflexive. Lessons from the UK Sustainable Development Commission's appraisal suggest that reflexivity in policy appraisal can be enhanced. This can be done through institutions and individuals responsible for the appraisal understanding the need for reflexivity; climate policy appraisals being conducted by organisations that sit on the boundary of knowledge and government; and the inclusion of a wide variety of stakeholders, so that a plurality of views are included in the analysis. It is concluded that opportunities for more reflexive appraisal may be enhanced by shocks to wider societal systems such as the 2008 global 'credit crunch'.

2.1 Introduction

Understanding and implementing the policy frameworks Europe needs to meet its climate change adaptation and mitigation goals is not simple. Climate change is a 'wicked issue' (Rittel and Webber, 1973; Rayner and Okereke, 2007) that defies

Making Climate Change Work for Us: European Perspectives on Adaptation and Mitigation Strategies, ed. Mike Hulme and Henry Neufeldt. Published by Cambridge University Press © Cambridge University Press 2010.

easy resolution. Indeed, climate policy is 'knowledge intensive' and involves complex scientific, technical, legal, policy and social issues (Fiorino, 2001: 322). Moreover, it spans many sectors with consequences for a diverse array of actors with widely different perspectives (Huitema *et al.*, 2008). Climate change is also a global and long-term issue raising difficult questions over intra- and inter-generational equity and the ultimate effectiveness of policy, which would not be observed for decades (Rayner and Okereke, 2007).

Developing responses to climate change entails dealing with socio-ecological systems that exhibit, among others, complexity, non-reducibility, spontaneity, variability and a collective quality (Dryzek, 1987: 28–33; Young *et al.*, 2006). Moreover, uncertainty surrounds the mitigation targets needed to avoid 'dangerous' climate change, the future of the international policy regime under which these will be adopted and the technology that will be most effective. Consequently, policy making in the field of climate change is confronted with a three-fold challenge: (i) radical changes in policy and policy goals are needed to address the issue of climate change, (ii) goals, however, are difficult to define due to the uncertainty and ambiguity involved, and (iii) goals need to be continually reassessed in the light of changing scientific, social, economic and political contexts. The challenge is enormous, because political systems often tend to be geared towards 'muddling through' and towards slow and incremental change on the back of existing policy (Lindblom, 1979). Indeed, a meta-analysis of the European climate policy evaluation studies suggests that current European climate policy is not sufficient to meet existing climate goals (Haug *et al.*, 2008). Neither is it flexible enough to respond to changes in the economics, politics, social processes and the understanding of the science surrounding climate change (Haug *et al.*, 2008). Despite difficulties in setting climate change goals, there is a need for quick and fundamental policy development for averting dangerous climate change.

'Learning our way out' of environmental problems has almost become the standard recommendation of the sustainability debate (see Dryzek, 1987; Tàbara and Pahl-Wostl, 2007). A widely advanced (e.g. Guba and Lincoln, 1989; Fischer, 1995; Armitage *et al.,* 2007) and used (European Commission, 2005; BERR, 2007) process to promote such learning is *ex ante* policy appraisal that informs policy development (Owens *et al.,* 2004). In this chapter, we argue that learning through policy appraisal can be used to induce reflexivity by determining the suitability of climate policy goals. We also develop an analytical framework to highlight some of the key characteristics of such an appraisal process.

We take a broad definition of appraisal as 'multiple methods of enquiry and argument to produce and transform policy relevant information that may be utilised in political settings to resolve public problems' (Dunn, 1981 cited in

Fischer 1995: 2). This definition includes the formal (e.g. official impact assessments, evaluations and consultation) and informal (e.g. ad hoc meetings with stakeholders) inputs of information to develop future public policy (i.e. it has an *ex-ante* or forward looking component). It can also include policy reviews by non-departmental and non-state actors. This definition incorporates many common features of policy development. Yet it is distinct from policy making as a whole, which can also exhibit additional processes such as path dependency, political discretion, informal rules, etc.

The next section in this chapter outlines an analytical framework contrasting two ideal-typical appraisals: one in which learning only allows for incremental changes in policy (instrumental appraisal) and one fostering learning that challenges existing policy frameworks, goals and underlying norms (reflexive appraisal). In so doing, we do not seek to develop a new type of appraisal system, but rather examine how appraisal processes – including existing formal appraisal systems in Europe such as impact assessment and sustainability assessment – can be made more reflexive. We then apply the framework to case studies of the appraisal of climate policies in the UK, Sweden and the European Commission in order to examine some of the factors that enabled and hindered reflexivity.

Finally, we give recommendations on how more reflexive policy appraisal may be used to help the Europe meet its climate goals.

2.2 A framework for the evaluation of the appraisal of climate policy

2.2.1 Reflexive vs. instrumental appraisal

Policy appraisal has been practised in the United States for decades. It has also become an increasingly popular tool for developing environmental policy in many European states since the publication of the Brundtland report on sustainable development (WCED, 1987; Jordan and Lenschow, 2008). However, studies on European appraisal systems suggest that their policy impact has been limited (Hertin *et al.*, 2009; Turnpenny *et al.*, 2008), with the primary focus of many appraisals often being on the instruments for the delivery and implementation of pre-determined policy objectives. In such cases, the capacity of the appraisal processes to re-assess policy goals for a complex domain such as climate policy (see above) is limited. For the purposes of this analysis, we label such an approach to policy appraisal as *instrumental*, i.e. one that focuses on the instruments of policy delivery rather than on broader goals.

By contrast, as we argue above, the shifting targets surrounding climate change require that policy is regularly re-appraised in a manner that is process- rather than goal-oriented, i.e. an appraisal process that identifies the goals and works in an iterative manner to reassess goals throughout the process. Given the right conditions and design, we propose that policy appraisal can act as a catalyst for more reflexive policy making to respond to the developing demands of climate change. We call such an approach *reflexive appraisal* as it focuses on whether current policy goals and overriding policy frameworks (e.g. better regulation) are appropriate in the current scientific, social, economic and political contexts. As such, reflexive appraisal can allow for the reframing of climate change goals and policy where necessary. The remainder of this section unpacks some of the possible elements of both instrumental and reflexive appraisal to better understand and contrast the two approaches.

2.2.2 Elements of instrumental and reflexive appraisals

There is a rich literature on learning in public policy and management focusing on the 'relatively enduring alterations of thought or behavioral intentions that result from experience and that are concerned with the attainment (or revision) of public policy' (Sabatier, 1998). For this chapter, though, we focus on the wider concept of social learning among a diverse range of stakeholders because it is increasingly being advocated as an approach to deal with critical environmental problems such as climate change (e.g. Dryzek, 1987; Tábara and Pahl-Wostl, 2007). The concept of social learning has been linked increasingly to the concept of policy appraisal (see Nilsson, 2005; Weaver *et al.*, 2006) and has also been applied to policy contexts more generally (e.g. Mostert *et al.*, 2006; Armitage *et al.*, 2007). Social learning is particularly promoted as a way to address difficulties in research uptake when uncertainty is high (as in the case of climate change) because it involves communication between key stakeholders, including policy makers and knowledge generators (Fiorino, 2001: 325). Moreover, it goes beyond mere consensus building or levelling differences in opinion, requiring genuine deliberative exchange of diverse viewpoints and values, paving the way for taking well-argued decisions.

Within the literature, three critical aspects of learning have been identified, namely: who learns, what is learnt, and to what effect? (Bennett and Howlett, 1992; Van de Kerkhof and Wieczorek, 2005). The literature is fairly imprecise over what exactly is meant by these three aspects (Armitage *et al.*, 2007). However, they provide a useful framework around which we can understand how learning in an appraisal process may lead to more reflexive critiques of policy goals.

Who learns?

Authors such as Hall (1993) largely focus on the lessons individual policy makers draw from their experiences. Others show greater interest in how groups of societal actors learn (e.g. Sabatier, 1998) within or outside formal policy making processes.

For the purposes of this chapter, we are interested in the role that societal actors or stakeholders play to facilitate learning in the appraisal process.

A meta-analysis of 262 *ex-post* European climate policy evaluations in the ADAM project (Huitema *et al.*, 2008; Chapter 6, Berkhout *et al.*) suggests that who conducts and participates in evaluations of climate policy can have a marked impact on the degree of reflexivity present in the analysis. We anticipate that the same will hold true for policy appraisal. An appraisal process only focusing on actors with a direct influence on policy ignores those who may have a legitimate stake in the issue, but are not included in existing decision making processes (Weaver *et al.*, 2006: 249). Studies on appraisal practice demonstrate that where stakeholder input was primarily limited to state and powerful non-state actors, opportunities for challenging policy goals were limited (e.g. see Russel and Turnpenny, 2009).

The impacts of climate change and related adaptation or mitigation policy are spread across society. There is a strong case therefore for the inclusion of a diverse range of actors in climate change policy to 'learn together to manage together' (Mostert *et al.*, 2006). Social learning literature (e.g. Tábara and Pahl-Wostl, 2007) suggests that opportunities for learning to induce more reflexive outcomes are higher in settings where a plurality of actors are included with varying views, values and interests and understandings of the problems at hand. In such settings, stakeholders can collectively discuss problems, potential solutions and the potential impacts of measures taken. This is not to propose that participation in itself will lead to reflexivity or that participation processes are not without problems (e.g. they can be captured by vested interests, they can derail and hold-up policy processes, etc.) (see Owens *et al.*, 2004). However, when participation is managed in the right way, the diversity of opinions and actors can facilitate opportunities for recursive deliberation and the consideration of radically different approaches to climate change.

What is learnt and how is it learnt?

Most writings on social learning distinguish between different types and degrees of learning. In this chapter, we focus on the concept of learning as first developed by Argyris and Schön (1996), namely single and double-loop learning. For them *single-loop learning* is 'when a [policy] mismatch is corrected without changing the underlying values and status quo that govern the behaviours' (Argyris, 2003). Single-loop learning is primarily related to the development of insights into policy options related to the policy problem in a given context (Van de Kerkof and Wieczorek, 2005: 736), i.e. it serves the objective of better goal attainment on the back of past policy and new information (Bennett and Howlett, 1992: 276). It is often incremental in nature, conforms to existing routines and does not challenge the overall terms of given policy goals (Hall, 1993: 279, 280; Lindblom, 1979). Thus it is more concerned with the implementation of existing policy goals and targets. In relation to our two types of policy appraisal, an instrumental appraisal is likely to be characterised by single-loop learning.

By contrast, *double-loop learning* is where a policy mismatch is 'corrected by first changing the underlying values and other features of the status quo' (Argyris, 2003; Argyris and Schön, 1996). Double-loop learning concerns not only problem solutions, but also insights into the problem itself and the context in which it takes place (Van de Kerkof and Wieczorek, 2005: 736). From a policy perspective, this includes insights into the suitability of existing policy goals. It is the type of learning that would be found in a reflexive appraisal. Such learning is more likely to be enhanced in an appraisal process that allows for a larger spread of stakeholder views, interests and values to be considered. Ultimately, it should promote new connections between actors to facilitate the reframing of the policy discourses surrounding climate change (Hall, 1993: 279) especially in relation to the changing economic, social, scientific and policy contexts. While double-loop learning is reflexive, it is not the only element in our definition of a reflexive appraisal (see below).

On a note of caution, while it is clear that anyone who has influence over policy choices can learn (Fiorino, 2001: 323), bombarding them with evidence and new information provides no guarantee that they *will* learn in a manner consistant with either single or double loop learning. Another point to consider is that while double-loop learning might help reframe policy actors' understandings of the suitability of policy goals, it may lead to outcomes that are beneficial to other policy arenas but to the possible detriment of climate change.

To what effect?

For many scholars, the end result of social learning is policy change, i.e. learning is only judged to have occurred where policies have been changed (Bennett and Howlett, 1992). Following Huitema *et al.* (2008), we apply more differentiated criteria. Firstly, policy change can occur for reasons unconnected to learning (e.g. political pressures, changes in government, etc.). Secondly, learning does not have to lead to change, as it might simply lead to better justifications and understandings of existing policy. Indeed, learning through a reflexive climate change policy appraisal might well find that current policy is suitable. For these reasons, our approach is to focus on the processes of the appraisal and the degree of reflexivity in its analysis, rather than whether it leads to policy or behavioural change.

So what does this mean for the appraisal of climate policy? The challenges of mitigation and adaptation span many sectors. Reflexive policy appraisal should therefore act as a conduit to feed the latest climate change considerations and variety of knowledge perspectives into sectoral policy processes (Weaver *et al.*, 2006: 242). To this end, reflexive policy appraisal processes can be geared towards a comprehensive consideration of trade-offs and synergies, knowledge integration and policy integration.

Policy appraisal is widely promoted and used as a tool for policy integration in two different ways (see Jordan and Lenschow, 2008). Firstly, it has been used to integrate cross-cutting issues such as climate change into sectoral policies (so-called single issue appraisals). Secondly, policy appraisal has itself become an integrated process where several cross-cutting issues are analysed in one process (so-called integrated appraisal – see Russel and Turnpenny (2009)). With both approaches, integration has primarily been state-focused with appraisal being one of set of approaches to improve knowledge exchange and co-ordination between ministries. To date, both approaches have been poor at integrating issues such as sustainable development (including climate change) into sectoral policy processes (see Jordan and Lenschow, 2008; Turnpenny *et al.*, 2008; Russel and Turnpenny, 2009). Moreover, synergies and trade-offs have been generally handled in a very incremental manner within a narrow sectoral framing of the policy problem at hand (see Turnpenny *et al.*, 2008; Russel and Turnpenny, 2009).

Whether a reflexive appraisal is a single issue or integrated is immaterial. More important is the fact that integration, however achieved, should transcend the boundaries of the state and have a wide scope that is sensitive to broader European sustainable development goals relating to the environment, society and the economy, e.g. those enshrined in the European Commissions Sustainable Development Strategy (Council of the European Union, 2006). Through inclusive stakeholder processes and double-loop learning, a reflexive appraisal should facilitate policy integration that connects non-state and state actors within the policy process and it should link to different governance scales (i.e. local, regional, national, EU and global). Moreover, we might expect integration to have a more social or deliberative aspect to it, i.e. integration actually occurs through interactions between stakeholders (see above) during the appraisal. Consequently, in a more reflexive appraisal, synergies and trade-offs are likely to be comprehensively dealt with within a wider framing of sectoral needs and constraints. Moreover, trade-offs would be placed within the context of sustainable development with an explicit understanding of ecological (as well as societal and economic) limits.

Formal policy appraisal as practised by many European Union states is the latest in a long line of attempts to better integrate knowledge (including stakeholder perspectives) into policy processes (Jordan and Lenschow, 2008). Despite a long history of attempts to develop policy with the aid of hard evidence, there is a growing body of literature suggesting that knowledge may be used for anything but learning. Some authors, for instance, argue that it may be symbolically used to bolster political positions and justify decisions already taken (e.g. Bulmer, 1980;

Majone, 1989; Fischer, 1995; Radaelli, 1995; Owens, 2005). By creating a more deliberative platform through which a diverse range of stakeholders can participate, a variety of different knowledge perspectives (i.e. science, economic analysis, social science, lay, etc.) can be better integrated into an appraisal process to make it more reflexive. Moreover, these different types of knowledge would help to critique existing climate policy goals and explore radical policy alternatives within an explicit understanding of sustainable development and ecological limits.

2.2.3 A framework of reflexive vs. instrumental appraisal

One of the initial tasks of the ADAM project was to develop a conceptual framework for the evaluation and design of climate change appraisals. The framework defined a set of five analytical themes (Weaver *et al.* 2006):

- *Participation and deliberation.* How and why is participation used in the appraisal and what are the interactions with other sources of knowledge?
- *Social learning and reframing.* To what extent does the appraisal stimulate social learning and reframing processes?
- *Trade-offs and synergies.* How does the appraisal deal with trade-offs and synergies over both short and long-term perspectives?
- *Knowledge integration and use of tools and methods.* What types of knowledge are used in the appraisal and how are they obtained, used and integrated?
- *Policy integration.* To what extent does the appraisal consider linkages across sectors?

Based on this framing and the above discussion, Figure 2.1 outlines the main characteristics of reflexive and instrumental appraisals. The distinction between the two types of appraisal is made on the basis of typical characteristics; most appraisals of climate policies will exhibit, to differing degrees, elements of both types. Crucially, no one element is indicative of reflexive (or instrumental) appraisal; for example a reflexive appraisal is more than just double-loop learning. A reflexive appraisal is represented by the presence and possible interaction between all of the elements so that the process encourages the revaluation of climate policy goals within the broader framing of sustainable development. We now examine the appraisal of climate policy in three different European jurisdictions in order to better understand the degree to which they display the various characteristics in Figure 2.1.

2.3 European appraisal of climate policy in different political contexts

This section presents empirical cases of the appraisal of major climate change public policy in different European jurisdictions: the European Commission, Sweden and the United Kingdom. In addition, a single appraisal in the form of an evaluation of

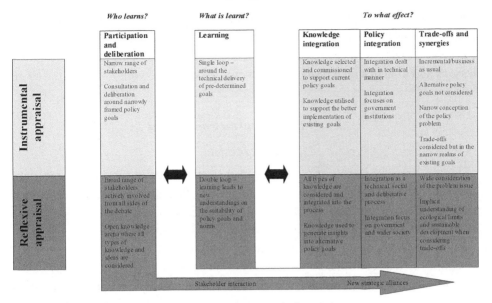

Figure 2.1. Reflexive versus instrumental appraisal.

UK climate change policy commissioned by the Sustainable Development Commission is examined. While this case is markedly different from the others, as it was not conducted as part of the formal policy process, it was one of the more reflexive appraisals identified in the ADAM project making it worthy of further investigation. The case studies draw on a combination of in-depth documentary studies, interviews with stakeholders within and around the appraisal process, and previous empirical work.

The appraisal systems in each of the cases are rather different in formal terms and have different histories. Moreover, many of the appraisal processes observed in these case studies are not necessarily aimed at redefining goals in a reflexive appraisal, being more about improving regulatory quality, and the transparency of policy making, etc. The European Commission's 'Impact Assessment' system and the UK's integrated Regulatory Impact Assessment procedure have both been in place since 2004 and are motivated by similar concerns for 'better regulation'. In Sweden, appraisals are carried out primarily within the context of Committees of Inquiry, a system that has been in place for much longer. Furthermore, the cases only represent a subset of many appraisal processes surrounding climate change.

That being said, the UK, Sweden and the European Commission have all committed in some way to 'evidence-based policy' and have institutionalised relatively advanced forms of appraisal. Moreover, each jurisdiction is regarded as an international front-runner when it comes to addressing climate change. Their performance on key areas of climate change public policy therefore represents a critical test of the potential for reflexive appraisal. Through examining these cases, we hope to gain some insight into current appraisal practice from which to draw lessons and to better understand the elements in our framework.

2.3.1 The European Commission

Background

In March 2007, the European Council agreed an ambitious 20 per cent cut in carbon dioxide equivalent emissions (rising to 30 per cent in case of a robust international treaty post 2012) to meet its long-term objective of limiting climate change-related temperature increases to no more than 2 °C. In January 2008, the European Commission's Climate Change and Energy Package (COM, 2008) was published to outline a strategy to meet these targets. This case study examines the appraisal processes surrounding European Union (EU) climate policy from the advent of its formal appraisal system (known as Impact Assessment) in 2003, until the publication of the 2008 Climate and Energy Package. In line with our appraisal definition above, the analysis not only includes the formal impact assessments but also stakeholder consultations and/or dialogue, Green Papers and Commission Communications.

Participation and deliberation

Public consultation is a formalised part of the European Commission's policy making processes, where the public and other groups send in written responses to proposals. In addition to formal consultations, there were a number of other forms of stakeholder engagement during the formulation of climate policy. This included four stakeholder forums established by Director Generals (DGs) Transport and Energy to discuss future energy policy; informal discussions with selected stakeholders; and discussions with Member States. The European Climate Change Programme also entailed establishing a number of stakeholder working groups to look at the implementation of international carbon reduction targets. These workshops were trumpeted as a major multi-stakeholder policy review exercise. While formal public consultation processes attracted a wide variety of actors, other types of stakeholder engagement appeared to be more exclusive. For example, the aforementioned stakeholder forums on energy policy were distinctly under-represented in terms of scientists and non-governmental organisations (Vasileiadou, 2008: 88–100). Moreover, interview data suggest informal dialogue also occurred with selected industry representatives and trade associations. Often, stakeholder input seemed to be more concerned with consulting on pre-determined policy directions rather than on deliberation and reflection on climate policy goals.

Learning

Many policy refinements seemingly resulted from quantitative analysis and from the discussions that this triggered. Ambitious targets on climate change and renewable energy consumption (20 per cent by 2020) were also based on quantitative analysis. Stakeholder processes were seen as an exercise of understanding the acceptability of pre-determined policy goals (e.g. improving the Emissions Trading Scheme). Thus, across the board any learning that occurred – at least by the commission – was of the single-loop type. This is not to say that double-loop learning did not occur elsewhere.

Trade-offs and synergies

Trade-offs were highlighted, but not addressed, in the quantitative analysis contained within the Impact Assessments over medium-term timescales (e.g. targets for 2020). On the whole, it appears that attempts to process potential trade-offs occurred through discussions and bargaining with stakeholders and Member States, during inter-service meetings between DGs and in the Council of Ministers (also see Turnpenny *et al.*, 2008).

Knowledge integration

European Commission climate and energy policy development incorporated quantitative modelling and stakeholder inputs. The setting of ambitious carbon dioxide equivalent reduction and renewable energy consumption targets were seemingly driven by systematic modelling processes. The importance of quantitative analysis was seen by some interviewees to lie more in 'defending policy' in negotiations with other DGs, member states, stakeholders and within the council of ministers, than in driving policy development. The integration of stakeholder knowledge was generally based on what was seen as practical in the light of discussions surrounding the modelling results.

Policy integration

Until work began on the Climate and Energy Package in 2007, energy and climate policy were being developed in separate domains. DG Transport and Energy had responsibility for the energy sector while Environment DGs, and DG Enterprise and Industry had responsibility for climate policy. Despite this institutional divide, appraisal processes did little to integrate energy and climate policy (see Vasileiadou, 2008). The catalyst for a more integrated approach in 2007 appeared to be heightened concern about climate change and energy security and a strong political lead by Commission head, José Manuel Durão Barroso rather than appraisal. Where appraisals considered integration, links to other sectors and high-level policy objectives (e.g. sustainable development) were explored but in an ad hoc rather than a systematic manner.

2.3.2 Sweden

Background

Sweden has not yet institutionalised a policy appraisal procedure within its ministries. Instead, 'Committees of Inquiry' are appointed for substantive investigations and appraisals. Committees come in different forms, ranging from one-person expert committees to broad parliamentary ones. They typically work for between one and three years, after which they report their findings. This analysis examines seven committees active between 2000 and 2006 working on policy related to climate, including: preparations for a national climate strategy; two committees preparing the implementation of the Emissions Trading Scheme; policies for public transport policy, sustainable consumption, biofuels and road taxation (also see Nilsson, 2006; Swartling and Nilsson, 2007; Nilsson *et al.*, 2008; Nykvist and Nilsson, 2008; Turnpenny *et al.*, 2008). The aim of these committees was to review and present policy options, detailed objectives and strategies to meet national climate change priorities, and to perform appraisals of the impacts of the proposals made.

Participation and deliberation

A relatively wide range of stakeholders were included throughout the committee processes: for example, representatives from different ministries, agencies, industry organisations, environmental groups and specialists. Stakeholder engagement was through direct membership of the Committees of Enquiry, as members of a committee reference group, or through ad hoc consultations. Participants were primarily selected on the basis of their established standing as a major interest group, personal contacts with committee leaders, or established technical expertise. Thus, despite the wide range of actors involved, there was a bias towards 'regime stakeholders', with limited representation for more politically marginal groups. Moreover, the stakeholder engagement was not a priority in the formal impact assessment stage, which tended to be a more technical desktop review exercise by the secretariat and commissioned expert.

Learning

Learning in the committee process was relatively strong. Several committee participants explicitly labelled the process an important learning experience (Nilsson, 2006). To keep abreast of issues and to provide constructive insights, stakeholders had to digest an array of information and understand different perspectives (Swartling and Nilsson, 2007). Learning was greatly facilitated by the arms-length distance from government ministries, which created the intellectual space for stakeholders to interact in a more learning-orientated rather than interest-based manner. At the same time, this arms-length distance limited the transfer of knowledge to the decision-making arenas, i.e. the ministry and the parliament

(Turnpenny *et al.*, 2008). Moreover, learning was very much of the single-loop kind as it was strongly constrained by existing political priorities, which the Committees were not mandated to challenge.

Trade-offs and synergies

Overall, there was relatively little consideration of trade-offs in a transparent, systematic and formal manner. When considered, they were often handled through economic analysis where social costs were internalised by the economic calculations of an appraisal. Trade-offs were also considered on an ad hoc and political basis without the aid of formal decision support such as multi-criteria analysis. The committee chair played the key role here, both in terms of negotiating positions between different interests and mediating with the ministry.

Knowledge integration

The knowledge gathered and considered was heavily dependent on mainstream economic modelling to assess economic impacts across sectors and the cost-efficiency of measures. Other impacts were typically handled rather summarily (Nilsson *et al.*, 2008). The committee processes displayed strengths and weaknesses with regard to knowledge integration. Because the appraisal was a process organised for producing insight into policy options and strategies, there was strong knowledge integration within the committee and the membership, including important policy actors such as bureaucrats and parliamentarians. Interactions with the sponsoring ministry, however, were more constraining. Ministries ensured that the workings of committees were strongly framed by existing policy goals and political priorities.

Policy integration

The committees were fairly strongly framed by sectoral priorities of the sponsoring ministry. Thus, the potential for policy integration was constrained. Nonetheless, we could see evidence of policy integration that was facilitated by learning processes whereby participant ministries came to understand each other's sectoral perspectives. This was a highly fluent and intuitive process and not something that was subject to formal process or procedure.

2.3.3 The United Kingdom

Background

The UK has recently published ambitious plans for a minimum 80 per cent cut in carbon dioxide equivalent emissions by 2050 in its Climate Change Bill (HMG, 2008). This surpasses commitments made by the majority of EU Member States.

This commitment follows a series of different policy initiatives to tackle climate change. This analysis examines the appraisal processes surrounding selected UK climate-related policies including energy policy, the implementation of the European Emissions Trading Scheme and the 2006 Climate Change Programme Review.

Deliberation and participation

UK central government has an official consultation process, supported by the publication of consultation notes and a policy appraisal. Consultation is usually a one-off process to allow stakeholders to provide a written comment on a policy proposal. On the whole, such consultation attracts responses from a large range of stakeholders. However, it was sometimes undermined by being conducted so late in the policy process as to make it meaningless (see Russel and Turnpenny, 2009). There were also instances of unofficial consultations with privileged interests prior to the official processes. Actors from large CO_2 intensive industries had a high level of access to government officials through which they could discuss climate policy when compared, for example, to environmental groups.

Learning

Appraisal processes in UK central government are supposed to provide information to help develop policy. However, analysis in some appraisals was so weak as to render them meaningless in terms of policy learning (see Russel and Jordan's (2009) account of the appraisal of energy policy). Moreover, some analysis was manipulated to bolster the credentials of pre-determined policy goals. Often, the appraisals were more an exercise of tidying up rather than policy development. That being said, there were some instances of appraisals being used to provide insights into policy options by UK central government, particularly in the 2006 Climate Change Programme Review (see NAO, 2007). In most cases, learning was based around whether policy would meet existing targets and goals and was thus of single-loop type (see NAO, 2007).

Trade-offs and synergies

On the whole, the examination of trade-offs in appraisals conducted by UK central government tended to be done implicitly (i.e. by listing cost and benefits), if at all. Trade-offs were treated as something that should be resolved through political bargaining between ministries (see Russel and Turnpenny, 2009). Such an approach reduces opportunities for analysing the implications of trade-offs in a more systematic manner as advised in the UK Treasury's appraisal guidance (HMT, 2003). The exception to this pattern can be found with the government's 2006 Climate Change Programme Review where trade-offs and synergies were considered in a quality assurance process that identified and packaged together overlapping policies (NAO, 2007).

Knowledge integration

With the exception of the 2006 Climate Change Programme Review, use of analysis in appraisals tended to be weak, if it was used at all. Much of the quantitative data did not fully consider uncertainties in the data and gave scant consideration to ecological limits. Apart from the 2006 Climate Change Programme Review, the integration of consultation data was not widely seen beyond using it to test the acceptability of a policy proposal. Many appraisals were more concerned with post hoc justification and rationalisation of policy rather than being treated as an opportunity to produce knowledge for policy development (also see Russel and Turnpenny, 2009).

Policy integration

Policy appraisal in UK government is meant to be a technical process to aid inter-ministerial coordination by providing information on potential policy spill-overs (Russel and Jordan, 2008). However, so poor was the analysis in many appraisals that it is inconceivable that they could have been used to inform inter-ministerial/sectoral discussions (also see Russel and Jordan, 2007). Moreover, the consideration of wider cross-sectoral impacts was generally limited.

2.3.4 The United Kingdom Sustainable Development Commission

Background

This case study is of an audit of the UK's climate change programme commissioned by the Sustainable Development Commission and conducted by the Edinburgh Centre for Carbon Management (ECCM, 2003). The Sustainable Development Commission is a non-departmental public body, which reviews the UK's performance on sustainable development and acts as a government watchdog (Russel, 2007).

Deliberation and participation

Stakeholder participation occurred primarily through a workshop towards the end of the appraisals. It seems that the stakeholder processes were included at the prompting of the Sustainable Development Commission to raise awareness. The workshop had about 50–60 participants from a wide variety of sectors including government, business and non-governmental organisations. The workshop allowed for deliberation on policies beyond 2010, between stakeholders in small breakout groups. Common themes were then extracted and integrated into the final report.

Learning

The appraisal was designed to be comprehensive in its evaluation of UK climate policy. It provided a strong and systematic critique of existing policy in terms of meeting international and national targets and actively considered future policy needs and opportunities. The stakeholder processes were seen to be a useful learning experiences in terms of understanding the opportunities and challenges faced by different sectors. Learning that did occur seemed to be more reflexive than would be expected from single-loop learning, i.e. it examined whether current policy and its goals were suitable enough to meet longer-term decarbonisation needs. Yet too many of the proposed alternative policies were simply the application of well-known policy instruments in new contexts (e.g. a carbon tax for the transport sector). Thus it is difficult to say that learning was of the double-loop type. In terms of wider learning by the political system, the appraisal had little influence on policy particularly in relation to the broader sustainable development implications of UK climate policy.

Trade-offs and synergies

Trade-offs were explicitly considered for each individual climate change policy against broader sustainable development criteria (e.g. whether a policy incorporated the precautionary principle). Information for each policy was explicitly laid out in a matrix and a reason for the score given was provided. Rather than compare the policies and make judgements about them, issues were flagged when it was felt they were important.

Knowledge integration

Robust, systematic analysis and modelling were crucial elements of the appraisal. The report was heavily geared towards quantitative modelling of carbon dioxide equivalent savings from policies. There was also some economic modelling. One interviewee, however, said that some of the more interesting findings came out of the more qualitative analysis on how the policies interacted with sustainable development. Wider stakeholder knowledge was not integrated into the whole process and was only drawn upon towards the end to draw up future policy options.

Policy integration

The integration of climate policy with broader sustainable development principles was actively and systematically examined. The appraisal also adopted a wide cross-sectoral analysis when considering the strength of existing policy measures and the direction needed for future policy. It implicitly viewed integration as a wider societal process through its framing of climate change as a problem that goes well beyond the boundary of the state, i.e. one that involves changing behaviour across society.

2.3.5 Overview – reflexivity in the appraisal of climate policies

In the above analysis, we have provided an overview of appraisal processes in three political jurisdictions, namely the European Commission, Sweden and the UK, as well as an appraisal commissioned by the UK Sustainable Development Commission. Notably, we found little evidence of true reflexivity in our analysis – except to a degree in the Sustainable Development Commission case – despite the fact that jurisdictions are seen as global leaders on climate change.

In terms of who learns (Figure 2.1), we generally found that stakeholder consultation included a range of actors from a variety of backgrounds. That being said, often privileged actors were informally consulted (as in the UK and European Commission cases) or participants were drawn from a narrow range of élites (as in the Sweden case). Moreover, formal participation, (with the notable exception of the UK Sustainable Development Commission appraisal) was essentially limited to information provision or consultation around the acceptability of policy (see also Leach and Pelkey, 2001; Olsson *et al.*, 2004; Sabatier *et al.*, 2005; Huitema *et al.*, 2008). Perhaps the most inclusive and discursive stakeholder process was found in the workshop that was part of the UK Sustainable Development Commission's appraisal.

The majority of the appraisal processes displayed some learning about climate policy. This does not necessarily mean that climate policy changed as a result, or that all those with a stake learnt through the appraisal processes. Learning does, however, imply that the appraisal processes generated some new insights into climate policy. The main type of learning we observed was more in the vein of the single-loop rather than the double-loop learning type, which limited opportunities for more reflexive insights into climate policy. The closest to double-loop learning came in the form of the appraisal commissioned by the UK Sustainable Development Commission which comprehensively examined the UK's existing policy and set out an alternative policy framework. However, whether it was actually double-loop learning is disputable, as goals were not explicitly re-examined and many of the policy recommendations were based around existing policy instruments.

When it comes to the question of 'To what effect?' the cases examined tended to display more instrumental than reflexive characteristics, with the exception of the UK Sustainable Development Commission appraisal. The consideration of trade-offs and synergies was often patchy. Knowledge integration was seemingly on the strong side (as is evidenced by presence of learning), although it was primarily framed around delivering existing policy goals rather than critically examining them. Policy integration was often considered through a sectoral lens.

2.4 Barriers and opportunities for reflexive appraisal

How can we account for the lack of reflexivity in the appraisals we analysed? Firstly, we could argue that there is a lack of understanding or willingness to acknowledge the aforementioned benefits that reflexive appraisal could bring to climate policy (see Argyris and Schön, 1996). Therefore, to encourage the revaluation of climate change goals, the institution concerned must understand the benefits of reflexive appraisal and actively seek to apply it.

Secondly, introducing the elements (Figure 2.1) of reflexive appraisal is not necessarily an easy task. For instance, organising a series of stakeholder processes that offer genuine opportunities for multi-way dialogue and learning is resource intensive in terms of actual and opportunity costs for the organisers and partici-pants alike (Mostert *et al.*, 2006). Thus, stakeholder groups and individuals with more resources (information, money, time, skills, etc.) are likely to be over-represented and could exert more influence. Unless under-resourced stakeholders are properly supported, there is a danger that public participation may augment power imbalances (see also Sabatier *et al.*, 2005; Huitema *et al.*, 2008) further reducing the scope for wider participation. Consequently, the process may be in danger of hijack by vested interests. Moreover, participatory processes can slow down the speed of decision making which may conflict with the political urgency to produce climate change policy (e.g. in preparation for international conven-tions, etc.).

Finally, with the exception of the UK Sustainable Development Commission appraisal, institutional actors working within political systems conducted the pro-cesses we examined. Studies show that where appraisals are primarily conducted by state institutions (e.g. Russel and Jordan, 2007; Turnpenny *et al.*, 2008; Hertin *et al.*, 2009), their impact on policy tends to be negligible and confined to supporting pre-determined policy goals, i.e. they tend to be more instrumental in character. Arguably, climate policy appraisal in Europe is characterised by incremental insti-tutionalism. It is not politically feasible to abandon existing policy and to reassess climate change goals because many resources have been invested in pursuing them. This is not to say that reflexivity cannot occur in such situations, but rather that opportunities for reflexivity are likely to be limited to longer term decision making cycles (see Sabatier, 1998) where external events (e.g. a new international agree-ment) provide opportunities to reframe climate policy and goals. There may be a greater scope for a more reflexive appraisal of climate policies within institutions with less direct investment in current climate policy (*cf.* the UK Sustainable Development Commission's appraisal). Conversely, being on the outside of the formal policy making system might mean that such institutions have less leverage to influence policy (*cf.* the Sweden case). Thus, there is a tension between the need for distance from the policy process to avoid the risk of policy capture, while at the same time ensuring that the appraisal process has a policy impact.

Before concluding on the policy implications of this chapter, it is useful to reflect on the value of appraisal framework that we developed (Figure 2.1). Based on this empirical work, it has not been possible to establish definitive links between the elements of the framework, but some interesting patterns were observed. For example, rather closed and one-way stakeholder dialogues were observed alongside more single-loop learning, as were more partial considerations of trade-offs and policy integration (all factors we associated with instrumental appraisal).

2.5 Conclusions – policy implications

In this chapter we argued that climate policy goals should not be fixed, due to changing social, economic, political and scientific contexts. We therefore proposed that climate policy and goals need to be regularly re-assessed in the light of these shifting contexts through a reflexive policy appraisal process. We developed a framework for investigating such appraisals and applied it to four appraisal case studies: the European Commission, Sweden, the UK central government and the UK Sustainable Development Commission. Broadly, we found an overall lack of reflexivity in the appraisal processes of these climate change policy leaders. What are the policy implications of these findings?

Firstly, as we argue in this chapter, the appraisal of climate change policy should aim to be reflexive. To this end, an appraisal should include a wide variety of stakeholders and display double-loop learning that critiques policy goals in the light of changing contexts. In addition, there should be an explicit consideration of trade-offs and synergies and a comprehensive understanding of policy integration framed around broader sustainable development goals. The whole process should be aided by the integration of a wide variety of analysis and knowledge types including lay perspectives. Incorporating each of these elements into the appraisal process is time intensive and therefore they need to be initiated at the earliest stages of policy development to enhance opportunities for reflexivity.

Secondly, in order to pursue such an approach, it is important that institutions and individuals responsible for the appraisal understand the need and have an appetite for reflexivity. Appraisal processes can then be framed in a way that maximises the opportunities for reflexivity (i.e. inclusive stakeholder processes that allow for multi-way dialogue). Because institutions may be unwilling to use an appraisal to conduct a reflexive critique on policies in which they have heavily invested and promoted, there is a strong case for such appraisals to be the responsibility of well-resourced organisations sitting on the boundary of knowledge and policy (see e.g. Pielke, 2007) (*cf.* the Sustainable Development Commission case).

Ensuring the effectiveness of such an organisation is, however, not easy. It has to be distant enough from the policy making process to avoid being bound by previous policy and institutional perspectives, while at the same time being powerful enough to influence government policy (*cf.* the UK Sustainable Development Commission and Sweden cases). Thus, there is a strong case for members of such a body to have limited terms of tenure (e.g. through secondment) to avoid policy capture. Moreover, such a body should to have the power to require a response to its appraisal analysis from policy making institutions. This is not to say that such an organisation will produce a reflexive appraisal, but that opportunities for reflexive appraisal will be higher. A good example of what such an organisation might look like can be found in the newly established Climate Change Committee in the UK, which currently has the remit to set carbon budgets and check the government's progress on meeting targets.

Thirdly, social learning theory, around which we develop our analysis, sees stakeholder input as key. Where stakeholder input is employed in appraisal it needs to go beyond simple consultation if it is to enhance opportunities for reflexivity. Stakeholder processes need to allow for multi-way dialogue. To encourage such dialogue, it is important to stress that reliable information must be made available to the participants in an understandable form (Gooch and Huitema, 2007). Ideally, it should also be an inclusive and iterative process, where new stakeholders can become involved as the appraisal develops. Moreover, workshops need to be designed with a clear focus and be facilitated in such a way as to provide all participants with a platform to speak. Finally, to enhance participation, the process should be meaningful and with a relevant input into the appraisal. This requires that the agenda is not unilaterally determined, but reflects the concerns of all stakeholders. Moreover it necessitates that the stakeholder process meaningfully interacts and contributes to other elements of the appraisal such as the consideration of trade-offs and synergies, etc. (Figure 2.1).

Fourthly, even though there are potential difficulties associated with reflexive appraisal, particularly stakeholder processes (for example opportunity costs, money and time delays and risk of capture by vested interests), reflexivity can still be achieved with more limited resources. Most important, is the desire to be reflexive in the analysis. Thus, if resource constraints do not allow for stakeholder engagement, other forms of analysis (e.g. modelling, desktop policy evaluations) can be geared towards critiquing existing goals, developing new goals and recognising the importance of issues such as trade-offs and policy integration. That being said, a diversity of stakeholders bring with them different experiences, ideas and perceptions of the problem. We therefore maintain, that good stakeholder interactions are likely to enhance opportunities for reflexivity if managed in the right way.

Finally, in order to maximise opportunities for more reflexive appraisals of climate policy, it might be necessary to connect appraisals to 'windows of opportunity for learning' caused by shocks or changes to wider societal systems (e.g. a period of economic downturn or extreme weather events (see, e.g. Birkland, 1998; Sabatier, 1998)). For example, it can be argued that the current global 'credit crunch' provided an opportunity for Barack Obama to campaign on changing American climate policy on the basis of stimulating the American economy through low carbon investment. In such situations, appraisals might be used at the right time to speak to several policy agendas and provide opportunities for more profound (double-loop) learning to occur.

Acknowledgements

The authors would like to thank all of the people who were interviewed during the research that fed into this chapter. They would also like to thank the two referees for their constructive comments.

References

Argyris, C. (2003) A life full of learning. *Organisation Studies*, **24**, 1178–92.
Argyris, C. and Schön, D. (1996) *Organisational Learning II: A Theory of Action Perspective*. Reading, UK: Addison–Wesley.
Armitage, D., Berkes, F. and Doubleday, N. (2007) *Adaptive Co-Management. Collaboration, Learning and Multilevel Governance*. Vancouver: UBC Press.
Bennett, C. J. and Howlett, M. (1992) The lessons of learning: reconciling theories of policy learning and policy change. *Policy Sciences*, **25**, 275–294.
Birkland, A. (1998) Focusing events, mobilisation, and agenda setting. *Journal of Public Policy*, **18**, 53–74.
Bulmer, M. (ed.) (1980) *Social Research and Royal Commissions*. London: George Allen & Unwin.
COM (2008) *Package of Implementation Measures for the EU's Objectives Climate Change and Renewable Energy*. Brussels: European Commission.
Council of the European Union (2006) *Renewed EU Sustainable Development Strategy*. Brussels: Council of the European Union, 26 June 2006.
Department for Business, Enterprise and Regulatory Reform (BERR) (2007). *Impact Assessment Guidelines*. London: BERR.
Dryzek, J. (1987) Ecological rationality. In *Environment and Political Economy*. New York: Basil Blackwell.
Dunn, W. (1981) *Public Policy Analysis*. Englewood Cliffs: Prentice-Hall.
ECCM (2003) Edinburgh Centre for Carbon Management. *Policy Audit of UK Climate Change Policies and Programmes: Rep-ort to the Sustainable Development Commission*. London: SDC.
European Commission (2005) *Impact Assessment Guidelines* SEC(2005) 791. Brussels: European Commission.

Fiorino, D. J. (2001) Environmental policy as learning: a new view of an old landscape *Public Administration Review*, **61**, 322–34.

Fischer, F. (1995) *Evaluating Public Policy*. Chicago: Nelson Hall.

Gooch, G. D. and Huitema, D. (2007) Participation in water management. In *The Adaptiveness of IWRM; Analysing European IWRM Research*. Timmerman, J. G., Pahl-Wostl, C. and Möltgen, J. (eds). London: IWA Publishing, pp. 27–44.

Guba, E. and Lincoln, Y. (1989) *Fourth Generation Evaluation*. London: Sage.

Hall, P. (1993) Policy paradigms, social learning and the state: the case of economic policy making in Britain. *Comparative Politics*, **25**, 275–96.

Haug, C., Rayner, T., Huitema, D. *et al.* (2008) *An appraisal of EU climate policies*. ADAM deliverable P2.4.

Her Majesty's Government (HMG) (2008). *Climate Change Act*. London: The Stationery Office.

Her Majesty's Treasury (HMT) (2003). *Green Book: Appraisal and Evaluation in Central Government*. London: HMSO.

Hertin, J., Turnpenny, J., Jordan, A. *et al.* (2009) Rationalising the policy mess? Ex ante assessment and the utilisation of knowledge in the policy process. *Environment and Planning A*, **41**(5), 1185–1200.

Huitema, D., Rayner, T., Massey, E. *et al.* (2008) *Climate Change Policy Evaluation Across Europe. Practices of climate policy evaluation in Finland, France, Germany, Poland, Portugal, the UK, and the European Union* Amsterdam: IVM.

Jordan, A. and Lenschow, A. (eds) (2008) *Innovation in Environmental Policy? Integrating the Environment for Sustainability*. London: Edward Elgar, 341pp.

Leach, W. D. and N. W. Pelkey (2001) Making watershed partnerships work: a review of the empirical literature. *Journal of Water Resources Planning and Management*, **127**, 378–85.

Lindblom, C. (1979) Still muddling through. *Public Administration Review*, **19**, 78–88.

Majone, G. (1989) *Evidence, Argument and Persuasion in the Policy Process*. New Haven, CT: Yale University Press.

Mostert, E., Pahl-Wostl, C., Rees, Y. *et al.* (2006) Social learning in European river basin management; Barriers and supportive mechanisms from 10 river basins. *Ecology and Society*, **12**, 19 [online] www.ecologyandsociety.org/vol12/iss1/art19/.

NAO (2007) National Audit Office. *Cost-Effectiveness Analysis in the 2006 Climate Change Programme Review*. London: NAO.

Nilsson, M. (2005) Learning, frames and environmental policy integration: the case of Swedish energy policy. *Environment and Planning*, **C23**, 207–26

Nilsson, M. (2006) The role of assessments and institutions for policy learning: a study on Swedish climate and nuclear policy formation. *Policy Sciences*, **38**, 225–49.

Nilsson, M., Jordan, A., Turnpenny, J. *et al.* (2008) The use and non-use of policy appraisal in public policy making: an analysis of three European countries and the European Union. *Policy Sciences*, **41**(4), 335–55.

Nykvist, B. and Nilsson, M. (2008) Are impact assessment procedures actually promoting sustainable development? Institutional perspectives on barriers and opportunities found in the Swedish committee system. *Environmental Impact Assessment Review*, **29**(1), 15–24.

Olsson, P., Folke C. and Berkes F. (2004) Adaptive co-management for building resilience in social–ecological systems. *Environmental Management*, **34**, 75–90.

Owens, S. (2005) Commentary: Making a difference? Some perspectives on environmental research and policy. *Transactions of the Institute of British Geographers*, **30**, 287–292.

Owens, S., Rayner, T. and Bina, O. (2004) New agendas for appraisal: reflections on theory, practice and research. *Environment and Planning A*, **36**, 1943–1959.

Pielke, R. A. Jr. (2007) *The Honest Broker: Making Sense of Science in Policy and Politics* Cambridge, UK: Cambridge University Press.

Radaelli (1995) The role of knowledge in the policy process *Journal of European Public Policy*, **2**, 159–183.

Rayner, T. and Okereke, C. (2007) The politics of climate change. In *The Politics of the Environment*. ed. Okereke, C. London: Routledge, pp. 116–135.

Rittel, H. and Webber M. (1973) Dilemmas in a general theory of planning. *Policy Sciences*, **4**, 155–159.

Russel D. (2007) The United Kingdom's sustainable development strategies: leading the way or flattering to deceive? *European Environment*, **17**, 189–200.

Russel, D. and Jordan, A. (2007) Gearing up governance for sustainable development: Patterns of policy appraisal in central government. *Journal of Environmental Planning and Management*, **50**, 1–21.

Russel, D. and Jordan, A. (2008) The United Kingdom. In Jordan, A. and Lenschow, A. (eds) *Innovation in Environmental Policy? Integrating the Environment for Sustainability*. London: Edward Elgar, pp. 247–67.

Russel, D. and Jordan, A. (2009) Joining up or pulling apart? The struggle to coordinate policy making in the United Kingdom. *Environment and Planning A*, **41**(5), 1201–16.

Russel, D. and Turnpenny, J. (2009) The politics of sustainable development in UK government: What role for integrated policy appraisal? *Environment and Planning C*, **27**(2), 340–54.

Sabatier, P. A. (1998) The advocacy coalition framework: revisions and relevance for Europe. *Journal of Public Policy*, **5**, 98–130.

Sabatier, P. A., Focht, W., Lubell, M. *et al.* (2005) *Swimming Upstream: Collaborative Approaches to Watershed Management*. Massachusetts: MIT Press, p. 343.

Swartling, A. and Nilsson, M. (2007) *Social Learning and EPI: Communicative governance in Swedish climate policy formation*. EPIGOV Paper No. 12, Ecologic, Berlin.

Tàbara, D. and Pahl-Wostl C. (2007) Sustainability learning in natural resource use and management. *Ecology and Society,* **12**, 3 [online] www.ecologyandsociety.org/vol12/iss2/art3/.

Turnpenny, J., Nilsson, M., Russel, D. *et al.* (2008) Why is integrating policy assessment so hard? A comparative analysis of the institutional capacities and constraints. *Journal of Environmental Planning and Management*, **51**, 759–75.

Van de Kerkhof, M. and Wieczorek A. (2005) Learning and stakeholder participation in transition processes towards sustainability. *Technological Forecasting and Social Change*, **72**, 733–47.

Vasileiadou, E. (2008) *Collaborative Policy-Making? Assessing stakeholder consultations in the European energy policy*. ADAM Deliverable D1.2b

Weaver, P., Haxeltine, A., van de Kerfhof, M. and Tabara, J. David (2006) Mainstreaming action on climate change through participatory appraisal. *International Journal of Innovation and Sustainable Development*, **1**, 238–57.

WCED (World Commission on Environment and Development) (1987) *Our Common Future*. Oxford, UK: Oxford University Press.

Young, O. R., Berkhout, F., Gallopin, G. C. *et al.* (2006) The globalisation of socio-ecological systems: an agenda for scientific research. *Global Environmental Change*, **16**, 304–16.

3

Scenarios as the basis for assessment of mitigation and adaptation

Lead authors:
DETLEF P. VAN VUUREN[1], MORNA ISAAC,
ZBIGNIEW W. KUNDZEWICZ

[1]Co-ordinating lead author

Contributing authors:
NIGEL ARNELL, TERRY BARKER, PATRICK CRIQUI,
NICO BAUER, FRANS BERKHOUT, HENK HILDERINK,
JOCHEN HINKEL, ANDRIES HOF, ALBAN KITOUS, TOM KRAM,
REINHARD MECHLER, SERBAN SCRIECIU

Summary

The possibilities and need for adaptation and mitigation depend on uncertain future developments with respect to socio-economic factors and the climate system. Scenarios are used to explore the impacts of different strategies under uncertainty. In this chapter, some scenarios are presented that are used in the ADAM project for this purpose. One scenario explores developments with no mitigation, and thus with high temperature increase and high reliance on adaptation (leading to 4 °C increase by 2100 compared to pre-industrial levels). A second scenario explores an ambitious mitigation strategy (leading to 2 °C increase by 2100 compared with pre-industrial levels). In the latter scenario, stringent mitigation strategies effectively reduce the risks of climate change, but based on uncertainties in the climate system a temperature increase of 3 °C or more cannot be excluded. The analysis shows that, in many cases, adaptation and mitigation are not trade-offs but complements. For example, the number of people exposed to increased water resource stress due to climate change can be substantially reduced in the mitigation scenario, but even then adaptation will be required for the remaining large numbers of people exposed to increased stress. Another example is sea level rise, for which adaptation is more cost-effective than mitigation, but mitigation can help reduce damages and the cost of adaptation. For agriculture, finally, only the scenario based on a combination of adaptation and mitigation is able to avoid serious climate change impacts.

Making Climate Change Work for Us: European Perspectives on Adaptation and Mitigation Strategies, ed. Mike Hulme and Henry Neufeldt. Published by Cambridge University Press © Cambridge University Press 2010.

3.1 Introduction

The future development of many factors that determine climate change and climate change policy is highly uncertain. This uncertainty includes, for instance, future man-made greenhouse gas emissions, the climate impacts and the capacity of societies to adapt to a changing climate. These factors determine both the need and the possibilities for mitigation and adaptation. Scenario analysis has been developed to explore different uncertain developments and their consequences. For example, the scenarios from the Intergovernmental Panel on Climate Change (IPCC) Special Report on Emission Scenarios (SRES) (Nakicenovic *et al.*, 2000) have been used in mitigation, climate analysis, impact and adaptation research, and have provided a means to compare information across the different research communities that are involved in these fields.

Both within the ADAM project and elsewhere in the scientific community, there is an interest in exploring the relationships between adaptation and mitigation based on consistent assumptions. This chapter briefly discusses how scenario analysis can contribute to an assessment of future adaptation and mitigation strategies. It also describes scenarios that are used throughout the ADAM project as a common basis of analysis. These scenarios are based on different combinations of adaptation and mitigation strategies[1] as illustrated conceptually in Figure 3.1. The first two cases include no mitigation. The first case is a *baseline scenario*, in which we also assume no explicit adaptation. This scenario is useful as an analytical point of reference, as it is rather implausible. The other non-mitigation case assumes an efficient *adaptation strategy*, reducing climate change impacts. A third scenario includes stringent *mitigation action*, and hence less adaptation. While adaptation and mitigation can partly be regarded as substitutes, they are certainly not mutually exclusive. Effective climate policy involves a portfolio of both adaptation and mitigation activities. For example, even with high levels of mitigation, some climate change impacts remain likely and will require considerable adaptation efforts. In contrast, a high degree of climate change could make effective adaptation impossible, which means that there is a need for some minimum level of mitigation.

For the baseline scenario, we assume a continuation of current trends. For this purpose, we developed a scenario consistent with the WETO-H_2 scenario recently published by the European Commission (EC, 2006). As energy use in such a scenario is mostly based on fossil fuels, this scenario will lead to considerable

---

[1] In this context, mitigation is defined as activity aiming to avoid impacts by constraining the level of climate change, whereas adaptation aims at avoiding, or reducing, adverse impacts (or exploiting opportunities) by adjusting human systems in response to observed or projected climate change.

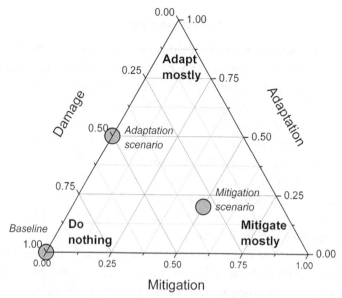

Figure 3.1. Climate policy leads to different combinations of three types of costs: mitigation costs, adaptation costs and residual damage (illustration). The figure also illustratively indicates the position of the ADAM scenarios. (Source: based on Klein *et al.*, 2007).

climate change. For the mitigation scenario we focus on the target of current EU climate policy: a maximum of 2 °C of temperature increase compared with pre-industrial levels. Using the most likely value for climate sensitivity given by the Intergovernmental Panel on Climate Change (IPCC) of 3 °C (Meehl *et al.*, 2007), this translates into a stabilisation level of 450 parts per million (ppm) carbon dioxide equivalent (CO_2e.).

While Figure 3.1 suggests that the costs and benefits of mitigation, adaptation and residual damages can be weighed against each other, there are several challenges in the appraisal of long-term mitigation and adaptation strategies (see also Section 3.4.8).

Spatial and temporal scales of proposed action are very important. Both mitigation and adaptation happen at various *spatial s*cales ranging from individual households to the global scale. For mitigation, benefits always occur globally – despite the fact that action is taken at national or local level. A critical factor in limiting mitigation costs is international co-operation (or competition) in technology development. For adaptation, in contrast, both costs and benefits occur at the local scale, though a supportive environment created at a larger spatial scale (e.g. in a multi-national entity, such as the European Union) can enhance adaptation at a smaller scale. Mitigation action involves some form of international co-operation, while adaptation is mostly explored at the local scale. From this perspective, Figure 3.1 is an enormous simplification of the real problem, as costs occur at different points in time and for different actors and stakeholders. The challenge of bringing adaptation issues into scenarios is discussed further in Chapter 4.

The *temporal scale* of mitigation and adaptation also varies over a wide range. Stringent mitigation scenarios typically require strong early reduction of emissions. The impacts in stringent mitigation scenarios typically only diverge from impacts in scenarios without mitigation after a few decades due to the high inertia of the climate system. Adaptation measures can often be implemented over a shorter time scale and become effective immediately, but some important exceptions exist, which may require decades to implement, such as changes in spatial planning or large scale engineering works for flood protection.

Other important factors are *risk and uncertainty*. The cause–effect chain of climate change (see Figure 3.2) is beset with risks (quantifiable as probability density functions with various, often asymmetrical, forms) and uncertainties. At each main stage of the chain, the uncertainties are due to different factors. Examples of factors affecting the various stages are: (i) emissions, affected by energy and land use; (ii) the climate system, affected by unknown climate sensitivity; (iii) adaptive capacity, affected by costs of infrastructure; and (iv) mitigation, affected by the wide range of costs of mitigation. Mitigation reduces the uncertainties, since it reduces the originating cause of climate change (Barker, 2003; Piani *et al.*, 2005). But mitigation and adaptation may both add to risks. For example, some geo-engineering options may compound risks of climate change by attempting to offset one set of risks (climate) while creating another set of different risks (e.g. ocean acidification). Exploring uncertainties should be part of a robust decision making process, but makes appraisal much more difficult. Among other issues, differences

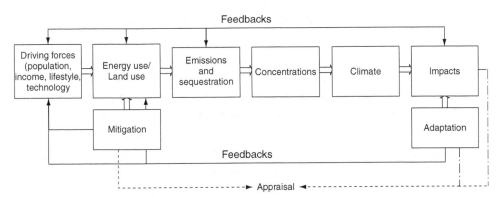

Figure 3.2. Driving force–pressure–state–impact–response framework for climate change. Thick lines indicate direct linkages. Solid small lines indicate potential feedbacks (many of which are not included in current scenarios). The dashed lines indicate categories that are generally used to explore the impact/costs/benefits of different scenarios.

in risk perception become relevant. One way of dealing with risks is to include assessments of probabilities. This is often done on the basis of past evidence, extrapolated to cover specific future circumstances. Uncertainties (unknown or unknowable shocks and surprises) are more difficult to represent numerically, but justify precaution and acknowledgement of ignorance. Scenarios can explore the potential for extreme events and the robustness of various policy portfolios to address the problem and to reduce the risks and uncertainties in the system (we come back to this in Section 3.4.2).

In this chapter we discuss two ADAM scenarios: a baseline scenario in which global mean temperature increases by 4 °C above pre-industrial levels, and a scenario with considerable mitigation efforts in which global mean temperature increase is limited to 2 °C. These scenarios have been developed as a basis of analysis for the whole ADAM project to enhance consistency in the analyses performed by various work packages. We describe the assumptions that have been made and indicate their major outcomes. We also discuss whether the mostly linear approach taken here in scenario development is warranted, given expected impacts, or whether a more integrated approach should be used for future analysis. In our descriptions, we focus on the global level (in view of the limited space). Clearly, adaptation decisions need more detail. Such detail is included in the underlying analysis not presented here; it would also need to come from more country or sector level analysis, for example, as presented in Chapter 8.

3.2 Scenario development

3.2.1 Types of scenarios

Based on the considerations about mitigation and adaptation described in the introduction, different types of scenarios can be defined. Firstly, we define a *baseline scenario*, as a trajectory of events assuming no major feedbacks from climate change and no specific policy efforts on either mitigation or adaptation (such scenarios are also sometimes referred to as 'business-as-usual' scenarios). The main purpose of this type of scenario is analytical, serving as a point of reference for other scenarios, and a starting point for both mitigation and adaptation analysis. Secondly, *adaptation scenarios* describe a world responding to climate change impacts. Their purpose is to explore the type of technologies and policies required to adapt to climate change, and the associated costs. Thirdly, *mitigation scenarios* describe a world with policies aiming to limit climate change.

In ADAM, a scenario has been developed with ambitious mitigation policies aiming to reach a 2 °C target (also called low-stabilisation scenarios). Its purpose is to explore the type of technologies and policies required to minimise climate change, and the associated costs of some of these technologies and policies.

In reality, future development may combine some of these elements (and aim for intermediate targets).

The scenarios can be used in different ways. Firstly, qualitative descriptions (storylines) and quantitative analyses can be used to describe the kind of conditions associated with certain development trajectories. Secondly, one may explore the implications of these scenarios, either in terms of physical impacts (e.g. change of climate or biodiversity loss) or in terms of costs associated with mitigation, adaptation or (residual) damages. In exploring a preferred mix of mitigation, adaptation and residual damage, two main approaches exist: (i) the risk-based approach that describes potential impacts as function of global mean temperature increase (and thus mitigation), and (ii) cost–benefit analysis that looks at the same impacts, but now in monetary terms. Most studies indicate that mitigation efforts, and associated costs, increase for scenarios aiming at lower greenhouse gas concentrations (Fisher *et al.*, 2007). At the same time, assessments of impacts indicate that the magnitude of impacts and adaptation costs (and, in particular, the sum of these two) increases for higher temperatures and concentration levels (see e.g. Parry *et al.*, 2007; Stern, 2006). The result is a virtual trade-off between mitigation costs on the one hand and climate change damages and adaptation costs on the other hand. However, for several reasons outlined in the introduction, such a trade-off cannot really be made. Moreover, uncertainties play a critical role in baseline development, technological development, climate impacts and climate sensitivity.

3.2.2 Further integration

In climate analysis, scenarios are generally developed in a manner consistent with the driver–pressures–state–impacts–responses (DPSIR) framework (Figure 3.2). Using this approach, the development of scenarios starts by describing changes in economic activities (income, energy use, agriculture, etc.) and estimating the resulting emissions. These emissions become inputs to climate models, whose outputs are estimates of climate impacts (as done for the SRES scenarios mentioned before) or, in case of integrated assessment models (IAMs), to a climate model included in the IAM. Next, changes in climate variables are used to assess possible impacts and, in some cases, adaptation opportunities.

The DPSIR framework is also used to develop scenarios for IPCC reports. These scenarios are developed by Working Group III researchers focusing on development of driving forces, energy system and land use parameters in baseline and mitigation scenarios. Subsequently, they are run by Global Circulation Models in Working Group I analyses to assess climate change. Finally, the scenarios are used for Impact, Adaptation and Vulnerability analyses by Working Group II researchers. In fact, many IAMs are built around a similar approach. As a result, possible feedbacks from climate change on driving forces, energy system or land use are typically not considered in the IPCC assessment, and very few models include any of these feedbacks. Ignoring these feedbacks might only be scientifically sound if they are not substantial enough to undermine the likelihood of the original scenario. For analytical reasons, there are major advantages to organising scenario development in a linear way along the DPSIR framework. It enables research to focus on elements of the chain, without the complication of interlinkages between the elements, uncertain feedbacks, etc. In the context of integrated analysis of both mitigation and adaptation needs and opportunities, this may, however, not be sufficient. Some examples of why an integrated approach might be necessary are as follows.

 (i) Climate impacts could be substantial in agriculture. In such cases, estimates of land-use related emissions not taking impacts into account might be wrong, and the mitigation potential of bio-energy could also be affected.
 (ii) Climate impacts may be so severe that they undermine the economic assumptions of the original scenario.
(iii) Land areas might be attractive for both mitigation and adaptation purposes.

An interesting question therefore is whether impacts are indeed so severe that a more integrated approach of scenario development should be preferred, or whether the impacts can be handled separately (as they mostly have been until now), simplifying the analysis framework. The few available studies that looked at the inter-relationships between adaptation and mitigation indicate that, in most sectors, the adaptation implications of any mitigation project are small and, conversely, the emissions generated by most adaptation activities are only a small fraction of total emissions (Klein *et al.*, 2007). In the scenarios presented here, based on the current state of the art in modelling and scenario development, we also ignore most feedbacks. However, we will discuss the impacts of different strategies, and by the end of the chapter come back to the question of whether more integrated (but also more complex) scenarios need to be developed.

3.2.3 *Approaches to scenario development and their use in ADAM*

The use of scenarios within ADAM illustrates different approaches to scenario application. One approach is to use scenarios to analyse the same climate goals with different models (parallel approach). Use of this approach in ADAM is described in Chapter 11; here, integrated models covering the overall global energy system have systematically analysed the implication of achieving stringent mitigation targets using the ADAM scenarios. Model assumptions were harmonised on the basis of the ADAM scenarios for income, population and as far as possible the energy system, in the case of the baseline scenario. For the mitigation scenario, modelling teams were free to follow the greenhouse gas emission profile, the cumulative emissions or the concentration target. The use of multiple models, following alternative approaches, provides insight into uncertainty. The results can, in fact, be seen as an improvement of the original scenario by providing alternative pathways for the same storyline.

The second approach deepens the original scenario by using more detailed models (serial approach). The more detailed model uses the default scenario descriptions as boundary conditions, such as annual temperature and precipitation profiles. The interesting aspect of this approach is that disciplinary strengths of many different models are used. This approach was employed here and in subsequent chapters in assessing the expected impacts of the two scenarios. This improves the original scenario by providing in-depth analysis of the consequences of the scenario. The shortcoming of this method is that possible interactions are often difficult to handle (as they require feedbacks and iterations).

In some cases, the coupled scenario development exercise may itself be highly integrated, or can be coupled in a more interactive way to the original scenario. In such a case, the models complement each other, providing an improved assessment by integrating various disciplinary models. To some degree, the coupled analysis within ADAM, assessing mitigation and adaptation of the European energy system, is an example of this. A description of this work is given in Chapter 7.

3.3 The ADAM scenarios

As explained in the introduction, in ADAM, a set of scenarios is used to explore different combinations of mitigation and adaptation. For analytical purposes, we define a baseline scenario without mitigation action leading to a global mean temperature increase of about 4 °C by the end of the century. Two variants of this scenario are used: one without adaptation (and thus high impacts) and one with adaptation. In terms of socio-economic projections the scenarios are considered to be the same (ignoring for analytical reasons some of the feedbacks). Finally, the mitigation scenario is based on a stringent mitigation strategy leading to an increase of about 2 °C above pre-industrial levels by 2100. Here, we briefly discuss the different assumptions and results. Various storylines may lead to these scenarios. Box 3.1 presents two different possibilities for each scenario. The scenarios were used for various purposes in the ADAM project (see Chapters 4, 7, 8 and 11 and Berkhout *et al.*, 2009).

Box 3.1. Storylines for climate policy

The development of emissions strongly depends on the development of international climate policy. In considering international climate policy, one may consider two important factors: (i) the ambition with respect to mitigation and (ii) the degree of international co-ordination (Berkhout *et al.*, 2009). Together these create four caricature storylines: (i) autonomous adaptation, (ii) co-ordinated adaptation, (iii) autonomous mitigation, and (iv) co-ordinated mitigation. The first two storylines would coincide with the high emission scenario, while the latter two storylines could possibly coincide with the low emission scenario. Table 3.1 provides the main characteristics of each of these storylines. The *autonomous adaptation* scenario does little mitigation action and organises adaptation at the local scale. In the *co-ordinated adaptation* scenario international climate policy fails to organise mitigation actions, but is still able to organise international mechanisms to finance adaptation costs (complying with responsibility and/or the polluters-pays-principle). The *autonomous mitigation* scenario starts from local, technology-focused mitigation policies without binding international commitments (an important challenge here is whether countries voluntarily invest enough in new technologies in the next one to two decades). Whether such a scenario may reduce emissions enough to limit global mean temperature increase to 2 °C is an open question. Finally the *co-ordinated mitigation* storyline emphasises a development pathway based on co-ordinated international climate policy with binding commitments. The fact that more than one storyline could lead to similar emission scenarios (as emphasised earlier by Nakicenovic *et al.* (2000) implies that even with a set target, various policy options exists. That does not mean that all pathways are equally likely to lead to similar outcomes. For instance, the stringent mitigation scenario described in this chapter will require early participation of developing countries in order to achieve the global emission reduction as described. Such an ambitious reduction is arguably more likely to be achieved under an international co-ordination framework than under a more locally orientated, more fragmented regime such as the *automonous mitigation* case.

3.3.1 *Population development and economic growth.*

The ADAM baseline scenario uses the 2004 revision of world population projections (UN, 2005) up to 2050, and the UN's long-range medium projections over the period 2050–2100 (Figure 3.3). The projections are based on the medium-fertility variant with global populations steadily increasing to reach a total of almost 9.1 billion people by 2050, levelling off and stabilising at about 9.2 billion people over the subsequent 50 years up to 2100. According to the UN's definition, medium-fertility rests on the assumption that total fertility in all countries converges towards a level of 1.85 children per woman (though not all countries reach this level by 2050).

The population growth patterns that are used in the ADAM baseline scenario take a middle ground within the range of population forecasting. This is because the UN

Table 3.1. *Four storylines for international climate policy (see Berkhout* et al.*, 2009)*

	Autonomous adaptation	Co-ordinated adaptation	Autonomous mitigation	Co-ordinated mitigation
Focus of climate policy	Adaptation	Adaptation	Stringent mitigation and adaptation	Stringent mitigation and adaptation
International co-operation	Autonomous	Co-ordinated	Autonomous	Co-ordinated
Characteristics of climate policy	Government policy to ensure access to cheap energy; abandonment of Kyoto-style approaches; adaptation at local scale.	No agreements on emission reductions leads to failure of mitigation; but developing countries are able to get an UNFCCC type agreement on sharing adaptation costs.	High technology orientation, but little international institutional arrangements; dedicated technology fixes; both co-operation and competitive relationships across different regions.	High level of mitigation based on high level of international co-ordination; universal and effective governance structure; adaptation mostly at local level.
Outcome	High level of climate change	High level of climate change	Low level of climate change	Low level of climate change

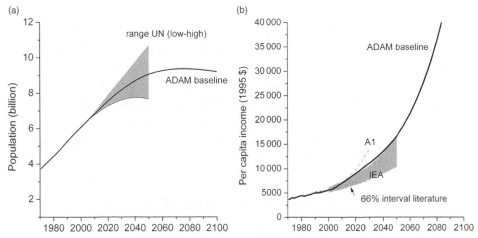

Figure 3.3. (a) Population development and (b) economic growth in the ADAM scenario compared with recently published scenarios (for economic growth the database of scenarios as compiled for IPCC (Fisher *et al.*, 2007, was used).

developed both low and high-fertility variants with projections varying from 7.8 billion (low series) to 10.6 billion (high series) by 2050 (Fisher *et al.*, 2007). The extent of uncertainties is illustrated by the wide range of projections available in the literature. For example the IPCC in its SRES report displays a range from just above 6 billion to over 15 billion of people to inhabit the planet by 2100. Higher or lower population numbers will have impact on both the demand-side through changes in consumption, and on the supply-side by affecting the availability of labour supply. These both in turn affect future growth prospects.

It should be noted that the economic growth projections presented in the ADAM baseline are only partially linked to the UN population projections. There is no direct feedback between the expected economic growth and changes in fertility, life expectancy rates and international migration flows, other than those taken into account in the UN's calculation of future population increases. Nevertheless, long-term growth rates underpinning the ADAM baseline do draw partly on population developments, particularly on likely future urban–rural migration flows across major economies. That is, the greater the availability of under-utilised labour resources and the higher the scope for rural–urban migration, the greater the perspectives for long-term sustained growth, as growth processes are arguably concentrated and dominating in large urban areas and city centres.

The projected economic growth assumptions are mainly the result of optimistic growth assumptions for China and India (*cf.* Fisher *et al.*, 2007). Outside these regions, ADAM assumptions are comparable to other more medium projections. The high income (Western) economies are projected to remain the richest in per capita terms. In terms of total economic activity however, the importance of developing regions grows rapidly, especially in much of Asia, particularly China and India, and in Latin America. In terms of gross domestic product (GDP) per capita, growth is between zero and two per cent per annum in Africa, the Middle East and Latin America. In Asia, it falls steeply from the current seven per cent per annum to four per cent per annum between 2010 and 2030 and to less than three per cent per annum by 2050. This largely reflects the end of the rapid catch-up process currently experienced by Asian economies and the economic slowdown as a consequence of the ageing population in China.

3.3.2 Energy use and greenhouse gas emissions

Baseline scenario

Energy use in the ADAM baseline scenario is based on the baseline published earlier by the European Commission (EC, 2006). The scenario shows energy intensity of the world economy in 2050 falls to about half of the 2001 value, and

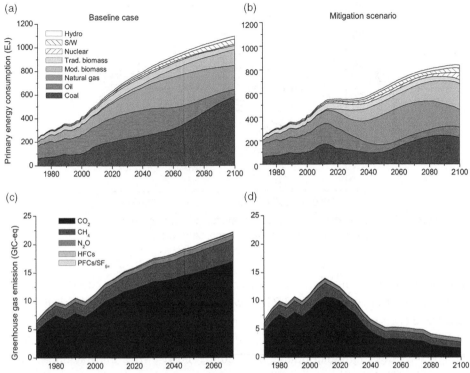

Figure 3.4. (a) Global primary energy use for the baseline scenario and (b) the 450 ppm scenario (right). (c) Global CO_2e. emissions for the baseline scenario and (d) the 450 ppm scenario (right). (Source: TIMER model). (See colour plate section.)

by 2100 it is halved again. World energy consumption, however, more than doubles in the 2000–2050 period and increases by another 25 per cent in the 2050–2100 period (Figure 3.4). Over the whole century, energy supply remains dominated by fossil fuels. In about 2040, conventional oil production peaks. Production of unconventional oil increases to compensate for reduced conventional oil production, so that total oil production does not decrease significantly before 2060. The peak of natural gas production occurs considerably later (around 2070). With an increase in coal use, the contribution of non-fossil energy sources to total primary production does not decrease significantly. The high degree of coal use (almost half of all energy used in 2100) reflects the relative abundance of coal, for which resource scarcity is not expected to play a role in limiting production or significantly increasing cost in any foreseeable future. The amount of non-fossil energy production also increases substantially. Nuclear energy use increases by a factor of two to three to 76 EJ over the period to 2100, the use of biomass increases strongly, while hydro-electricity production increases by about 60 to 80 per cent. All these resources grow further after 2050. The largest relative increase is that of wind and solar energy;

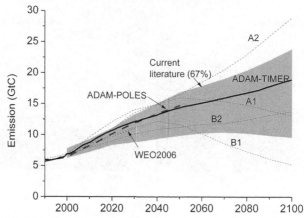

Figure 3.5. World carbon dioxide emissions from energy production and use in the baseline scenario compared to emissions according to other scenarios.

this rises from less than one per cent of all non-fossil energy to between 10 and 14 per cent in 2050. Between 2050 and 2100, wind and solar energy double again. Total renewable energy use in 2050 is 120 to 140 EJ, and 190 EJ in 2100.

As a result of the trends described above, emissions of carbon dioxide from energy activities more than double until 2050, and rise by a third again between 2050 and 2100 (see Figure 3.5). This scenario is consistent with a large range of other scenarios in scientific literature (Fisher *et al.*, 2007).

Land-use-related emissions of greenhouse gases other than carbon dioxide (in particular methane) increase steadily in the period 2000 to 2050 (driven by increasing agricultural production), but at a slower rate than energy-related carbon dioxide. In the second half of the century, a stabilising population also leads to a stabilisation of agricultural emissions. The ADAM baseline scenario lies at the low end of the range of similar scenarios that have recently been published (Rose *et al.*, 2006). Carbon dioxide emissions from land-use fall back to zero during the first half of the century, because of a stabilisation of agricultural area.

Mitigation scenario

The ADAM mitigation scenario corresponds to the ambition of the EU to limit global mean temperature increase to maximum 2 °C compared to pre-industrial levels (using a best-guess value for climate sensitivity). This scenario aims at stabilising greenhouse gases at around 450 ppm CO_2e after an initial overshoot to about 510 ppm CO_2e (den Elzen and Van Vuuren, 2007).

The emission reduction is achieved in various ways. One element is to increase energy efficiency, which reduces the total amount of energy use. By 2050, energy use is reduced by more than 20 per cent in this scenario compared to the baseline

2100 Baseline

2100 Mitigation

☐ Agriculture
■ Bio-Energy
■ Afforestation
☐ Other

Figure 3.6. Land use in the different scenarios. Geographical details are for illustration only. The figure shows the visible impact of land-use related mitigation options on future global land use. (See colour plate section.)

(see Figure 3.4). Another measure in this scenario is a switch away from coal to natural gas, especially during the first half of the century, when other technologies for emission reduction are still underdeveloped. Oil use is also reduced, so that 'peak oil' due to depletion is not reached, as in the baseline scenario, and unconventional oil resources are minimally exploited. The scenario also shows an increasing use of energy from non-fossil sources, which account for most of the growth in total energy use. Non-fossil energy use increases from about 15 per cent of total primary energy use in 2010 to more than 30 per cent in 2050 and is over 40 per cent of the total by the end of the century. Most of this growth is due to an increase in bio-energy use. Finally, carbon capture and storage is applied in most remaining stationary uses of fossil fuels. In addition, also non-carbon dioxide greenhouse gas emissions are reduced. As a result, global emissions peak around 2020, and reduce further with time. By 2050, emissions are reduced by more than 70 per cent compared to the baseline and more than 80 per cent by 2100. The consequences of the mitigation policies are not only obvious for energy, but also for global land use. Substantial land areas are used for afforestation and bio-energy (see Figure 3.6).

3.3.3 Climate change

The atmospheric greenhouse gas concentration and temperature change resulting from the emissions of the two scenarios is shown in Figure 3.7. There is some uncertainty in the concentration levels, and temperature change due to uncertainties in the carbon cycle. The solid and dashed lines indicate the outcome for best-guess assumptions (such as a climate sensitivity of 3 °C). The shaded area indicates the uncertainty due to the carbon cycle and climate sensitivity. Global mean temperature under the baseline case increases almost linearly to 2.1 °C

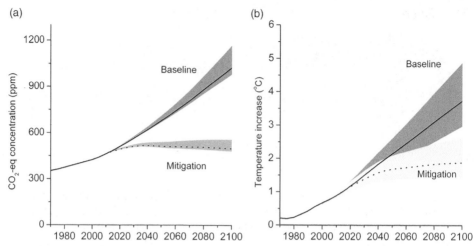

Figure 3.7. Atmospheric carbon dioxide equivalent concentration (taking into account the Kyoto gases) (a) for the baseline scenario (4 °C, no climate policy) and mitigation scenario (2 °C) (left) and (b) global temperature change (since the pre-industrial age) for the same two scenarios (right). Ranges are based on van Vuuren *et al.*, 2008.

above the pre-industrial levels in 2050 and to 3.7 °C in 2100. The uncertainty in these values, however, ranges from 3 to 5 °C. In the mitigation scenario, the global mean temperature increase by 2100 is reduced to 1.9 °C for the best-guess assumptions. Again, there is considerable uncertainty, and in fact, during the first decades the temperature ranges of the baseline case and the mitigation scenario strongly overlap. By the end of the century, however, there is a clear difference. Even so, Figure 3.7 indicates that the mitigation case could also lead to a temperature increase of 2.6 °C compared with pre-industrial levels.

The mitigation scenario presented here is among the most stringent in scientific literature, and two important conclusions can be drawn. Firstly, global warming can be mitigated but not be stopped; the most stringent scenarios still lead to an increase of about 2 °C above pre-industrial levels, whilst assuming there is global co-operation to reduce emissions from about 2015/2020 onwards. Secondly, because this stringent scenario could also lead to considerably greater climate change, adaptation policies could be hedged against the higher numbers. For example, such policies may be to 'aim for 2 °C, but prepare for 3 °C'. In the assessment of impacts below, we focus on the central climate change projections.

Models agree that the level of temperature and precipitation change will not be the same at different locations. The patterns of change are, however, very different across the models. Some general trends can be observed; for example, the change in annual mean temperature is larger at high latitudes than at low latitudes. In terms of

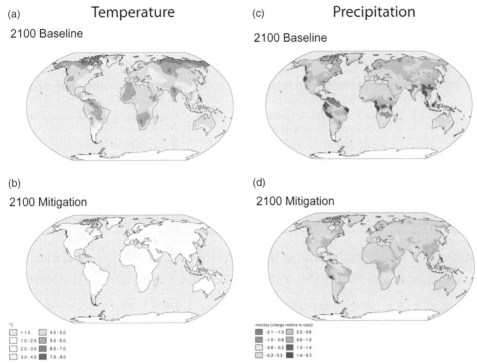

Figure 3.8. Map of change of annual mean temperature and precipitation in 2100 relative to 1990; (a) temperature; scenario: baseline (leading to 4 °C warming in 2100); (b) temperature; scenario: mitigation (2 °C); (c) precipitation; scenario: baseline (leading to 4 °C warming in 2100); (d) precipitation; scenario: mitigation (2 °C). (See colour plate section.)

precipitation, however, differences across models are even greater; there are only a few places where the majority of models agree on whether it gets wetter or drier (e.g. all models expect Southern Europe to be become drier). The patterns used for the ADAM scenarios are shown in Figure 3.8.

3.4 Impacts and adaptation in the different scenarios

3.4.1 Introduction

The Synthesis Report of the IPCC Fourth Assessment Report (IPCC, 2007) gives an overview of impacts associated with climate change. Some of these impacts result from changes in climate averages, but other impacts (and arguably more severe ones) may result from changes in extreme events. Table 3.2 summarises some of the main impacts. There are six broad categories of important impacts: health, agriculture, water availability, coastal flooding, urban areas and energy system, and large-scale disruptions of the climate system. In this section, we sketch some impacts

Table 3.2. *Possible impacts of climate change*

	Impacts associated with global average temperature change	Impacts due to changes in extreme events
Health	Increasing burden from malnutrition, diarrhoeal, cardio-respiratory and infectious diseases. This will affect particularly populations with low adaptive capacity.	1. Reduced mortality from cold exposure, increased risk of heat related morbidity and mortality (heat waves) 2. Risks related to heavy precipitation events (landslides, flooding of rivers etc.) 3. Food and water shortage and increased risk of water- and food- borne diseases as a result of drought 4. Risks related to sea-level rise
Food	Negative impacts on vulnerable groups. Region specific changes (both positive and negative) in cereal crop productivity.	1. Changed yields in agriculture (due to extreme temperatures, droughts, heavy precipitation) 2. Land erosion and degradation (due to heavy precipitation events, droughts) Increased livestock deaths (due to drought)
Water	Increased availability in some areas, decreased availability and increasing drought and water stress in other areas. Effects are both through changes in rainfall + evapotranspiration and through changes in snow and ice melt. This will affect agriculture.	1. Effects on water resources relying on snowmelt and glaciers (due to changed extreme temperatures) 2. Effects on water supplies (due to changed extreme temperatures, changed seasonality, droughts, heavy precipitation events) 3. Increased water demand (due to heat waves, droughts) 4. Changed (reduced or increased) hydropower generation potentials due to changing droughts
Coasts	Increased damage from floods and storms due to sea level rise. This will affect low-lying coastal systems.	Increased risk and costs of coastal protection from extreme weather events.
Industry, settlements and society	Affected by impacts in all of the above categories, compounding pressures associated with rapid urbanisation, industrialisation and ageing in some societies.	Affected by impacts in all of the above categories. Specific impacts include: 1. Changes in energy demand for space conditioning

	The most vulnerable are generally those in flood plains, those whose economies are closely linked with climate-sensitive resources and the poor.	2. Reduced quality of life due to heat waves for people without appropriate housing 3. Disruption due to flooding caused by heavy precipitation 4. Water shortages due to drought 5. Disruption due to cyclones 6. Increased costs of coastal protection from extreme high sea level
Large-scale disruption	1. Partial loss of ice sheets on polar land implies metres of sea level rise. Rapid sea level rise on century time scales cannot be excluded. 2. Large-scale and persistent changes in the meridional overturning circulation (MOC) of the Atlantic Ocean could cause various changes to ocean behaviour.	

and adaptation requirements under the two main ADAM scenarios (baseline and mitigation) within these categories. The descriptions are not intended to be exhaustive, but instead provide some indication of the magnitude of impacts and adaptation challenges. We also use the results to discuss the possibilities and the need to develop more integrated scenarios.

3.4.2 Human health: temperature-related mortality and malaria

Climate change influences human health in various ways. It is likely, however, to remain a relatively minor factor compared to other drivers that impact human health (such as life-style-related factors) (Hilderink *et al.*, 2008). We focus here on climate change impacts on temperature-related mortality and malaria.

Temperature-related mortality

The effect of heat, in the form of heatwaves, on mortality has been described in various studies. The strongest physiological evidence for additional mortality levels is available for cardiovascular disease. The impact of climate change may occur via changes in extreme temperatures, changes in average temperatures or in seasonal variation of temperature and the literature shows varying results. McMichael *et al.* (1996) made an estimation of temperature-related mortality using relative risk ratios, showing that there is an optimum temperature at which the death rate is lowest (also know as the U-shaped dose-response relation). This study also shows that for higher temperatures heat stress-related mortality increases, whereas cold-related mortality decreases. Tol (2002) concluded that, in monetary terms, the reduction in cold-related mortality due to climate change will outnumber the increase in heat-related mortality. This conclusion is, however, strongly influenced by the approach used to value a life and also subject to the large uncertainties in the relationships between temperature and health. Adaptation may occur both by the adjustment of the human physiology to higher temperatures (McMichael *et al.*, 1996) and an increase of air conditioning use (Kinney *et al.*, 2008). Given the complexities in using dose–response relationships between temperature and mortality, we have not directly related them to the two ADAM scenarios.

Malaria

Annually more than one million people, mostly African children, die from malaria, a vector-born infectious disease. Malaria vectors spreading the infection, i.e. the anopheles mosquitoes, can only survive in climates with high average temperatures, no frost and sufficient precipitation. The MARA/ARMA malaria suitability model (Craig *et al.*, 1999) incorporates these factors to determine climatically suitable areas. It was run for the ADAM scenarios (Figure 3.9) (see Appendix for details). The number of deaths from malaria is predominantly influenced by factors such as access to preventative measures, such as insecticide-treated bed nets, and access to health care, both of which are linked to income and urbanisation. This is demonstrated by comparing a hypothetical scenario without GDP growth and the baseline scenario: the assumed income growth reduces malaria deaths by around 50 per cent and suggests that adaptation strategies might be similarly effective. The difference between the mitigation scenario and the baseline case however, is much smaller; mitigation reduces malarial health risks by about 2 per cent.

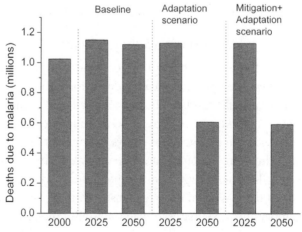

Figure 3.9. Death due to malaria in the baseline scenario, adaptation scenario, and in the mitigation scenario.

3.4.3 Agriculture: Impacts on yields

For the IPCC, Easterling *et al.* (2007) have synthesised a large amount of research on the impacts on crops of incremental temperature change, with or without adaptation. Their information represents an overview of the best knowledge currently available on the impact of climate change on crop yields (both with and without adaptation) and therefore we have compared their results to our scenarios. Easterling *et al.* (2007) have summarised the results using best-fit polynomials of the percentage yield change as a function of mean local temperature change[2]. These results can be used to estimate the potential global impacts of the ADAM baseline and mitigation scenarios, with and without adaptation, for maize, wheat and rice. Figure 3.10 shows the yield change according to Easterling *et al.* (2007) at low latitudes (tropics) and at mid to high latitudes (temperate zones) at temperature changes equal to the global mean temperature change in 2100 according to the ADAM scenarios. Although the results are highly uncertain, some preliminary conclusions seem to be possible from this figure. Firstly, the baseline scenario, with high climate change and no adaptation, causes a very substantial decrease in yields for all cases shown. Climate impacts, especially in the tropics, may reduce yields by 10 to 35 per cent for the crops studied. Secondly, engaging in either mitigation or adaptation limits the decrease in yields and, in some cases, may enable an increase in yields. In the tropics, impacts remain negative and typically in the order of a 10 per cent loss. Thirdly, the mitigation scenario with adaptation may result in an improvement in yield compared to no climate change.

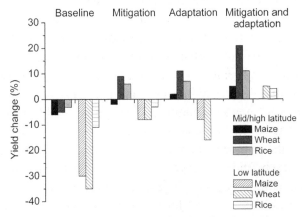

Figure 3.10. Indicative results for the sensitivity of maize, wheat and rice yield change at low and mid to high latitudes to climate change for each of four scenarios (following Easterling *et al.* (2007).

---

[2] We have in each case taken the global mean temperature change for a scenario and used that as an indication of the average local temperature change to be expected.

The importance of mitigation is illustrated by the analysis of Tubiello and Fischer (2007), who found that when comparing a non-mitigated scenario with a mitigation scenario, global costs of climate change in agriculture were reduced by 75 per cent to 100 per cent with mitigation, and the number of additional people at risk of malnutrition was reduced by 80 to 95 per cent. The importance of adaptation may be seen in the work of the same authors (Fischer *et al.*, 2007) on climate change impacts on water irrigation requirements, where they found that mitigation reduced the impacts of climate change on agricultural water requirements by about 40 per cent.

These results underline the need to look at both mitigation and adaptation. While mitigation limits the damages of climate change, adaptation is still necessary, since even with 2 °C mean climate change, impacts can be significant.

3.4.4 Water resources: potential water availability

The effects of the two ADAM scenarios on water resources are assessed using a global-scale water resources impact model (Arnell, 2003)[3]. Figure 3.11 shows the percentage change in average annual runoff by 2100 (relative to the 1961–1990 mean) under the baseline scenario and the mitigation scenario. The differences between the scenarios in change in runoff are similar to, but slightly larger than, the relative difference in change in global average temperature under the two scenarios. For example, the change in temperature by 2050 is approximately 20 per cent smaller under the mitigation than the baseline scenario (1.7 °C compared to 2.2 °C), and the change in runoff is generally between 20 and 25 per cent smaller.

The human implications of the difference between the baseline and mitigation scenarios can be assessed by examining changes in runoff in watersheds exposed to water resource stress (here defined to have runoff less than 1000 m^3/capita/year). The results show quite substantial differences in exposure to increased water resource stress from 2050 onwards between the mitigation and baseline scenarios. In 2020 there is little difference in runoff and this water stress between the two scenarios. Figure 3.12 shows the numbers of people exposed to an increase in water resource stress due to climate change under the two scenarios; the mitigation scenario reduces the numbers exposed by 135 million and 457 million in 2050 and 2100, respectively (the numbers are sensitive to the assumed pattern of climate change). Mitigation, however, does not eliminate the impact of climate change, and adaptation will clearly be required for the remaining billion people exposed to a situation of water resource stress due to climate change. Underlying results show that the effects of mitigation are regionally variable, and in some circumstances, mitigation appears to increase the numbers of people exposed to increased stress.

---

[3] We used pattern-scaling to obtain maps of temperature and rainfall change, on the basis of the HadCM2 run.

2100 Baseline   2100 Mitigation

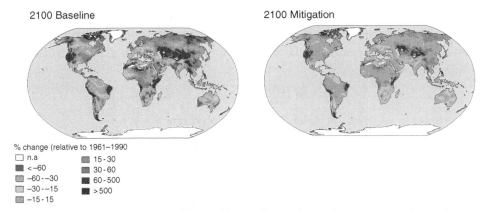

% change (relative to 1961–1990)

☐ n.a	■ 15 - 30
■ < –60	■ 30 - 60
■ –60 - –30	■ 60 - 500
☐ –30 - –15	■ > 500
■ –15 - 15	

Figure 3.11. Change in runoff by 2100. The figure shows the percentage change in average annual runoff by the 2050s, relative to 1961–1990, under the baseline and mitigation scenario. (See colour plate section.)

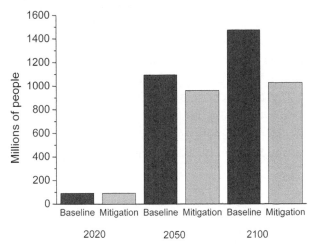

Figure 3.12. Numbers of people exposed to increase in water resource stress due to climate change, under the baseline and mitigation scenarios. The simulations are based on the HadCM2 climate model pattern.

This occurs because of the non-linear response of runoff in some circumstances to gradually changing rainfall and temperature.

The indicator used is that of exposure to water resource stress, and does not represent the actual water stress (which is determined by water management structures and practices). In a sense, it can be viewed as an index of 'adaptation demand', defining adaptation as the introduction of management practices to reduce the increased exposure to water resource stress. Adaptation may include measures to increase water storage, transport of water, or reduction of water demand by increasing efficiency. Another way of characterising adaptation is to assume that the aim of adaptation is to eliminate water resource stress. The effect of climate change on this demand for adaptation is indexed by the difference in exposure with and without climate change.

3.4.5 *Coasts: sea level rise*

Another important impact of climate change results from increasing sea levels. For this we used the DIVA model (version 2.0.3; see model appendix) to explore the vulnerability of coastal zones to sea level rise. The expected global sea level rise in 2025 is 19 cm (since the pre-industrial age) in both scenarios. In later years the scenarios start to diverge, with sea level rise in 2050 being 31 cm in the mitigation scenario and 35 cm in the baseline scenario, and in 2100 the rise is 49 cm and 71 cm, respectively.

The model simulates costs of damage caused by sea level rise and associated storm surges as well as costs of adaptation in terms of dyke building and nourishing beaches, assuming either no adaptation, or alternatively optimal adaptation (DINAS-COAST Consortium, 2006; Hinkel and Klein, 2007). Figure 3.13 shows that the sum of damages and adaptation costs are highest for the baseline case and lowest for the mitigation scenario with adaptation. The figure also shows that adaptation reduces overall costs rather effectively. The necessity of engaging in adaptation even under an ambitious mitigation effort is underlined by the fact that costs in the mitigation-only scenario are much larger than the costs in the

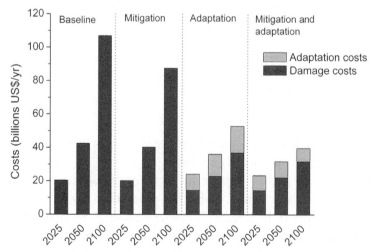

Figure 3.13. Global total annual adaptation costs and damages up to 2100 as a result of sea level rise, as modelled in the DIVA model using the ADAM scenarios.

adaptation-only scenario. For sea level rise, damages can be avoided more effectively through an adaptation-only strategy than through a mitigation-only strategy, although the combination of the two has the strongest positive impact. The causes of the increasing costs over time are the gradually rising sea level and rising GDP. While the costs involve substantial investment flows (10s of billions of US$ worldwide), they are a relatively small fraction of global GDP, even for sea level rise at the level of the baseline scenario. However, for individual countries or regions (particularly small island states) these costs can be a much larger fraction of income.

3.4.6 *Industry, settlements and society: Heating and cooling demand*

Since the demand for space cooling and heating is linked to climate, it is expected to be influenced by climate change. We have used some simple relationships to describe heating and air conditioning demand in the residential sector, and explore the impacts of climate change on this simulated energy demand (Isaac and van Vuuren, 2009). It should be noted that changes in population and income are projected to lead to a considerable growth in the energy demand for heating and air conditioning in the coming century (see Figure 3.14, no climate change case). Changes in cooling and heating practices are examples of that part of adaptation to climate change which is expected to occur 'autonomously' (i.e. without policy intervention). As a result, we do not have a separate adaptation scenario, but include adaptation in all scenarios. This is not perfect adaptation, however, since the extent to which a population is able to fulfill demand for space conditioning depends on income. Unfulfilled demand for heating and cooling can lead to health impacts

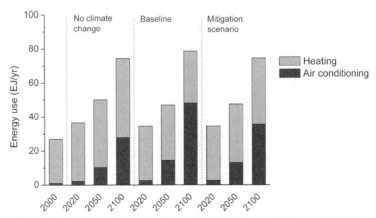

Figure 3.14. Global annual energy demand for heating and air conditioning in the residential sector in the year 2000 and during the coming century for two scenarios (baseline and mitigation) and if no climate change at all is assumed (TIMER model).

(as described in Section 3.4.2) and to loss of labour productivity. In addition to these effects, comfort is reduced when indoor temperatures are not optimal.

Figure 3.14 shows that globally, the increase in energy demand in time is much larger than the difference between the energy demand in the baseline scenario, the mitigation scenario, and the values calculated without taking climate change into account. The effect of climate change on the combined energy demand is also smaller than the effect on heating and air conditioning individually, since increases in air conditioning compensate for decreases in heating. On the regional and country level, impacts can be far more substantial: for example, in India, a large increase in energy demand is expected due to increased cooling, and in Western Europe and the USA there is a substantial decrease in energy demand expected due to reduced heating.

3.4.7 Changes in extreme weather events

Climate change is expected to include changes in frequency and intensity of some weather-related extreme events. Extremes like floods, droughts, storm surges and 'warm' extremes are projected to become more frequent, widespread, and intense, while cold-extremes, such as cold spells, are likely to become less frequent and weaker. Assessing risks on the basis of changes in average conditions may result in averaging out risks over temporal and spatial scales. A more risk-based, geographically explicit method is therefore preferable (see Chapter 8, Mechler *et al.*). Substantial progress has yet to be made, however, as comprehensive knowledge on disaster impacts and risks is limited and heterogeneous in nature, and uncertainties, particularly for future projections, are very large. For example, climate models do not always agree whether precipitation increases or decreases. Furthermore, flood risk analysis is hampered by little available information on modelled variability in precipitation events and coarse resolution of events, often only allowing assessments in terms of sensitivity analyses.

There was no global assessment of extreme events in the ADAM project, but we provide here a few examples of other work in the context of the ADAM scenarios. The global average number of people affected by flooding each year is currently about 50 million. Hirabayashi and Kanae (2008) compared changes in the number of people affected by floods for different degrees of global warming. For 2 °C warming above pre-industrial levels the number is projected to be 142 million, while for 4 °C warming it rises to 453 million. Impacts of flood disasters on human welfare are likely to occur disproportionately in countries with low adaptation capacity. Projected flooded area in Bangladesh, the most flood-vulnerable country in the world, is expected to increase by at least 23 to 29 per cent with a global temperature rise of 2 °C (Mirza *et al.*, 2003). However, the uncertainty of socio-economic factors and adaptation still leads to a wide range of estimates for the costs of future flood damage. With respect to drought, the projections for the 2090s made by Burke *et al.* (2006), using scenarios comparable to our baseline case, show a net overall global drying trend. The proportion of the global land surface suffering from extreme drought is predicted to increase by a factor of 10 to 30; from one to three per cent for the present day to about 30 per cent by the 2090s. The number of extreme drought events per 100 years and mean drought duration are likely to increase by factors of two and six, respectively, by the 2090s.

Rising costs due to weather-related extreme events are already increasing the need for effective economic and financial risk management. The costs of major events are expected to range from several percent of annual regional GDP in very large regions with big economies, to more than 25 per cent in smaller areas (Parry *et al.*, 2007).

3.4.8 Economic evaluation of impacts

Cost–benefit analysis (CBA) can be used to express the costs and benefits of climate change of different strategies (see Section 3.1) in terms of a common monetary unit. These costs are discounted into net present value calculations. The costs include: (i) real, measurable, economic costs (so-called market costs); and (ii) other impacts expressed in monetary terms on the basis of an 'assumed' value, such as the loss of biodiversity cost based on the willingness-to-pay concept. In the past, damage functions have been published as part of work on various IAMs (see Hof *et al.*, 2008). More recently, damage estimates for the DICE model were extended with explicit adaptation cost estimates (De Bruin *et al.*, 2009). We have used the FAIR model (see model appendix) to develop economic costs estimates under these scenarios, especially to assess the impact of relevant uncertainties (Hof *et al.*, 2008, 2009). Some important observations are that most models typically assess the costs of mitigation to be between zero and three per cent of GDP, for optimal implementation of 450 ppm CO_2e stabilisation scenarios at a global scale (Fisher *et al.*, 2007). Regional costs can be considerably higher, for example, greater than 10 per cent for oil-exporting countries. A few models report net economic gains even for very stringent stabilisation targets such as in the case of Barker *et al.* (2006; 2008). At the same time, estimates of the costs of impacts of climate change vary over a very wide range. While the damage curves for a baseline scenario included in most models typically lead to costs in the order of a few per cent of GDP, under extreme assumptions, these costs may be up to 25 per cent or higher. Finally, adaptation investments are mostly assessed to be smaller than mitigation investments and residual damages. However, they are very important in limiting residual damages. While uncertainties imply that CBA cannot be used to provide quantitative results on optimal mitigation and adaptation levels, the outcomes can very well be used to explore the impacts of different assumptions.

Under default settings of the FAIR model, the discounted costs as share of GDP due to climate change impacts for the period 2005–2200 at 2.5 per cent discount rate amount to nearly 4.5 per cent in the baseline (Figure 3.15). These costs rise sharply over time, reaching 17 per cent in 2200. Adaptation or mitigation reduces these costs substantially to around 2.5 per cent. The adaptation scenario results in relatively low costs due to discounting, as in the long-run this scenario still leads to 8 per cent costs of GDP in 2200. By comparison, the mitigation scenario leads to less than 2 per cent costs of GDP in 2200, but follows a completely different time profile, with mitigation costs early in the century. The combination of mitigation and adaptation leads to the lowest discounted costs, namely 2 per cent.

A crucial caveat needs mentioning at this point with regard to the economic evaluation of climate change impacts. First of all, calculations cannot be regarded as reliable for the extreme tails of risks (i.e. low probability, high impact events). As a subjective assessment on how to handle such risks is involved, Weitzman (2008) questioned the usefulness of CBA for policy makers. Secondly, the value of the discount rate to account for time preference and risk is currently heavily debated, with arguments relating to subjective time preference and risk perception (Nordhaus, 2008; Price, 2005; Stern, 2006). Finally, irreversible changes, for example a warming of the oceans leading to the loss of coral reefs, need subjective quantification of damages (Ackerman and Heinzerling, 2004).

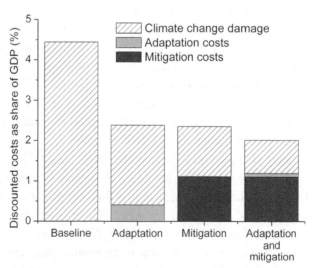

Figure 3.15. Mitigation costs, adaptation costs, and residual damages due to climate change as share of GDP according to the FAIR model (Hof *et al.*, 2009).

3.4.9 Uncertainties in climate change, impacts and adaptation

There are many sources of uncertainty in projections of future climate change and its impacts. Uncertainties are associated with every step in the causal chain shown in Figure 3.2: emissions, climatic drivers (e.g. the carbon cycle), climate (mainly climate sensitivity and pattern of climate change), and impacts (including adaptive capacity). The initial uncertainty, relating to future human development, is considerably amplified along this chain. This is illustrated by the fact that, under the same emission scenario, different models lead to different impacts, due to model differences in the later steps in the chain. This difference is often larger than that arising in one model with different emission scenarios. For example, for precipitation changes until the end of the twenty-first century, the multi-model ensemble mean exceeds the inter-model standard deviation only at high latitudes (Kundzewicz *et al.*, 2007). Uncertainties in climate change projections increase with the length of the time horizon. In the near term (e.g. the 2020s), climate model uncertainties play the most important role; while over longer time horizons (e.g. the 2090s), uncertainties due to the selection of emissions scenario become increasingly inportant (Jenkins and Lowe, 2003).

The impact of future climate change on extreme events is particularly uncertain. This is partly due to a mismatch between the larger spatial and temporal scale of coarse-resolution climate models, and the local occurrence and short life of some weather extremes (e.g. cloudburst precipitation and flash floods). Impacts and adaptation are most relevant at the local scale, as people experience events in a particular time and place. Resolving the mismatches at both the spatial and the temporal scale requires downscaling, giving rise to another source of uncertainty, no matter which method of downscaling is used.

Uncertainty has implications for adaptation. The large range of projections observed in different climate model-based scenarios (*cf.* ENSEMBLES Project of the EU) suggests that planning for adaptation should not be based on a limited number of scenarios, since the range of simulations obtained might not represent the full range possible. Robust adaptation procedures, which do not rely on precise projections of changes, therefore need to be developed.

3.5 Conclusion

In this chapter, we have discussed how scenario analysis may contribute to the assessment of mitigation and adaptation strategies. We have also presented two scenarios used in the ADAM project as a starting point for analysis. We specified impacts in those scenarios for a selected number of parameters, focusing mainly on mean climate changes. Further improvements can be made by focusing on extreme

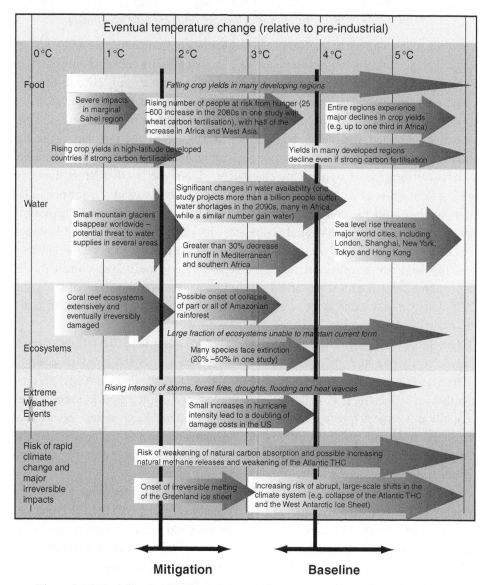

Figure 3.16. Stabilisation levels and probability ranges for temperature increases (Stern, 2006), with lines marking the temperature change for the baseline and mitigation scenarios.

events and by exploring the implications of various risk levels. Both the IPCC (Parry *et al.*, 2007) and the Stern Review (Stern, 2006) made a more comprehensive overview of impacts as a function of global mean temperature rise, but did not couple impacts to specific scenarios. Figure 3.16 shows the impacts as assessed by the Stern review, indicating the position of the two ADAM scenarios on the climate change axis.

(i) *We described two sets of possible climate change trajectories for the world for analysis of mitigation, adaptation and impacts.* The first set, including the so-called baseline and adaptation scenario without climate policies, is expected to lead to a global mean temperature increase by the end of the century of around 4 °C (for the most likely values for climate parameters, and current economic trends). This scenario has high adaptation needs as explicitly described in the adaptation scenario. The second scenario assumes stringent mitigation and limits global mean temperature change to 2 °C, with a probability of 50 per cent, using known techniques and costing one to two per cent of GDP. Even under this scenario, adaptation measures will still be needed.

(ii) *While it is possible to explore different consequences of scenarios (including uncertainties) it is not practical to scientifically determine an optimal mix between mitigation, adaptation and residual damages.* As discussed in this chapter, the weighing of the consequences of climate change and the various policy responses is complicated by large differences in scale, space and time; large uncertainties; and clear differences in interest between individual actors. As a result, subjective interpretation of risks will always play an important role.

(iii) *Effective climate policy includes both adaptation and mitigation.* Even the most stringent climate policy scenarios can still result in a global mean temperature increase of more than 2.5 °C and at best a temperature increase of 1.5 °C. The need for a combination of mitigation and adaptation has been shown for most of the impacts explored in this chapter. For example, adaptation can be more effective than mitigation in dealing with sea level rise, but mitigation still has a role to play in reducing damages and costs of adaptation. Agriculture presents an example where adaptation and mitigation are both clearly necessary. Crop yields in agriculture are projected to be negatively impacted by climate change in the absence of both adaptation and mitigation action. Without stringent mitigation, adaptation could limit negative impacts, but not remove them.

(iv) *While impacts of climate change can be severe and, depending on subjective choices, may warrant stringent climate policy, these impacts are not necessarily of the order of magnitude that significantly undermine assumptions of population and economic growth at a global scale.* Current 'middle-of-the-road' estimates of mitigation costs and climate damage are likely to be in the order of a few per cent of GDP. While climate change may have an impact on millions of people, other challenges are likely to influence global population growth more significantly. Mitigation studies are often optimistic in assuming global participation, while studies on damages are almost exclusively focused on changes in average climate. In an analytical context, the severity of impacts is relevant; although there might be merit in improved integration of impacts and adaptation, low costs negate the need for global analysis to include all feedbacks on main drivers based on the consistency of the storylines. Clearly, at the local scale, the situation is likely to be very different; impacts for individual countries can be far more substantial than at the global scale.

(v) *Impacts may differ very much between locations.* Sea level rise is very important for some low-lying island states and countries. These countries could be significantly affected, or even destroyed, by large adaptation costs and/or damages. For agriculture, positive and negative impacts are projected to occur in different places and at different times, with low-income countries often experiencing greater negative impacts. The wealthier north, where agriculture is currently temperature-limited, would benefit.

(vi) *Important focus areas for further analysis include variability and extreme events and the role of governance.* In our work, we focus on changes in mean values. There is strong evidence that changes in probability of extreme events may be of more relevance than mean values, but information on this is still scarce. The role of different actors is another issue; some forms of adaptation require active governmental involvement; other measures, such as installation of space cooling systems, are likely to be implemented by private investors. The differences between these two adaptation protagonists are relevant for scenario development.

References

Ackerman, F. and Heinzerling, L. (2004). *Priceless: On Knowing the Price of Everything and the Value of Nothing*. New York: The New Press.

Arnell, N. (2003). Effects of IPCC SRES emissions scenarios on river runoff: a global perspective. *Hydrology and Earth System Sciences*, **7**(5), 619–41.

Barker, T. (2003) Representing global climate change, adaptation and mitigation. *Global Environmental Change*, **13**, 1–6.

Barker, T., Pan, H., Köhler, J., Warren, R. and Winne, S. (2006). Decarbonising the global economy with induced technological change: scenarios to 2100 using E3MG. *Energy Journal*, Special Issue, 143–60.

Barker, T., Scrieciu, S. S. and Foxon, T. (2008) Achieving the G8 50% target: modelling induced and accelerated technological change using the macro-econometric model E3MG. *Climate Policy*, Special issue **8**, S30–45.

Berkhout, F., Haug, C., Hildingsson, R. and Stripple, J. (2010). Exploring the future: the role of scenarios and policy excercises. In A. Jordan, D. Huitema, F. Berkhout and H. van Asselt (eds.) *Climate Change Governance in Europe: Confronting Dilemma of Mitigation and Adaptation?* Cambridge, UK: Cambridge University Press.

Burke, E. J., Brown, S. J. and Christidis, N. (2006) Modelling the recent evolution of global drought and projections for the 21st century with the Hadley Centre climate model. *Journal of Hydrometeorology*, **7**, 1113–25.

Craig, M. H., Snow, R. W. and le Sueur, D. (1999) A climate-based distribution model of malaria transmission in Africa. *Parasitology Today*, **15**(3), 105–11.

de Bruin, K. C., Dellink, R. and Tol, R. S. J. (2009). AD-DICE: An Implementation of Adaptation in the DICE Mode. *Climate Change*, **95**(1–2), 63–81.

den Elzen, M. G. J., and van Vuuren, D. P. (2007). Peaking profiles for achieving long-term temperature targets with more likelihood at lower costs. *PNAS*, **104**(46), 17931–6.

DINAS-COAST Consortium (2006). *DIVA 1.5.5*. Potsdam: Germany. Potsdam Institute for Climate Impact Research.

Easterling, W., Aggarwal, P., Batima, P. *et al*. (2007) Food, fibre and forest products. In Parry M. L., O. F. Canziani, J. P. Palutikof, van der Linden, P. J. and Hanson, C. E. (eds.). *Climate Change 2007: Impacts, Adaptation and Vulnerability. Contribution of Working Group II to the Fourth Assessment Report of the Intergovernmental Panel on Climate Change*. Cambridge, UK: Cambridge University Press.

EC (2006). *World Energy Technology Outlook 2050 (WETO H2)* Brussels: European Commission.

Fischer, G., Tubiello, F. N., van Velthuizen, H. and Wiberg, D. A. (2007) Climate change impacts on irrigation water requirements: effects of mitigation, 1990–2080. *Technological Forecasting and Social Change*, **74**, (7), 1083–107.

Fisher, B., Nakicenovic, N., Alfsen, K. *et al*. (2007) Issues related to mitigation in the long-term context. In Metz, B., Davidson, O., Bosch, P., Dave, R. and Meyer, L. (eds.). *Climate Change 2007 – Mitigation*. Cambridge: Cambridge University Press.

Hilderink, H., Lucas, P. L., ten Hove, A. *et al*. (2008) *Towards a Global Integrated Sustainability Model*. Bilthoven: Netherlands Environmental Assessment Agency.

Hinkel, J. and Klein, R. J. T. (2007). Integrating knowledge for assessing coastal vulnerability. In Fadden, L. M., Nicholls, R. J., and Penning-Rowsell, E. (eds.). *Managing Coastal Vulnerability*. London: Earthscan.

Hirabayashi, Y. and Kanae, S. (2008) First estimate of the future global population at risk of flooding. *Hydrology Research Letters*, **3**, 6–9.

Hof, A. F., den Elzen, M. G. J. and van Vuuren, D. P. (2008) Analysing the costs and benefits of climate policy: value judgements and scientific uncertainties. *Global Environmental Change*, **18**(3), 412–24.

Hof, A. F., de Bruin, K., Dellink, R., den Elzen, M. G. J. and van Vuuren, D. P. (2010) Costs, benefits and inter-linkages between adaptation and mitigation. In Biermann, F., Pattberg, P. and Zelli, F. (eds.). *Global Climate Governance after 2012: Architecture, Agency and Adaptation*. Cambridge: Cambridge University Press.

IPCC (2007) *Climate Change 2007: Synthesis Report. Contribution of Working Groups I, II and III to the Fourth Assessment Report of the Intergovernmental Panel on Climate Change*. Core Writing Team, ed. Pachauri, R. K. and Reisinger, A. Geneva: IPCC.

Isaac, M. and van Vuuren, D. P. (2009). Modeling global residential sector energy demand for heating and air conditioning in the context of climate change. *Energy Policy*, **37**(2), 507–521.

Jenkins, G. and Lowe, J. (2003) *Handling Uncertainties in the UKCIP02 Scenarios of Climate change*. Exeter: Met Office.

Kinney, P. L., O'Neill, M. S., Bell, M. L. and Schwartz, J. (2008) Approaches for estimating effects of climate change on heat-related deaths: challenges and opportunities. *Environmental Science Policy*, **11**, 87–96.

Klein, R. J. T., Huq, S., Denton, F. *et al*. (2007) Inter-relationships between adaptation and mitigation. In Parry, M. L., Canziani, O. F., Palutikof, J. P., van der Linden, P. J. and Hanson, C. E. (eds) *Climate Change 2007. Impacts, Adaptation and Vulnerability. Contribution of Working Group II to the 4th Assessment Report of the Intergovernmental Panel on Climate Change*. Cambridge, UK: Cambridge University Press, pp. 745–77.

Kundzewicz, Z. W., Mata, L. J., Arnell, N. *et al.* (2007) Freshwater resources and their management. In Parry, M. L., Canziani, O. F., Palutikof, J. P., Hanson, C. E. and van der Linden, P. J. (eds), *Climate Change 2007: Impacts, Adaptation and Vulnerability. Contribution of Working Group II to the Fourth Assessment Report of the Intergovernmental Panel on Climate Change.* Cambridge, UK.: Cambridge University Press.

McMichael, A., Haines, A., Slooff, R. and Kovats, S. (1996). *Climate Change and Human Health.* Geneva: World Health Organization.

Meehl, G. A., Stocker, T. F., Collins, W. D. *et al.* (2007) Global climate projections. In Solomon, S., Qin, D., Manning, M. *et al.* *Climate Change 2007: The Physical Science Basis. Contribution of Working Group I to the Fourth Assessment Report of the Intergovernmental Panel on Climate Change.* Cambridge, UK: Cambridge University Press.

Mirza, M. M. Q., Warrick, R. A. and Ericksen, N. J. (2003) The implications of climate change on floods of the Ganges, Brahmaputra and Meghna Rrivers in Bangladesh. *Climatic Change,* **57**, 287–318.

Nakicenovic, N., Alcamo J., Davis G. *et al.* (2000) *Special Report on Emissions Scenarios: A Special Report of Working Group III of the Intergovernmental Panel on Climate Change* Cambridge, UK: Cambridge University Press.

Nordhaus, W. D. (2008) *A Question of Balance: Weighing the Options on Global Warming Policies.* New Haven & London: Yale University Press.

Parry, M. L., Canziani, O. F., Palutikof, J. P., van der Linden, P. J. and Hanson, C. E. (2007) *Climate Change 2007: Impacts, Adaptation and Vulnerability. Contribution of Working Group II to the Fourth Assessment Report of the Intergovernmental Panel on Climate Change.* Cambridge: Cambridge University Press.

Piani, C., Frame, D. J., Stainforth, D. A. and Allen, M. R. (2005) Constraints on climate change from a multi-thousand member ensemble of simulations. *Geophysical Research Letters,* **32**, (L23825), doi: 1029/2005 GL024452, 2005.

Price, C. (2005) An intergenerational perspective on effects of environmental changes: discounting the future's viewpoint. In Innes, J. L., Hickey, G. M. and Hoen, H. F., (eds.). *Forestry and Environmental Change: Socioeconomic and Political Dimensions.* Vienna: International Union on Forestry Research Organisations (IUFRO).

Rose, S., Ahammad, H., Eickhout, B. *et al.* (2006) *Land-based Mitigation in Climate Stabilisation.* Energy Modeling Forum Report, Stanford University. www.stanford.edu/group/EMF/projects/group21/EMF21sinkspagenew.htm2006.

Stern (2006). *Stern Review on the Economics of Climate Change.* Cambridge, UK: Cambridge University Press.

Tol, R. (2002) Estimates of the damage costs of climate change, Part II. Dynamic estimates. *Environmental and Resource Economics,* **21** (2), 135–60.

Tubiello, F. N. and Fischer, G. (2007) Reducing climate change impacts on agriculture: Global and regional effects of mitigation, 2000–2080. *Technological Forecasting and Social Change,* **74** (7), 1030–56.

UN (2005) World Population Prospects: The 2004 Revision. *CD-ROM Edition-Extended Dataset.* United Nations, Department of Economic and Social Affairs, Population Division.

van Vuuren, D. P., Meinshausen, M., Plattner, G.-K. *et al.* (2008) Temperature increase of 21st century mitigation scenarios. *Proceedings of the National Academy of Sciences,* **105**, (40), 15258–62.

Weitzman, M. L. (2009) On modeling and interpreting the economics of catastrophic climate change. *The Review of Economics and Statistics,* **91**(1), 1–19.

4

National responsibilities for adaptation strategies: lessons from four modelling frameworks

Lead authors:

ASBJØRN AAHEIM[1], THERESE DOKKEN, STEFAN HOCHRAINER,
ANDRIES HOF, EBERHARD JOCHEM, REINHARD MECHLER,
DETLEF P. VAN VUUREN

[1]Coordinating lead author

Contributing authors:

HENRY NEUFELDT, TAOYUAN WEI

Summary

Most of the literature about adaptation and vulnerability deals with decision making on a local level. This chapter examines the possible relevance of developing national or global adaptation strategies, and sorts out possible challenges. Firstly, the challenge in balancing mitigation and adaptation is discussed. Adaptation is very efficient, but substantial damages will remain even after optimal adaptation. Large regional differences in adaptation costs indicate the importance of establishing an international adaptation regime. Secondly, a closer examination of the energy system identifies innovations required to utilise the full potential for adaptation, and emphasises that successful R&D strategies may turn challenges into a competitive advantage. Thirdly, the possible market barriers to adaptation are identified. It is shown that climate change is likely to increase migration of labour and capital to urban areas. This adaptation is, however, hampered because capital, labour and natural resources are immobile to a certain extent. Finally, depending on the economic and financial vulnerability of an economy and its key actors, and the extent and frequency of disasters, countries may exhibit differential economic follow-on effects after a disaster. It is shown that increased frequencies and intensities of disasters, such as floods, may have substantial fiscal and macroeconomic consequences. An identification of governments' roles in adaptation strategies is important, but is far from comprehensive when it comes to potential issues. Public goods are used as a common denominator for the identification of subjects to which governments should pay attention. A lot of research remains before an extensive overview of adaptation options, which can be characterised as public goods on a national scale, can be provided.

Making Climate Change Work for Us: European Perspectives on Adaptation and Mitigation Strategies,
ed. Mike Hulme and Henry Neufeldt. Published by Cambridge University Press © Cambridge University Press 2010.

4.1 Introduction

The literature on adaptation frequently reminds us that 'mitigation is global, while adaptation is local' (see e.g. Tol, 2003). Policy makers have also adopted this statement (Robinson *et al.*, 2008). It is probably meant to point to the fact that policies aiming at adaptation can be distinguished from policies aiming at mitigation because their dependency on actions of others differ widely; it is useful to adapt even if nobody else does, but mitigation is meaningless unless it is as a part of collective global effort. The statement is therefore an intuitive one, although the implications are unclear. By surveying the literature on adaptation, and to some extent on mitigation, one nevertheless gets the impression that 'mitigation is global' implies that all decisions will have to be taken by national bodies and co-ordinated with other nations, while 'adaptation is local' implies that it is the decisions made on the local level that matter.

However, there is no necessary link between the policy issue and relevant level of decision making in this case. Even though the atmosphere is a public good, and co-ordinated action among countries of the world is required to control emissions, the act of cutting emissions is, in the end, something individual agents will have to decide about on their own. It is of high importance to understand better how agents respond to policies initiated by national governments in response to the choice of policy instrument, or whether public policies, aiming at a change of local structure, can contribute to reduce emissions.

Similarly, the fact that adaptation depends on local impacts and vulnerabilities, and therefore will have to take widely different forms depending on local conditions, does not imply that adaptation is irrelevant for the national level of decision making. How agents adapt is subject to frames set by national authorities, such as institutions and economic incentives. Adaptation may, moreover, lead to structural change that policy makers on the national level are concerned with. After a survey of the literature, the Intergovernmental Panel on Climate Change (IPCC) nevertheless concludes that decisions on adaptation and mitigation are taken on different levels (Klein *et al.*, 2007). If this conclusion is drawn on the basis of the emphasis of decision levels in the literature, it indicates that the national responsibilities for adaptation are addressed rarely.

This chapter identifies issues related to adaptation that have to be dealt with by national bodies and on a national level. It aims at indicating the benefits of developing national strategies of adaptation, but is not comprehensive in the description of subjects that could, or should, be included in such a strategy. Common to the issues discussed in this chapter is that adaptation can be considered a public good; if taken on an individual basis, insufficient adaptation will take place because the benefits of individual actions are shared by many.

Four approaches that support national adaptation strategies are discussed. The first addresses the implications of including adaptation in policies aiming at mitigation. It has been recognised for some time that impacts may be reduced significantly, or benefits enhanced, if adaptation is taken into account. Early contributions by Nordhaus and Boyer (2000) and Tol (2002) show the economic implications. The purpose of including adaptation was first and foremost to improve existing damage estimates and to evaluate implications for mitigation. It is assumed that adaptation occurs more or less autonomously, with little input from policy. This approach has recently been challenged. Based on a disaggregation of climate change impacts into adaptation costs and residual damage costs, we argue that there is indeed a strong role for adaptation policy. This is in line with other results from the ADAM project and is also supported by the EU White Paper on Adaptation (Aaheim *et al.*, 2008; EC, 2008). In this chapter we discuss the impact of adaptation strategies in Europe and demonstrate a good adaptation strategy, given the uncertainty in climate change.

The second approach outlines research and development required to fully utilise future adaptation options. In this part, we focus on the energy system, where the full potential for adaptation is still in need of research and development before it is ready to be implemented. The outcome of research and development is, however, uncertain. If successful, the investment will be beneficial, not only to the investor, but also to others. Research has, in other words, the characteristics of a public good, with insufficient incentives for private agents to invest. Governments will therefore have to play an active role to encourage development of means for adaptation. We outline potential technologies for adaptation in the energy system, and point out benefits for Europe. These include the 'private' part of the benefits of research, which are related to the improvement of the competitiveness of the inventor.

The third example of national responsibilities relates to the market barriers to adaptation. Analyses of the impacts of climate change by means of macroeconomic modelling tools are based on national aggregates, with strictly limited capability to deal with variabilities of impacts and vulnerability. At the same time, variability on its own represents potential challenges for adaptation to climate change. The ability to adapt depends on available alternatives, which differ considerably between the local and the national scale. A full utilisation of the potential for adaptation may require vast migration of natural resources, labour and capital. Migration is, however, costly and sometimes impossible. We present an illustration of the potential implications of these market barriers.

Finally, we address the financial implications of extreme events. Damages in the wake of extreme events represent the greatest worries for future climate change. Floods and windstorms may be restricted to regions, but can, at the same time, have transboundary effects. By means of an experiment applied to a number of Central European countries, it is shown that flood events may result in financial constraints over a long period of time. This affects economic growth and enhances the vulnerability to future events. Moreover, the ability to recover from the impact depends notably on the extent to which neighbouring countries are affected. As an adaptation strategy, national or EU funds may be established to prepare for an increase in the frequency of extreme events under climate change.

4.2 Adaptation costs in Europe compared to other world regions

Integrated assessment models that provide monetary estimates of climate change damages form an excellent tool to analyse relations between monetary impacts of adaptation and climate change damages. In the past, damage estimates generally consisted of both adaptation costs and damages that remain after adaptation, the so-called residual damages. As an example, the integrated model DICE (Nordhaus and Boyer, 2000) estimates sectoral damages, which are summed to arrive at a total damage projection for the whole economy. In agriculture, for example, it is assumed that farmers will change their crops according to a change in climate. Thus, adaptation is implicitly included in the damage estimates. More recently, some models provide explicit estimates of the impact of adaptation on damages (Hope, 2006; de Bruin *et al.*, 2009a; de Bruin *et al.*, 2009b). It is still hard to calibrate these models based on empirical data, but they allow us to analyse adaptation strategies by addressing the extent of adaptation required. In this section, we use such a model.

The amount of adaptation needed depends on the damages expected from climate change. Therefore, it is useful to first consider some of the different damage projections for Europe if no mitigation takes place (Section 4.2.1). In Section 4.2.2, we apply the recently developed damage and adaptation functions of AD-RICE (de Bruin *et al.*, 2009a) to shed some light on how much adaptation would be needed in Europe under different (global) mitigation scenarios. Section 4.2.3 compares the European estimates of adaptation costs with other world regions.

4.2.1 Damage projections for Western Europe

Damage estimates of climate change involve wide ranges of scientific uncertainties (e.g. impact of climate change on the number of storms or change in mortality) as well as value judgements (e.g. how to monetise non-market damages and how to weigh the welfare of future generations compared to the current one).

Figure 4.1 gives an indication of these uncertainties by comparing the damage estimates for Europe in a medium development baseline (IPCC SRES B2 scenario) without any mitigation activities according to different integrated models. The

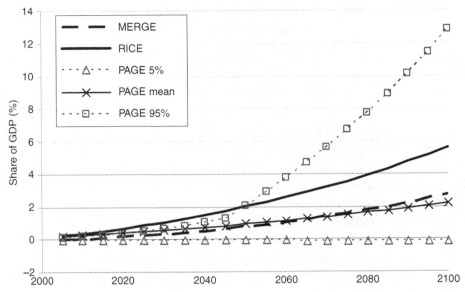

Figure 4.1. Western European climate change damage projections according to different models in a B2 baseline scenario.

global temperature increase in this baseline reaches 3.5 °C above the pre-industrial level in 2100. The damage estimates of MERGE (Manne and Richels, 2004) and PAGE (Hope, 2006) are based on a literature review of aggregated climate change impacts, while the damages of RICE (Nordhaus and Boyer, 2000) are based on sectoral impact estimates of climate change.

The damage functions taken from the three models all include both market and non-market costs. MERGE is the only model that does not take into account the risk of large-scale disruptions of the climate system (e.g. collapse of the thermohaline circulation) in the damage projections. Figure 4.1 shows the 5th and 95th percentiles of damage costs in the PAGE model, which allows for probabilistic calculations,

The damage projections of MERGE and the mean estimate of PAGE are relatively similar, with damages exceeding 2 per cent of gross domestic product (GDP) at the end of the century in Europe. MERGE projects slightly lower damages at the beginning of the century, but higher damages from 2070 onwards. According to RICE, damages are much higher (5.6 per cent of GDP in 2100) and even exceed the damage projections of the PAGE 95th percentile damage function up to 2050. This is mainly because the PAGE 95th percentile damage projection is relatively low for a small increase in temperature level, but increases sharply for temperatures higher than 2 °C above the pre-industrial level.

Overall, climate change damages in Western Europe are likely to be substantial in the current century. However, the uncertainty of the monetary impacts of damages is large, as shown by the 5th and 95th percentiles of the PAGE damage estimates.

4.2.2 Inter-linkages between damages, adaptation costs and mitigation costs in Western Europe

The need for adaptation measures depends on how large the potential negative effects of climate change will be. These potential negative effects decrease with increasing mitigation efforts, as climate change should be reduced. To assess quantitatively the interaction between mitigation, damages and adaptation costs, a model that includes these components explicitly is needed. Such models are scarce, however. The most recent attempt to explicitly model adaptation costs and damages of world regions is the AD-RICE model (de Bruin *et al.*, 2009a), which is based on the RICE model of Nordhaus. The damage function of RICE is a combination of the optimal mix of adaptation costs and residual damages. AD-RICE disaggregates these two components so that adaptation costs and residual damages can be quantified. Adaptation is defined as the share of gross damages avoided. It is assumed that adaptation costs are a function of the level of adaptation and that these costs increase at an increasing rate, as cheaper and more effective adaptation options will be applied first. With optimal adaptation, the sum of adaptation costs and residual damages is minimised (and is the same as the damages in the RICE model). Quantitatively, the AD-RICE model has been calibrated against insights into relative costs as reported in the literature.

We have integrated the AD-RICE damage and adaptation functions described above within the FAIR framework. FAIR is a policy support tool aimed at integrating information from detailed energy, climate, and socio-economic models. It describes the interactions between multi-gas emissions, greenhouse gas concentrations and the climate system, as well as the interaction between the climate system and the economy through climate change damages and mitigation costs. Integration of the AD-RICE damage and adaptation functions with FAIR allows the reporting of adaptation costs and residual damages separately and, moreover, modelling adaptation as a policy variable (see Section 4.2.3).

Figure 4.2 shows the projected adaptation costs and residual damages for different mitigation targets using the AD-RICE damage and adaptation functions, assuming optimal adaptation[1]. Firstly, it is shown that, in the current AD-RICE settings, residual damages are much higher than adaptation costs. This indicates that only a small part of the damages can be abated through adaptation. Recent top-down estimates of adaptation costs confirm this picture. The United Nations Framework Convention on Climate Change (UNFCCC, 2007) estimates adaptation costs to be potentially more than US$ 100 billion per year. The global DICE model estimates

---

[1] Optimal adaptation means that the marginal costs of adaptation are equal to the marginal benefits of avoided damages, implying adaptation up to the point where the adaptation costs start to exceed the benefits of adaptation.

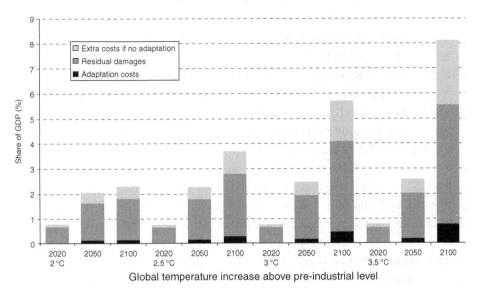

Figure 4.2. Adaptation costs, residual damages and the extra costs if no adaptation measures are implemented in Western Europe according to AD-RICE under different global mitigation scenarios. Costs are plotted as a function of scenarios aiming to stabilise temperature increase below certain targets in the long-run.

total damages,equal to residual damage, and adaptation costs, to be considerably higher at US$ 1.3 to 1.8 trillion in 2050, depending on the mitigation scenario.

Figure 4.2 also shows that, in the short term, the effect of mitigation efforts on the optimal level of adaptation is negligible, as can be seen by the equal adaptation costs and residual damages for the different mitigation scenarios in 2020. Residual damages in Western Europe are projected to amount to 0.6 per cent of GDP in 2020, while adaptation costs are only 0.04 per cent of GDP. This means that investment in adaptation measures amounts to over US$ 5 billion[2] in Western Europe by 2020, regardless of the mitigation target; in the short-term temperature increase is hardly influenced by mitigation. If these adaptation measures are not implemented, an increase in damages of US$ 22 billion is projected. Some adaptation is therefore worthwhile in the short term.

By 2050, temperature levels across the different mitigation targets have diverged with a corresponding divergence of adaptation costs and residual damages. If greenhouse gas emissions are cut drastically to limit climate change to 2 °C above the pre-industrial level, adaptation costs could be limited to 0.13 per cent of GDP. If, on the other hand, no mitigation efforts are made, leading to a global temperature increase of 3.5 °C above the pre-industrial level by the end of the century, adaptation

---

[2] We use constant 2000 US$ throughout.

costs would amount to 0.18 per cent of GDP. In absolute terms, this means that Western Europe would have to invest $10 billion more in adaptation measures for the B2 baseline scenario, than the strong mitigation scenario, by 2050. The absolute difference in residual damages is even higher. Residual damages in Western Europe are projected to be US$ 320 billion in 2050 if climate change is limited to 2 °C above the pre-industrial level and US$ 390 billion in the baseline scenario. Residual damages and adaptation costs really start to diverge between different mitigation scenarios by the end of the century. AD-RICE projects that limiting climate change to 2 °C would save more than US$ 200 billion annually in adaptation costs and a trillion US$ in residual damages by the end of the century. Adaptation could reduce damages by US$ 200 billion at the cost of US$ 50 billion for the 2 °C target and by US$ 1.1 trillion at the cost of US$ 250 billion for the baseline scenario.

We conclude that relatively small investments in adaptation can substantially decrease climate change damages in Western Europe, as shown by Figure 4.2. However, even with optimal adaptation, a large share of the residual damages cannot be avoided, and these increase with greater global warming.

4.2.3 Comparing adaptation costs in different world regions

The last section showed that adaptation costs in Western Europe are in the order of US$ 5 billion in 2020, rising to US$ 35 billion in 2050 (assuming strong mitigation measures leading to a 2 °C temperature increase in 2050). Figure 4.3 shows how these numbers compare with other world regions, again using the adaptation cost estimates of AD-RICE. In absolute terms, Western Europe is expected to face the highest adaptation costs in 2050, followed by South Asia and South East Asia. The global adaptation costs are estimated at US$ 115 billion in 2050, with non-Annex I countries incurring more than half of this world total.

East Asia and Japan are expected to have particularly low adaptation costs. The reason for this is that RICE projects low damages due to positive impacts for agriculture and non-market amenity value, such as climate-related time use, for small increases in temperature. Note, however, that for larger increases in temperature these regions will also face high climate damages and therefore adaptation costs.

The adaptation costs as a share of GDP is a better measurement for analysing the relative burden of adaptation costs to the economy. This shows that most low income regions such as South Asia and especially South East Asia and Western and Eastern Africa face high adaptation costs as a share of their GDP compared to the other regions. In the latter two regions, this ratio is four times the global average; South Asia has high adaptation costs due to their expected large impacts on health and the agricultural sector; for Western and Eastern Africa this is mainly due to health impacts.

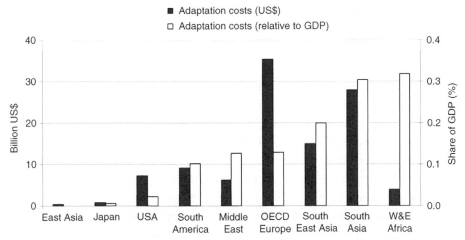

Figure 4.3. Adaptation costs as share of GDP and in US$ for several world regions in 2050, according to the adaptation cost curves of AD-RICE and assuming a mitigation scenario leading to a maximum temperature increase of 2 °C above the pre-industrial temperature.

The large regional differences in the projected adaptation costs, with higher costs especially in most non-Annex I regions, together with the importance of adaptation in reducing potential damages, indicate the need for an international adaptation regime. The Adaptation Fund established during the UNFCCC conference in Bali in 2007 could form the basis of such a regime.

As a final note, we have to keep in mind the caveats involved in this analysis, most importantly the large uncertainty in the monetary estimates of climate change impact. Our conclusions are based on only one estimate of climate change damages and adaptation costs and do not take into account the uncertainty range of these estimates shown in Figure 4.1. Therefore, our estimates have to be interpreted with more than the usual care.

4.3 The European energy system – multi-national adaptation policies

To be specific about possible adaptation options, there is a need for quantitative assessments of the expected climatic changes. The options discussed in this section refers to the hybrid model system briefly described in Chapter 7 (Eskeland *et al.*) and more extensively presented in Jochem *et al.* (2009). The quantitative results are based almost exclusively on the expectations of rising average ambient air and river temperatures in Europe until 2050, given the fact that the predictability of changes due to extreme events (storms, drought, floods, heatwaves) is still in its infancy. While adaptation to extreme events is still an open issue, the question of how much multi-national action is needed to develop adaptation policies for the European energy sector remains.

The level of necessary adaptation of energy use and production will differ by country and sector, and strongly depends on future global greenhouse gas emissions. It also depends on present adaptability of the capital stock and resources as well as their future options[3]. The predicted adaptation needs for energy use and conversion are modest for the next 20 to 30 years, but increase significantly thereafter, depending on the speed of temperature rise in Europe. Despite the time lag, early action is needed, given the long lifetimes of buildings, industrial plants, transport infrastructure and electricity generation, transmission and distribution. It is not just a concern and responsibility of governments at the local or national level, but a policy issue at the EU level at least, if not at the Organisation of Economic Co-operation and Development (OECD) and emerging countries level. It is also an issue for investment in good producers, planners, architects, or institutions offering education and professional training.

4.3.1 *Adaptation in energy demand sectors*

For Europe south of the Alps, it is unlikely that increase in electricity demand for cooling will be outweighed by reductions in the need for heating energy. The net result in an Adaptation scenario[4], for instance, is about 1.7 EJ less final energy demand by 2050, or − 3.3 per cent for all 29 European countries studied (Jochem *et al.* 2009). However, the decline is only 1.7 per cent in southern Europe compared with 4.3 per cent in western Europe, and more importantly, the additional electricity demand of 7 per cent to be expected in southern Greece, Malta, Cyprus, southern Italy and Spain, Bulgaria and Romania. Electricity used for space cooling is presently far more carbon-intensive than energy used for heating (e.g. gas, heating oil, wood fuels) in most European countries. As electricity is more expensive than heat from fossil fuels or district heat, adapting to higher temperatures by the residential or service sector implies additional energy cost to maintain the same comfort in southern Europe.

This difference in adaptation among European countries gives rise to distributional issues between northern and southern countries. It may also lead to a greater

---

[3] E.g. existing types of buildings, cooling systems of buildings, industrial processes, and thermal power plants, potentials to adapt within the re-investment cycle and by new investments, geographical location, or socio-economic resources including know-how, technical expertise and capital of investors involved.

[4] The Adaptation scenario (also called a 4 °C scenario) assumes that global average temperature will rise by 4 °C compared to pre-industrial levels (Van Vuuren *et al.*, 2007).

need to balance summer electricity flows via the trans-European electricity transmission grid, particularly during extreme heatwaves.

Adaptation due to reduced heating supports mitigation measures that increase heat protection of buildings and facilitate other heating systems such as ventilation systems (instead of warm water systems) with heat exchange and driven by heat pumps. These new systems would easily allow cooling with the same ventilation system. Air conditioning is very ineffective today, using much electricity. Substantial improvements of these systems have to be made by intensive research and development (R&D). Also required are re-designed education of planners and architects, additional professional training, and changing technical regulations for new types of low energy buildings and passive houses that include air conditioning in southern European countries. R&D as well as the new technical standards have to be initiated by directives of the European Commission and have to be partially pursued in R&D programmes of the Commission. Left to private investors or national governments alone, adaptive changes may lead to maladaptation in the form of technological 'lock-in' situations due to underinvestment (e.g. in insulation) or wrong decisions (e.g. regarding the heating system) with only marginal benefits arising. In addition, if there is no adaptation policy at the European level, markets for highly efficient heating and cooling equipment will disperse, mass production and related economics of scale will not be adequately used, and hence heating and cooling would be unnecessarily costly.

4.3.2 Adaptation in electricity supply and water systems

Similar conclusions can be drawn from electricity supply, which is very likely to increase in importance in the European energy system. Decreased precipitation and warmer temperatures in summer (particularly heatwaves) will force additional investments in cooling systems of thermal power plants, which form the majority of the European power generation system. To avoid risks of electricity supply disruptions, an integrated adaptation strategy will involve the investment in additional power lines, reinforcement of transmission lines, and decentralisation of power generation.

There are other cross-sectoral policy opportunities for European adaptation technology, such as water supply, with which southern European countries again have to face the challenge of decreasing precipitation. Investments in irrigation systems, re-use of secondary water, long distance water transfers, or even in costly desalination may have to be made. This foreseeable challenge should be taken up by R&D at the European level in order to develop highly efficient water supply and re-use systems in time. Such a strategy will also provide important economic benefits when exporting technologies to many emerging and developing countries in Asia, Africa and Latin America facing similar problems.

To conclude, the example of the energy system shows that low cost adaptation requires a future-orientated and integrated R&D strategy at the European level. Highly efficient cooling and water systems are not only important for southern European countries, but also offer an excellent export potential for the European industry. Improved long distance transmission connectivity, smart grids, and electricity storage facilities have important linkages to mitigation options. The new technologies, in particular ventilation and cooling of buildings, need technical standards induced by directives from the European Commission. They also require changes in the curricula for education and professional training of planners, architects and craftspeople. The issue is urgent, given the long lifetimes of buildings, heating and cooling systems, power plants and transmission lines, and so policy action, at least at the European level (or at the OECD level by the International Energy Agency (IEA)), is absolutely necessary in the next few years.

4.4 Market barriers to adaptation

The previous section shows that adaptation may reduce the impacts of climate change considerably if the potential is fully utilised. However, climatic conditions, vulnerabilities and adaptive capacity vary from one local community to another, which may explain why Klein *et al.* (2007) find that most of the adaptation literature focuses on local cases. When national markets are considered as a supermarket, where all transactions take place without costs, as usually considered in computable general equilibrium (CGE) models, possible implications of variability are ignored.

For instance, it can be assumed that climate change affects the natural resources in a community to such an extent that people have to move to a new place and start work in a different sector of the economy; CGE models are designed to explain this shift from one sector to another, but assume that it is frictionless. In reality, there are barriers to moving, at least over longer distances: people depend on family and other social relations, real estate cannot be moved, and the value may fall. Moreover, real capital cannot always be transformed into something useful in a different sector, at least not without costs. A limitation of the negative consequences of these structural impacts could constitute an important part of a national strategy of adaptation to climate change. One challenge, however, is to develop appropriate analytical tools.

This section suggests a way by which these market barriers to adaptation may be identified in CGE models. The approach has two steps. Firstly, impacts of climate change are attached to specific economic activities of the model, such that the behaviour of economic agents becomes subject to the climatic conditions to which they adapt. Secondly, the national or regional economies are divided into provinces in order to capture variabilities of climatic changes. Barriers to adaptation are taken into account by limiting the ability to move production factors between provinces. We start with a survey of climate impacts in Europe, which is based on available studies. Then, we show how the results are used and integrated into a CGE model for the Iberian Peninsula.

4.4.1 Impacts of climate change

Adaptation is driven by the impacts of climate change, which are integrated into the CGE model GRACE (Aaheim and Rive, 2005) by adjusting affected activities as described by Aaheim and Schjolden (2004). For example, losses in resource based sectors, such as agriculture, are interpreted as a reduction in the availability of natural capital, and extreme events imply loss of real capital. Changes in the demand for commodities and services directly caused by climate change are represented by shifts of demand. The external estimate is, in other words, interpreted as the change in output as if the composition of all input factors, including primary input, remains constant after climate change has taken place. The model then re-allocates input according to market responses. This can be interpreted as autonomous adaptation.

The adjustments are based on existing literature, but major revisions had to be made to fit the results from different studies to a consistent change of climate across Europe. This increases the uncertainty of the estimates, which are highly uncertain from the outset. For this study, the estimated impacts are based on the SRES A2 scenario (Nakicenovic and Swart, 2000), which expects a global warming of +3.1°C for the period 2071–2100 compared to the control period 1961–1990. Adjustments for the European regions represented in the model are presented in Table 4.1. The variations across regions are based on maps for changes in temperature and pre-cipitation in PESETA (2007). For some impacts, the background for the assessment is extremely poor. Fish stocks are assumed to move north, and the changes may be substantial, but there are no available estimates. Extreme events are estimated by a simple extrapolation of a climate related trend of current events. A closer study of market responses to extreme events is presented in Chapter 8.

In general, southern parts of Europe are likely to be more negatively affected than northern parts. Agricultural productivity is expected to decrease in the south, but this is outweighed by an increase in the north. The total effect is assumed to be positive (PESETA, 2007). A similar pattern of change is expected for forests (Alcamo *et al.*, 2007; Fronzek and Carter, 2007). Impacts to fisheries will probably be large (Alcamo *et al.*, 2007): negative in some regions and positive in others, but in any case highly uncertain. Energy demand for cooling will increase as tempera-ture rises, leading to an increased demand for electricity, while the demand for energy for heating will decrease (de Cian *et al.*, 2007; Tol, 2002). The potential for the most important renewable electricity source in Europe, hydropower, is expected to decline by about 6 per cent by the 2070s (Lehner *et al.*, 2005) because of a drier

Table 4.1. *Estimates of direct impacts of climate change by sector and region in Europe. Percentage change under climate change compared with a situation without climate change*

Expected changes		The Baltics[1]	British Isles[1]	Centre East[1]	Centre North[1]	South. Eur[1]	Centre West[1]	Iberia[1]	Nordic Count.[1]
Climate change	Temperature (°C)	+3.0	+2.5	+3.5	+3.5	+4.0	+3.5	+4.0	+3.5
	Precipitation (per cent)	+13	0	−5	+5	−15	0	−25	+10
Impact on economic sectors	Agricultural land	−0.5	−8.3	−1.5	3.1	−2.1	−5.1	−11.0	15.6
	Mass of forest	3.1	0.2	−1.3	9.5	−3.7	−7.2	−15.5	29.5
Percentage	Stock of fish	25.3	−4.8	−4.6	7.3	−50.0	−12.4	−51.5	14.1
	Electricity generation	1.3	0.2	−3.8	−2.0	−3.5	−1.2	−1.6	15.1
	Energy demand	5.4	−10.1	−8.3	−6.9	−7.1	−5.8	−0.3	−2.0
	Industry	–	–	–	–	–	–	–	–
	Transport demand	2.2	4.1	1.0	5.5	−2.3	1.1	−3.4	2.8
	Services demand	1.6	1.8	0.7	1.4	−0.9	0.3	−1.5	2.1
Extreme events (percentage capital loss per year)		0.7	0.6	1.2	0.5	0.7	0.5	1.2	0.4

1. The regions are: Baltics: Estonia, Latvia, Lithuania, Poland; British Isles: Ireland, United Kingdom; Central East: Bulgaria, Czech Rep., Hungary, Romania, Slovakia; Central North: Austria, Germany, Switzerland; Southern Europe: Cyprus, Greece, Italy, Slovenia, Central West: Belgium, France, Luxembourg, the Netherlands; Iberia: Portugal, Spain; Nordic: Denmark, Finland, Iceland, Sweden, Norway.

south. The direct impacts to the industrial sector are assumed to be zero. This sector will, however, be affected indirectly through market effects.

Impacts to the transport sector are difficult to estimate, but infrastructure is likely to be damaged by more extreme events. Modal shifts in passenger preferences are also likely (Wooller, 2003). Tourism is an important sector in the European economy and, in the model applied below, tourism is part of the service sector. Changes in tourism also have an impact on demand for transport services. The total amount spent on tourism is not expected to change, but there may be a significant change in destination (Berrittella *et al.*, 2006) as climate has an important bearing on the decision. With hot summers in the Mediterranean and drier summers in the north, tourism is expected to flow northwards (Hamilton *et al.*, 2005).

The general picture of climate change in Europe is that temperature will increase and humidity decrease in the warm and dry areas of the south. The colder and more humid areas of the north however, face an increase in temperature close to, or a little lower than, the average, and higher humidity. This has major negative consequences for Southern Europe, but a positive impact in the north. The Baltics and the Central East of Europe have less capacity to take advantage of the possible benefits. This is partly related to the fact that resource-based activities such as agriculture, forestry and fisheries, where production technologies substantially differ, are affected most by climate change.

4.4.2 Implications of variability across sub-regions

To take into account the variability across provinces within the eight European regions in Table 4.1, an estimate of each impact in each province is needed. This information is available only for countries, while the provinces of the model represent a further level of disaggregation. The alternative is to specify relationships between estimated impacts and climate parameters, based on available information. One option is to calibrate a chosen relationship for each country by the estimate for that country, and then calculate a relationship for the whole region. Another is to assume that the same relationship applies to all of Europe, and estimate this relationship on data from all the countries. The latter alternative is chosen for this study. The strength in this approach is that the relationship is based on multiple studies, even though each single study provides a very uncertain estimate. The weakness is that applying a common relationship between a few climate indicators and economic impacts to all of Europe is, indeed, making a strong assumption.

The justification of this choice is illustrated in Figure 4.4, which shows the difference between 'observed' (one dot for each country) and estimated changes (line) for agriculture. The relationship assumes that the productivity of land in agriculture is

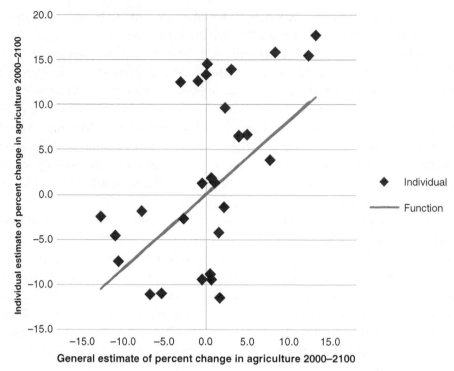

Figure 4.4 Observed and estimated impacts estimates for agriculture. Percentage change of productivity in agriculture as a result of climate change. Estimated change from individual studies compared with estimates of a general impacts function.

a function of temperature level and level of precipitation. From changes in temperature and precipitation, the change in productivity is estimated. For example, higher temperature leads to a reduction in productivity in warm regions, and an increase in cold regions. The observations admittedly are spread out in the diagram, but it may be noted that, for most countries, there is a correspondence between the observations and estimations when comparing positive and negative impacts.

Similar relationships for electricity production were based on a combination of changes in temperature and precipitation, in addition to the share of hydropower. Impacts to forestry, energy demand, tourism and fisheries refer to temperature only. Having estimated these impact relationships, variability of impacts across provinces may be taken into account. The variability depends partly on the variability of climate change, and partly on the different contributions from different sectors in provinces. It is assumed that the same 'technology' applies for all sectors in all provinces of a region. Barriers to adaptation occur as a result of constraints on the equilibrium conditions. These are imposed by province for labour and real capital, and by province and sector for natural capital.

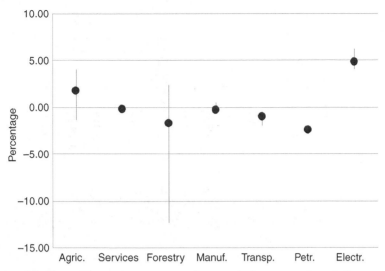

Figure 4.5 Change in sector output under equal changes and variations under differentiated impacts.

Table 4.1 shows that the Iberian peninsula may be affected significantly by climate change. The PESETA study gives a higher than mean expected increase in temperature with a much lower annual precipitation. The immediate impact is a loss in all sectors, particularly the resource-based sectors. However, a reduction of supply increases the prices, and the economic results are more positive. The dots in Figure 4.5 show the economic impacts after market responses if all sub-regions are affected equally. Agriculture output is higher, which is explained by an increase in prices. The increase in electricity production is due to a combination of higher demand and higher prices. The output in other sectors is reduced and some sectors are excluded from the figure, because of highly uncertain impact estimates.

If the impacts are differentiated according to expected variability of climate change and to the relative importance of each sector in the ten provinces into which Iberia was divided, the picture changes. The variability is indicated by the bars in Figure 4.5, which show the minimum and maximum change by sector. As expected, the main variabilities occur for the resource-based sectors. For some provinces, the loss of productivity in agriculture is too large to be compensated for by the increase in prices. For forestry, there is an increase in output in some regions, even though the total output from forestry reduces for the whole region. The total effect on the gross domestic product for the entire Iberian region is slightly more negative when incorporating the variability. This indicates that markets represent an important driver for adaptation within the provinces, although the losses are differentiated across regions.

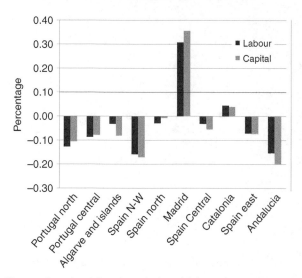

Figure 4.6. Change in price of real capital and labour by province in the case of differentiated impacts.

As mentioned earlier, variability may be a motivation for migration of labour and capital. To identify vulnerable regions, the change of factor prices resulting from climate change under differentiated impacts can be compared. Figure 4.6 shows the change in capital price and wages by province. Higher wages and higher return on capital occur in the Madrid province and Catalonia, which includes Barcelona. The factor prices decrease in all the other provinces. Climate change thus seems to reinforce the pressure on already loaded areas. Possible negative implications of market barriers to adaptation occur in the north west of Spain and in the south (Andalucia).

4.5 Risk-based economic planning for extreme events

To some extent, the management of disaster risks falls within the realm of local and regional authorities, yet central governments are also key actors. Generally, they assume responsibility for replacing damaged infrastructure, providing post-event relief and ensuring rapid recovery of the economy overall. The associated planning problem in this subset of the economics of disaster risk management and adaptation, is one of contingency liability planning, and it is generally within the realm of central governments to take the appropriate measures.

The literature on the economic modelling of extremes is only now emerging. Extreme events have been included in some modelling studies of adaptation, with global or regional resolution. They use add-on damage functions based on average

Table 4.2. *Government liabilities and disaster risk*

Liabilities	Direct: obligation in any event	Contingent: obligation if a particular event occurs
Explicit government liability recognized by law or contract	Foreign and domestic sovereign borrowing, Expenditures by budget law and budget expenditures	State guarantees for non-sovereign borrowing and public and private sector entities, reconstruction of public infrastructure
Implicit; a 'moral' obligation of the government	Future recurrent costs of public investment projects, pension and health care expenditure	Default of subnational government and public or private entities, disaster relief

Source: *Modified after Schick and Polackova Brixi*, 2004.

past impacts and are contingent on gradual temperature increase (Nordhaus and Boyer, 2000; Pielke and Sarewitz, 2005; Hope, 2006). Also, with respect to the economics of disaster risk management, there is a substantial, yet heterogeneous body of research. Existing approaches utilise a wide variety of models such as input–output, computable general equilibrium, economic growth frameworks and simultaneous-equation econometric models. One issue with the modelling undertaken in this line of research is the deterministic nature of the modelling approaches. In essence, disaster risks have been represented as averages (expected annual losses), or as singular events based on event information on actual disasters (see Mechler, 2004; Hochrainer, 2006). As an alternative, we suggest a probabilistic approach, which links up with concrete adaptation decisions at hand. It is, moreover, based on the direct relationship between disaster risks and key vulnerabilities, which reflects the nature of low probability/high impact of natural disasters.

Disaster risk emanates from explicit and implicit contingent public sector liabilities, classified in Table 4.2. The explicit liability consists of rebuilding damaged or lost infrastructure, which is due to the public sector's allocative role in providing public goods. Implicit liabilities are related to the commitment of providing relief due to the distributive function in reallocating wealth and providing support to the needy (see Table 4.2).

The case of Austrian large-scale flooding in 2002 (estimated to have been a 100–150-year event) provides an interesting example underlining the need for planning for extremes (for the Austrian case and the modelling framework used see Chapter 8, Mechler *et al.*). To cover part of a total loss of €3 billion, the government had to provide reconstruction and relief support of about €500 million with another

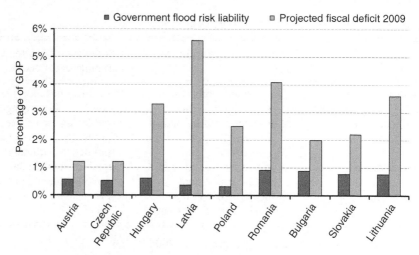

Figure 4.7. Government liabilities and disaster risk in selected flood-prone European countries. (Source for economic data: European Commission, 2008.)

€140 million provided by the EU solidarity fund. The money essentially had to be diverted from other planned spending leading to disruptions in the coalition government and eventually its collapse. In contrast, addressing liabilities and potential fiscal risk in advance is likely to lead to reduced costs and less fiscal volatility.

The above example shows how important it is to ensure sufficient flexibility in a government's budget to fund unexpected projects without weakening the sustainable fiscal position (see Heller, 2005 and Heller, 2007). This flexibility in the budget can be called fiscal space which, as an analytical concept, has been receiving increasing attention in the wake of the financial crisis with constrained public sector access to financial capital markets. EU countries, such as Hungary and Latvia, have been unable to get access to funds, and Hungary recently had to approach the World Bank for a US$20 billion credit line.

Limited fiscal space coupled with substantial contingent liabilities may be unsustainable. There are a number of emerging economy countries, for example, the new Eastern European member countries, which have joined the EU and need to adapt to more rigorous and changed fiscal and economic conditions. As shown in Chapter 8 (Mechler *et al.*) most of these countries are also subject to large scale flood risk. Given the government's role in absorbing flood risk in terms of implicit and explicit fiscal liabilities, we can compare this to the fiscal position. Figure 4.7 compares flood risks as a proportion of GDP to projected deficits for the flood prone Eastern European countries, with Austria as a reference point.

Hungary, encompassing the Tisza River basin to the east (see Chapter 9, Werners *et al.*), is an interesting case for several reasons. In Europe, Hungary ranks only behind the Netherlands with respect to geographical area exposed to floods. Over half of the country's territory, two-thirds of its arable land and a third of its railways are exposed to riverine, ground water and flash floods. Adding to the geographical scale of the problem, floods appear to be worsening in their intensity and frequency (Feyen *et al.*, 2008). With increasing losses, the Hungarian Government is concerned about continuing its tradition of taking almost full responsibility for flood risk management, including flood prevention, response, relief and public infrastructure repair. With membership in the European Union, Hungarians have committed to a programme of fiscal austerity.

Using the risk-based catastrophe simulation (CATSIM) modelling framework, we address the risk planning problem for Hungary. The model estimates for a country the fiscal and economic consequences of disasters that occur as stochastic events (see Hochrainer, 2006). The events are generated based on the risk estimation discussed in Chapter 8 (Mechler *et al.*). The model shows how monetary disaster risks may be absorbed by the economy in the short and in the long term and assesses a government's contingent disaster obligations and the potential shortfalls for financing (*financial vulnerability*), as well as the costs and benefits of vulnerability-reducing options. The model incorporates rare disasters explicitly as probabilistic events. Decisions on adaptation are thereby based on the whole range of possible future scenarios.

In order to fund spending needs if contingent liabilities become manifest in terms of disaster losses, governments may raise funds *ex post* by, *inter alia*, diverting from the budget, raising taxes or acquiring additional capital in domestic or international markets. Alternatively, in the *ex ante* approach, financial planning mechanisms such as reserve funds or public insurance instruments may be used. Figure 4.8 shows schematically how losses and financing sources interact.

Financial vulnerability results from the difference between the contingent post-disaster liabilities of the government and the sources of funding available to the government. It can be assessed by simulating the risks to public assets and estimating the government's ability to cover these risks as well as to provide private sector assistance. The shortfall in financing for a given event is the *resource gap*. Incorporation of disaster risks and potential resource gaps into macroeconomic projections leads to an analysis of the consequences for variables, such as economic growth or a country's fiscal position. Because the fiscal position in the future is dependent on the occurrence and magnitude of past disaster events (as well as the financing ability), the whole range of possible futures has to be simulated (see Chapter 8, Mechler *et al.*). In Hungary, for example, according to our analysis, government disaster liabilities can lead to important fiscal

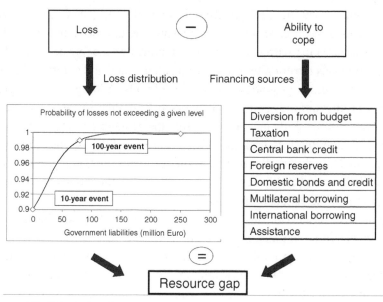

Figure 4.8. Calculating the resource gap.

consequences. While the fiscal space in Hungary without disaster events would increase by approximately 70 per cent in 2018, including disaster shocks, it would be substantially reduced. Taking all scenarios into account, the mean reduction would be approximately 50 per cent in 2018 compared with 2009. Hence, the analysis suggests that disasters may significantly add to fiscal volatility both today and in a future climate with increasing flood (and other) losses. Planning may be necessary to avoid these impacts.

There are several options the government can consider in pro-actively reducing its exposure to such risks, including public assets insurance, implementing or upgrading a national reserve fund, or limiting its relief provision. In the Hungarian case, given large disaster liabilities and faced with severe economic and fiscal challenges, more private responsibility for the reduction of losses from flood disasters may be considered an appropriate strategy, for example, by purchasing flood insurance. This enhances individual responsibilities and may prevent additional losses due to a reduction of moral hazard. On the other hand, shifting the liability of flood risks from the realm of public responsibility to individuals puts a heavy load on the most vulnerable, which, in the Hungarian case, are those people living in the poorest areas (Linnerooth-Bayer *et al.*, 2006). Moreover, flood losses, including relief to flood victims, can be considered as public goods, which will have to be a public responsibility. This highlights the importance of clarifying the role of the public sector in disaster risk management and adaptation strategies and the need for finding a balance between efficiency and fairness.

4.6 Conclusions

Most of the research on adaptation addresses options and obstacles to adaptation for individuals or local stakeholders. When considering adaptation in a context of national policy making, for example, as a part of a coherent strategy for climate policy, both the means and the targets are different from those on a local level. The responsibilities of national authorities are related, in particular, to initiatives that will not be taken, or be insufficient, if left to the local level. We have associated these responsibilities with public goods; in an attempt to identify and assess the potential of developing national strategies for adaptation, we have addressed four issues.

Firstly, the choice of adaptation strategy should be considered a part of a coherent effort to mitigate climate change. Adaptation reduces the damages of climate change significantly, even though there are limits as to how far it is possible to go. Substantial so-called residual damages remain after beneficial adaptation has taken place. There is however, more to lose if means to adapt in advance of expected changes are neglected, than if adaptation is exaggerated in advance.

Secondly, the example of the energy sector shows that an early adaptation policy in Europe opens up large opportunities for the associated industries and services to deliver low cost adaptation technologies to other parts of the world that have to face similar, or even more, adaptation to climate change.

Thirdly, a national strategy for adaptation may be useful to limit negative impacts of variability. These may occur as a result of market barriers related to the costs of moving production factors from place to place. The example from Iberia indicates that variabilities tend to reinforce the pressure on already loaded provinces, while labour and capital will move out from less loaded provinces.

Fourthly, extreme events represent particular challenges as their financial impacts may spread to other countries and into later periods, thereby affecting fiscal stability and economic growth. The example from Hungary shows that flooding today, and in a future with increased flood risk, may have severe fiscal and macroeconomic consequences, which depend profoundly on global economic performances.

Although the distinction between national and local responsibilities for adaptation is rarely addressed in the literature, issues discussed in this chapter, such as the benefits of adaptation, have been dealt with on several occasions. The purpose here has been to contribute to a clarification of the responsibilities at the national and supernational levels. This may facilitate implementation of adaptation measures, for example, in the energy system, or it may limit the consequences of extreme events. It may also help to identify objectives for national adaptation strategies, such as to limit the impacts of climate change in order to relax the need for mitigation, or to address potential losses due to variability within countries and regions. This chapter suggests the importance of the identification of governments' roles, but is far from comprehensive. We have used public goods as the common denominator for the subjects to which governments should pay attention. Much remains, however, before an extensive overview of adaptation options, which can be characterised as public goods on a national scale, can be provided.

References

Aaheim, A., Berkhout, F., McEvoy, D. *et al.* (2008) Adaptation to Climate change: why is it needed and how it can be implemented, *CEPS Policy Brief* no. 161, May 2008.

Aaheim, A. and Rive, N. (2005). *A Model for Global Responses to Anthropogenic Change in the Environment (GRACE)*, Report 2005:**5**. Oslo: CICERO.

Aaheim, A. and Schjolden, A. (2004) An approach to utilise climate change impacts studies in national assessments. *Global Environmental Change*, **14**(2), 147–60.

Alcamo, J., Moreno, J. M., Nováky, B. *et al.* (2007) Europe. In *Climate Change 2007: Impacts, Adaptation and Vulnerability. Contribution of Working Group II to the Fourth Assessment Report of the Intergovernmental Panel on Climate Change*, ed. Parry, M. L., Canziani, O. F., Palutikof, J. P., van der Linden, P. J. and Hansen, C. E. Cambridge, UK: Cambridge University Press, pp. 541–80.

Berrittella, M., Bigano, A., Roson, R. and Tol, R. S. J. (2006) A general equilibrium analysis of climate change impacts on tourism. *Tourism Management*, **27**, 913–24.

de Bruin, K. C., Dellink, R. B. and Agrawala S. (2009a). *Economic Aspects of Adaptation to Climate Change: Integrated Assessment Modelling of Adaptation Costs and Benefits*. Paris: OECD.

de Bruin, K. C., Dellink, R. B. and Tol, R. S. J. (2009b) AD-DICE: an implementation of adaptation in the DICE model. *Climatic Change* (in press): doi 10.1007/s10584–008–9535–5

de Cian, E., Lanzi, E. and Roson, R. (2007) *The Impact of Temperature Change on Energy Demand: A Dynamic Panel Analysis*. FEEM Working paper no. 46. 2007, Fondazione Eni Enrico Mattei.

EC European Commission (2008) *General Government Data. General Government Revenue, Expenditure*, Brussels: Balances and Gross Debt.

Feyen, L.; Barredo, J. I.; Dankers, R. (2008) Implications of global warming and urban land use change on floodings in Europe. In *Proceedings of The International Urban Water Conference, Water and Urban Development Paradigms Towards an Integration of Engineering, Design and Management Approaches*, ed. Feyen, J. *et al.*, Heverlee, Belgium, pp. 217–25.

Fronzek, S. and Carter, T. (2007) Assessing uncertainties in climate change impacts on resource potential for Europe based on projections from RCMs and GCMs. *Climatic Change*, **81**, 357–71.

Hamilton, J. M., Maddison, D. J. and Tol, R. S. J. (2005) Climate change and international tourism: a simulation study. *Global Environmental Change Part A*, **15**, 253–66.

Heller, P. (2005) *Understanding Fiscal Space*. IMF Policy Discussion Papers Wahington DC: International Monetary Fund. <http://ideas.repec.org/p/imf/imfpdp/05–4.html>

Heller, P. (2007) *Fiscal Policy for Growth and Development: the Fiscal Space Debate*. Istanbul: Background paper prepared for the G-20 workshop on fiscal policy.

Hochrainer, S. (2006) *Macreoeconomic Risk Management Against Natural Disasters*. Wiesbaden, Germany: Deutscher Universitaetsverlag (DUV).

Hope, C. (2006) The marginal impact of CO_2 from PAGE2002: an integrated assessment model incorporating the IPCC's five reasons for concern. *Integrated Assessment*, **6**, 19–56.

Jochem E., Barker T., Scrieciu S. *et al.* (2009) *EU-Project ADAM: Adaptation and Mitigation Strategies: Supporting European Climate Policy*. Deliverable M1.2: Report of the Reference (4 °C) Scenario for Europe. Fraunhofer ISI, Karlsruhe.

Klein, R. J. T., Huq.S., Denton. F. *et al.* (2007) Inter-relationships between adaptation and mitigation. In *Climate Change 2007: Impacts, Adaptation and Vulnerability. Contribution of Working Group II to the Fourth Assessment Report of the Intergovernmental Panel on Climate Change,* ed. Parry, M. L., Canziani, O. F., Palutikof, J. P., van der Linden, P. J. and Hansen, C. E. Cambridge, UK: Cambridge University Press. pp. 745–77.

Lehner, B., Czisch, G. and Vassolo, S. (2005) The impact of global change on the hydropower potential of Europe: a model-based analysis. *Energy Policy*, **33**, 839–55.

Linnerooth-Bayer, J., Vári, A. and Thompson, M. (2006) Floods and fairness and Hungary. In Verweij, M. and Thompson, M. (eds) *Clumsy Solutions for a Complex World: Governance, Politics and Plural Perceptions*, pp. 181–204. Houndmills, Basingstoke, UK: Palgrave-Macmillian.

Manne, A. S. and Richels, R. G. (2004) *MERGE: An Integrated Assessment Model for Global Climate Change*. Stanford University.

Mechler, R. (2004) *Natural Disaster Risk Management and Financing Disaster Losses in Developing Countries*. Karlsruhe, Germany: Verlag fuer Versicherungswissenschaft.

Nordhaus, W. D. and Boyer, J. (2000) *Warming the World: Economic Models of Global Warming*. Cambridge, MA: MIT Press.

Nakicenovic, N. and Swart, R. (eds) (2000) *Special Report on Emissions Scenarios*. Cambridge, UK: Cambridge University Press.

PESETA (2007) Limiting global climate change to 2 degrees celsius. The way ahead for 2020 and beyond. In *Communication from the Commission to the Council, the European Parliament, the European Economic and Social Comittee and the Committee of the Regions. Limiting Global Climate Change to 2 degrees Celsius. The way ahead for 2020 and beyond. Impact Assessment*. Commission of the European Communities, 57 pp.

Pielke, R. A. J. and Sarewitz, D. (2005) Bringing society back into the climate debate. *Population and Environment*, **26**(3), 255–68.

Robinson, M., Wallström, M. and Brundtland, G. H. (2008) Global justice must be part of the emission debate. *Canberra Times*, 12 December 2008.

Schick, A. and Polackova Brixi, H. (eds) (2004) *Government at Risk*. Washington, DC: World Bank and Oxford University Press.

Tol, R. S. J. (2002) Estimates of the damage costs of climate change, Part II. Dynamic estimates. *Environmental and Resource Economics*, **21**(2), 135–160.

Tol, R. S. J. (2003) *Adaptation and Mitigation: Trade-Offs in Substance and Methods*.
    Working paper FNU-33, Hamburg Vrije Universität.
UNFCCC (2007) *Climate Change: Impacts, Vulnerabilities and Adaptation in Developing
    Countries*. New York: United Nations.
Wooller, S. (2003) *The Changing Climate: Its Impact on the Department for Transport*. UK
    Department for Transport. http://www.dft.gov.uk/print/pgr/scienceresearch/key/the
    changing climateitsimpact01909H.

5

Learning to adapt: re-framing climate change adaptation

Lead authors:

JOCHEN HINKEL[1], SANDY BISARO, THOMAS E. DOWNING,
MAREEN E. HOFMANN, KATE LONSDALE, DARRYN MCEVOY,
J. DAVID TÀBARA

[1]Co-ordinating lead author

Summary

While research on climate change adaptation is still framed predominantly from an impact modelling and decision-analytical perspective, several limitations with this perspective have become apparent: climate change impacts are often difficult to project; information on impacts often does not meet the needs of decision makers; and institutional barriers continue to hinder the implementation of adaptation options. This chapter offers a different perspective by re-framing adaptation to climate change as a process of social learning involving scientific, policy and practitioner communities. We argue that this framing is valuable, because adaptation situations are complex and diverse, with multiple actors at different scales, differing perceptions and a lack of consensus as to what constitutes effective responses. Hence, in many instances, adaptation is not a matter of a single decision based on certainty about impacts and adaptation options, but a continuous learning process between actors and institutions at all levels of decision making. From this perspective, we present four distinct adaptation situations: (i) national adaptation policy; (ii) local risk management; (iii) regional trans-boundary co-operation under increasing water resource pressures; and (iv) local resource management in a developing world context. For each situation, we discuss what enabled or hindered adaptation. Our findings suggest that scientific evidence is an important enabling factor at higher levels of decision making. At more local levels, however, other factors such as awareness brought about by the direct experience of climate-related shocks, social networks and space for learning are influential. Higher level institutions play an important role in providing these incentives, but current institutional arrangements tend not to be directly connected to the climate change issue; a redistribution of institutional responsibilities is often crucial. We conclude that more research and space for social learning is needed to find out what kind of adaptation response is effective for a given adaptation situation.

Making Climate Change Work for Us: European Perspectives on Adaptation and Mitigation Strategies, ed. Mike Hulme and Henry Neufeldt. Published by Cambridge University Press © Cambridge University Press 2010.

5.1 Introduction

Research on climate change adaptation is framed predominantly from an impact modelling and decision-analytical perspective. Climate scenarios, models and data are used to derive information on potential impacts, adaptive capacities and vulnerabilities. Then, based on this information, adaptation measures are identified, evaluated and compared via multi-criteria, cost-effectiveness or cost–benefit analysis (see e.g. Carter *et al.*, 1994; 2007; Adger *et al.*, 2007). This is reflected in the majority of current research activity, which seeks to compare model results and map out model uncertainties, as well as improving the estimation of costs, benefits and the effects of adaptation measures.

Several limitations of this research framing have become apparent. Impacts and effects of adaptation options are often difficult, sometimes impossible, to project. Regional climate models offer no coherent picture of what changes to expect, and impact models, embracing similar uncertainties, are only available for a limited number of the relevant sectors. Even when it is possible to assess impacts with some confidence, the information produced often has little influence on adaptation decisions, because decision makers were not involved in the assessment process and the information produced is often not perceived to be credible, relevant or salient (Cash *et al.*, 2003; Haas, 2004). Moreover, the literature increasingly acknowledges institutional barriers to adaptation, i.e. institutions can actually prevent the implementation of options identified as being effective, or affect the translation of adaptive capacity into action (Adger *et al.*, 2007). Indeed, evidence from institutional and policy analysis shows that successful adaptation often depends on the interplay between formal and informal institutions as well as policy processes operating at multiple levels of decision making (e.g. Pelling and High, 2005; Pelling, 2006; Young, 2006; Berkes, 2007).

These difficulties illustrate that our understanding of adaptation, in many instances, must be framed differently. Adaptation involves many actors with different interests at various levels of decision making. There is often no clear argument about what exactly the adaptation problem is; there is uncertainty and ambiguity as to how improvements might be made; and the problem is potentially unlimited in terms of the time and resources it could absorb. These kinds of problems have been called *wicked* (Rittel and Webber, 1973; Darwin *et al.*, 2002) or *unbounded* (Chapman, 2004). Wicked adaptation problems require different approaches, away from the 'predict and provide' approaches of the impact modelling and decision analytical framing described above, to a more enquiry-based and reflexive approach that enables us to question our underlying assumptions about what is happening and what needs to change (Senge, 1990).

Social learning is gaining recognition as a concept that offers such a reflexive approach (Armitage *et al.*, 2008). It is said to occur whenever people are successful in implementing conscious changes in cognitive and normative frameworks of action (i.e. in re-framing), as well as tailoring institutional arrangements in a way that allows them to achieve collective goals that cannot be achieved individually (Folke *et al.*, 2005; Pahl-Wostl 2007). A social learning process explicitly acknowledges different framings of the problem and fosters the possibility to re-frame as new experiences are made. In the context of climate change adaptation, a 'successful' social learning would be collaboration between actors from the scientific, policy and practitioners communities to develop new ways of thinking and (inter-)acting, together with new institutional arrangements to better cope with climatic risks (Tàbara, 2005; McEvoy *et al.*, 2008a; Tàbara *et al.*, 2009; see also Chapters 1 and 2).

We suggest that better integrating the concept of social learning within the field of climate change adaptation requires two developments. Firstly, more attention must be placed on understanding the requirements for social learning, i.e. what institutional arrangements can be useful platforms for learning to occur (Pahl-Wostl, 2007). Learning processes can be blocked by, amongst other factors, power or perceived power relationships, disciplinary jargon, 'black boxed' technologies and attitude (Pasteur and Scott-Villiers, 2004). Essentially, if individuals are scared, confused, bored, and too busy, etc., the potential for learning will be reduced. That said, evidence also suggests that conditions for social learning can be promoted by clear and collaborative leadership; polycentric and transparent institutional designs; and the creation of spaces for interaction (Mostert *et al.*, 2007). Secondly, researchers, policy makers and the wider stakeholder community then need to engage in a collaborative learning process aimed at identifying useful framings for a given adaptation situation. Particular attention should therefore be paid to improving our understanding of what acts to facilitate or hinder adaptation. This is necessary as climate risks and associated adaptation responses are often context specific, differing in scale and the type of actors and institutions involved. In each situation, different framings of adaptation may be appropriate.

Addressing this emerging agenda, this chapter attempts to identify and analyse some of the key factors that either enable or hinder climate change adaptation and associated social learning in four adaptation situations:

(i) national adaptation policy,
(ii) local risk management and early warning,
(iii) regional trans-boundary co-operation under increasing water resource pressures, and
(iv) local resource management in a developing world context.

Each of these situations is presented in two parts with different formats. The first part presents the situation in the form of a stylised narrative, 'telling' the adaptation story chronologically without using technical terms or providing names of geographical locations, organisations or individuals. We have chosen this format for several reasons. Firstly, it grounds knowledge on adaptation in ordinary language; a narrative tells a story to which it is easy to relate. Secondly, adaptation is a continuous process, which must be seen in the context of its history; in our opinion, current literature on adaptation often fails to reveal key insights because situations are not described chronologically. Thirdly, adaptation situations are diverse and complex. In order to be able to understand what hinders or promotes adaptation, it is necessary to present the relevant issues with as much clarity as possible; omitting unnecessary details such as references, names, etc. helps to concentrate on the necessary detail. The second part then returns to more normal scientific parlance. We embed the adaptation story into its 'real world' context, introduce relevant literature and analyse the issues more thoroughly. Subsection headers highlight factors that proved to be important for promoting or hindering adaptation and social learning.

5.2 National adaptation policy

Narrative

In the late 1980s a notably strong and influential prime minister, with a degree in chemistry, first highlighted the potential seriousness of the threat of a changing climate on this island nation. A decade of interdisciplinary research collaborations had been building up a sound body of evidence suggesting that the climate was changing and that it was having a discernable influence on the ecology with the potential for a loss of species and habitats. This caught the attention of the wildlife-loving people and, due to the visual nature of the story, was often dramatically portrayed in the media. Supporting this was the development of global climate models by the national meteorological office, which were taken up as the de facto *scenarios for climate impact studies around the world. Another significant development in the late 1990s was the founding of a national, government-funded organisation to communicate the scientific evidence of climate change and, through sectoral and regional networks, to support stakeholders to respond to the challenges and opportunities.*

Severe, nationwide floods in 1998 and 2000 and a record-breaking European heatwave in 2003 further promoted the urgency of the climate change on the political agenda. The tone of the media coverage shifted from a debate between climate change sceptics and believers to an enquiry into realistic responses. In 2006, a review of the financial implications of coping with climate change was published. The report, overseen by an eminent economist, was taken seriously by individuals and sectors that had previously been unconcerned by the issue. This pushed the national government to

> *make further commitments to support the process of adaptation. A national 'Adapting to Climate Change' programme was created to co-ordinate and drive forward the work of the government and the wider public sector, raising awareness of the need to act now, and embedding this in government policies at all levels. At the local level, there was a commitment of further funding for the stakeholder support organisation and a local government process indicator on adaptation to climate change.*

Context

The above narrative describes the development of a national adaptation policy framework in the UK. Planning for adaptation is well on its way in the UK[1]. In July 2008 the Government published 'Adapting to Climate Change in England: a framework for action', which summarised the Government's Programme and sets out a vision for a UK that is adapting well to the impacts of climate change. The local government process indicator 'National Indicator 188' on adaptation to climate change is one of a set of indicators that measure performance in local government in the UK. Local authorities will use the indicator to measure progress in assessing the risks and opportunities from climate change across their area and in developing an action plan to address their priority needs to adapt.

Influential key players: Champions and leaders, the 'movers and shakers'

The attention brought to the issue by the Prime Minister, Margaret Thatcher, was likely to be influential. That she was also an Oxford University-trained chemist would also have added to the gravitas of her concerns. She was not simply responding to popular science scare stories but could be expected to understand the detail of what was purported to be happening in the atmosphere. Her speeches on the subject suggest as much; for example, her speech to the Second World Climate Conference in 1990 covered the absorption of atmospheric carbon and the need for precaution given the uncertainty in the available data[2]. Other key players in the national government were influential at a number of different levels: for example, in the creation of the stakeholder support organisation; in developing and actively participating in global climate change negotiations; and in the promotion of an adaptation programme for all levels of government[3].

---

[1] Nationally, the UK (Adaptation Policy Framework) has perhaps the most advanced and comprehensive programme, though other countries such as Finland (FINADAPT) and the Netherlands (ARK) are also developing strategic planning policy responses.

[2] For a transcript of the speech, see: http://www.margaretthatcher.org/speeches/displaydocument.asp?docid=108237

[3] For more information, see: http://www.defra.gov.uk/Environment/climatechange/adapt/index.htm

The importance of shocks – making climate change visible

Direct experience of floods, wind storms, sea surges and droughts can dramatically bring the issue into popular focus. The visibility of such events and the disruption and misery caused to ordinary people are captured and communicated quickly and vividly though the media, sparking debates about responsibility and calls for urgent action by the government. Once the connection is made that a single flood is not a one-off 'freak' event and could be part of a trend, it is difficult for the government to not take this seriously.

Rigorous scientific evidence

The accumulation of detailed observations of changes in species distribution attributable to changes in the climate over a number of decades significantly contributed to making the story believable. The evidence was clearly not just anecdotal accounts of one-off weather events, there was strong, scientifically rigorous evidence that there were measurable changes in the climate and that this was having a discernable effect on species distribution and crop yields. Progress was reinforced by having one of the world's leading centres for climate change research, the Meteorological (Met.) Office Hadley Centre, develop the main global and regional climate models used to build up a picture of likely climate changes up to 100 years into the future. To take the science seriously, despite the uncertainty, there needs to be a development of trust in the projections and it helps to have scientists who understand the local culture and media, and who can explain what models can (and cannot) tell us.

Involving the 'decision makers' early on

The UK Climate Impacts Programme (UKCIP)[4] was established in 1997 to help organisations understand and adapt to unavoidable impacts. UKCIP liaises between scientific research, policy makers and stakeholders, communicating the science in an accessible way and providing tools and 'space' to discuss the implications of the science for the stakeholders. It also co-ordinates research programmes to ensure that they are relevant to the needs of those making decisions. UKCIP climate change scenarios, developed with the Meteorological Office on behalf of the Government, show how the UK's climate might change in this century and are widely used in research into the impacts of climate change. UKCIP has always understood the value of meeting stakeholders face to face, facilitating dialogues and supporting regional and sectoral coalitions to explore the implications of the changing climate.

---

[4] For more information, go to: http://www.ukcip.org.uk/.

Talking about money

Sir Nicholas Stern, Head of Britain's Government Economic Service, released the 700-page report on the economics of climate change on 30th October 2006[5]. It suggested that climate change was 'market failure on the greatest scale the world has seen'. This was not the first economic report on climate change, but it is probably the largest and most widely known report of its kind. Despite being severely criticised by some quarters for the economics, particularly the overestimation of the present value of the costs of climate change and underestimating the costs of emission reduction, even many critics agreed that the conclusions were right (if for the wrong reasons). This report brought industry and business sectors into the debate and thus renewed urgency for the government to take action.

Conclusion

The combination of factors such as scientific evidence, the experience of shocks, media coverage and influential personalities have progressed national adaptation policy in the UK. It has needed a relatively well-resourced national government to take notice of its duty of care towards its citizens and to protect the nation's assets. Compared with other countries, the adaptation policy framework in the UK is in a relatively advanced state (Massey, 2007). This reflects the magnitude of the observed and projected risks, and the human capacity and resources with which the country can respond. Furthermore, the UK has the advantage of being an island nation; other countries often face the additional challenge that adaptation strategies need to involve international negotiations because the affected regions cross national boundaries (e.g. the Alps, the Mekong delta).

5.3 Local risk management: responding to heat stress in cities

Narrative

Inhabitants of a major city in northern Europe are used to the distinctive micro-climate that the city provides, with temperatures sometimes as much as 7 degrees higher than surrounding rural areas. This brings benefits such as more comfortable evening temperatures; however, the experience of recent hot summers has shown that there are also significant negative consequences. Heat stress not only has implications for human comfort (both indoors and outdoors) but, even more critically, it can have tragic consequences for human health, especially when combined with the enhanced ozone formation that occurs in metropolitan areas during periods of high temperature.

[5] Available at: http://www.hm-treasury.gov.uk/sternreview_index.htm.

The problems associated with the intensity of the urban heat island were brought into stark relief by the heatwave of 2003. This unprecedented extreme event took the city authorities by surprise, causing hundreds of heat-related deaths especially amongst the more elderly sections of the population. Those living in poorer neighbourhoods, with limited social networks, were also disproportionately affected, highlighting some of the key socio-economic influences that underpin vulnerability to heat stress. Although the loss of life was dramatic, critical infrastructure was also adversely impacted by the heatwave. Particularly badly hit was the transport network; responsible authorities had to cope with melting roads and buckled railway lines, with uncomfortable travel conditions for those using the underground tube system.

The experience of this extreme event resulted in heat stress becoming a critical issue in the minds of those living in the city, with its high profile reinforced by considerable media attention at the time. With scientific evidence suggesting that such conditions are likely to become the 'norm' by as early as the 2040s, adapting the city to heat has now moved to the top of political agendas. This is evidenced by the implementation of a national heatwave plan, which has been tailored for use by the local authorities in the city. A key component of this is a new heatwave warning system which was activated (and tested) in 2006.

Context

Heat is not a societal issue confined to the southern regions of Europe but is one that is likely to become increasingly problematic for urban areas in more northern latitudes. Whilst the above narrative is based on London in the UK, there are concerns that the traditional form and functioning of many large cities of northern Europe are not well designed for periods of sustained higher temperatures. For example, France was particularly badly hit by the heatwave of 2003 (with around 15 000 heat-related deaths) and other countries such as Portugal, Germany, Italy, and the Netherlands also reported an increase in mortality.

Experiencing shocks

The 2003 heatwave event is considered a pivotal moment in raising awareness of the need for adaptation in cities (South East Health Protection Agency, personal communication, 2007). Indeed, the health dimension (and the headline message that 'heat kills') has been an important influence on adaptation activity in London, with this specific heat event not only illustrating the increasing climate risks being faced by modern cities and their inhabitants, but also starkly exposing a lack of institutional preparedness for responding to heatwaves. Post-2003, this combination of environmental, social and political factors has not only been instrumental in altering our perception of heat-related risks, but has led to health protection moving up political agendas in England.

Responsibility for adaptation response

The most directly responsible agencies are those operating in the public realm with remits relating to the promotion of human health and well-being. These include individual actors (such as doctors acting as outreach facilities to local communities) and organisations which operate at differing scales; firstly, at the level of policy implementation (health sector bodies, local authorities, and emergency planning units) and secondly, at national government level. Governments have an important enabling role to play at the macro-level, for example by inducing adaptive behaviour through regulatory and fiscal instruments, and by promoting local capacity building through information campaigns and other awareness raising activity (McEvoy *et al.*, 2008a). One important consequence of the 2003 heatwave is that the roles and responsibilities of different actors have since become much more clearly defined (London School of Hygiene and Tropical Medicine, personal communication, 2007).

Scientific evidence: what do we need to know for risk management?

Whilst responding to heatwaves is clearly now on mainstream policy agendas, the newness of this issue has led some stakeholders to highlight the importance of 'evidence-based' responses (Greater London Authority, personal communication, 2007). International empirical evidence has highlighted that those at highest risk from heat stress are groups that are already considered the most vulnerable in society: examples include the elderly; the very young; people with chronic diseases or on medication; those that are in occupations that heighten exposure to heat stress; and finally, those with limited access to social networks (WHO Europe, 2008). Other urban variables further increase heat risk; even though other areas of England experienced hotter temperatures than London, the number of deaths was greatest in the capital. Analysis has suggested that factors such as access to working air conditioning, and even whether the vulnerable person is located in a top floor apartment, can affect risk (Kovats and Ebi, 2006).

The need for comprehensive responses

The most obvious tangible outcome of changing risk perceptions and political priorities was the introduction of a national heatwave action plan in 2004, and the introduction of a heatwave warning system first activated (and tested) in 2006. Similar responses have been adopted by many other European cities post-2003. However, whilst action plans and early warning systems are necessary responses, they are only a partial solution. Warnings need to be reinforced by a portfolio of accompanying actions: these include the general promotion of health, active checking regimes on vulnerable people; and outreach to the homeless (London School of Hygiene and Tropical Medicine, personal communication, 2007)., Effective communication, integral to these actions, is seen as particularly important; autonomous adaptation will happen by those most able, whereas the more vulnerable may find this problematic. This illustrates the importance of social networks to adaptation efforts.

A more comprehensive response to heat also requires the adaptation of our physical surroundings: the form and functioning of the built environment itself. For example, options such as the use of shutters, cooling parts of buildings, can be valuable adaptation measures. This has been adopted through the introduction of new legislation requiring care homes for the elderly to provide 'cool rooms' and to ensure that room temperatures are checked regularly (NHS, 2008). Even more fundamentally, a changing climate will require strategic action at the scale of neighbourhoods, and even the whole conurbation, in order to reduce the urban heat island effect. In the case of a densely populated city like London, this will require significant 'retro-fitting' of the city in order to be better prepared for a changed future climate.

Conclusions

The experience of the 2003 heatwave has acted as a major stimulus in changing local risk perceptions and behaviour, the most practical consequence being the introduction of a heatwave action plan, which is now being adjusted by London authorities to accommodate local conditions. However, there is concern that this plan can only address heat stress to a limited degree, because the responses remain largely compartmentalised within the health sector; more comprehensive answers actually lie with policy mechanisms outside their immediate control, for example changes to national building regulations (London School of Hygiene and Tropical Medicine, personal communication, 2007). A significant problem is that many buildings are still being designed based on climate guidance of the 1970s. Influencing future urban form will be a challenging task. Whilst there are concerns that lessons from 2003 may be quickly forgotten, there are encouraging signs that key actors in London have begun to respond to this new agenda; climate change science, and the wider considerations of sustainable development, are now being passed to different London stakeholder communities in more accessible formats. A checklist for developers (Greater London Authority, 2005), an adaptation guide for sustainable communities (TCPA, 2007), and the 2008 climate change adaptation consultation document for London (Greater London Authority, 2008) are all recent examples of this impetus for adapting to change.

5.4 Regional transboundary co-operation under increasing water resource pressures

Narrative

Antonio lives in a small inland town on the border between two arid regions. He runs a small consultancy, which provides legal and financial advice to small businesses, including farm co-operatives, local town councils and tourist brokers. His services also embrace transferring knowledge of possible EU subsidies and programmes, finding local partners to develop strategic projects on both sides of the border, and supporting other regional development agents, including environmental NGOs and universities. Managing the EU funding programmes for regional co-operation has begun to occupy an increasing amount of his working time. Very recently, climate policy has become the next business opportunity for him. Antonio knows that, during the last decade, over five European research projects have used his area as a case study to learn about the effects of climate change and the possible strategies to cope with it, and that new regional climate change strategies are now being developed in both countries. At the same time, he is aware that regional and local authorities have not yet developed enough management capacities, regulatory procedures and networks to inform and to support small businesses, farms and policy agencies about what to do in this field. With the funds obtained through the European Union's INTERREG programme, he is now part of an ad hoc transboundary agency aimed at supporting cross-border co-operation and development, and his work now begins to deal with transboundary climate change co-operation activities. Nevertheless, he feels that both regional and national agencies working on regional development do not really take seriously 'all this talk' about climate change and do not properly consider the needs of the local people.

Context

This narrative is situated in the Guadiana river basin (see also Chapter 9, Werners *et al.*), which is one of the three main international river basins in Iberia and spreads over a total area of 66 800 km². It is characterised by an arid environment with a low annual precipitation of about 440 mm/year. Due to mounting pressures from agriculture, new urban developments and increasingly from industry, competition for water resources is becoming an acute problem. Progress towards climate adaptation in this region depends mainly on four basic factors: (i) awareness of the problem; (ii) the existence of sufficient motives and incentives to take action; (iii) individual possibilities and capacities for agents' transformation in the face of climate threats; and (iv) the extent to which new institutions or changes in the existing ones redistribute responsibilities to sustain action in the long term.

Awareness raising through the media and scientific knowledge

Awareness of the risks associated with climate change is relatively recent in the area. Nevertheless, in some sectors such as agriculture, mounting water scarcity has been recognised for a long time as one of the main factors constraining its development. Besides the media, scientific knowledge is now playing an important role in raising awareness, as the Guadiana river basin is one of the most studied river basins in Europe regarding climate change and integrated resource management. In particular, in the case of agriculture, scientific knowledge about how different crops, technologies and practices could better respond to climate change is now available. This is being considered by regional authorities to develop new strategies to support local farmers.

Role of national and regional adaptation plans in stimulating action

National and regional climate adaptation programmes have now been put in place both in Portugal and Spain (MMARM, 2008; Junta de Andalucía, 2009). They provide regional climate scenarios of expected changes in precipitation and temperature until the end of the twenty-second century, and impacts are being assessed for the case of water availability, biodiversity and coastal areas. These programmes, together with the EU INTERREG funds and the emergence of new transboundary agencies and policy entrepreneurs (Perkmann, 2007) are beginning to play an important role in mainstreaming climate change into the Guadiana river basin development strategies (McEvoy *et al.*, 2008b; Cots *et al.*, 2009). However, with the exception of agriculture, and to some extent tourism, such plans do not meet the demands of local agents, because they neither specify concrete options and opportunities for adaptation, nor are they integrated into the existing local development activities.

Individual capacities and social networks

In the Guadiana farming sector, climate change adaptation processes have been observed mainly in the form of changing some agricultural practices, modifying working time budgets, and embracing non-farming activities such as rural tourism. New co-operatives have also been created, which secure a more regular income for farmers and increase their capacity to negotiate better prices for their products with retailers. However, these strategies, which are also aimed at increasing economies of scale, have also stopped some of the less productive traditional varieties of crops, such as olives, being grown. Thus, the local agricultural biodiversity is reduced and the farming structure becomes potentially more vulnerable to climate risks and biophysical change. Furthermore, a widespread response by farmers to the increased aridity has been to extend irrigation to crops for which this was traditionally not considered necessary, worsening competition for water.

Cross-border collaboration

At the institutional level, there has been a notable increase in the number of cross-border legislative agreements for regional co-operation, and the use of international river basins between Spain and Portugal. However, climate change has not yet been sufficiently mainstreamed into transboundary co-operation and regional development plans so as to secure a long-lasting adaptation process and redistribute the necessary responsibilities between the agents living in the area. Climate change is still largely perceived as an external issue, which can be dealt with separately from the core of development goals, such as expanding transport and irrigation infrastructures.

Conclusions

In the case of the Guadiana river basin, some steps towards social learning in climate change adaptation have been achieved, although this is not an example of full success. An increased awareness of the risks of climate change is now widespread. A growing number of local and regional agencies and private actors are beginning to show real interest in developing new policies and business, which integrate climate concerns and opportunities into their strategic goals. Some farmers have begun to transform their productive patterns and now embrace other activities such as rural tourism. Likewise, climate policies and programmes are beginning to be seen as new business opportunities for small and medium enterprises, with a rising interest in mainstreaming climate into development policies. However, when it comes to implementing enduring institutional structures that take climate concerns into account, progress is definitively scant.

5.5 Local resource management in a development context

Narrative

Going back several decades, rural livelihoods, based mainly on agriculture and livestock, have been threatened in this country by a number of factors, including climate variability, soil degradation and land scarcity. Overstocking, poorly managed grazing practices, and an absence of commercial markets were seen in the eyes of donors and government ministries as the main barriers to increased productivity of the pasture, reduction of soil erosion and degradation. However, this was not the only view; for local users, land use and livestock ownership continue to be of vital importance to local culture. In the early 1990s, upon return to democratic rule, the party that was to form government campaigned on a platform opposing grazing fees, which had been championed by donor agencies as a measure to increase productivity and conserve the pasture resource; the land was the property of the people, who should not have to pay for it.

By the mid 1990s a large hydrological development project was under way, with sales of water from the project making up about 20 per cent of the country's gross domestic product (GDP). This internationalised the resource degradation issue: a local commons problem on the pasture affected water of the country's larger and economically dominant neighbour. By 2003, the country had signed the Ramsar convention, signalling that the government had recognised the local, regional, and international services supplied by its wetlands and pasture resource. However, awareness of wetlands issues varied in rural communities where their conservation largely depended on a wetland's proximity to the village: those located closer being better managed. Efforts to raise awareness in the communities regarding wetlands conservation were stalled by a lack of resources and capacity in the relevant ministry. Many local workshops never took place and the planned co-ordination across the government and private sectors at a national level never materialised.

In the past 5 years, more frequent droughts, followed by hails and rains of increasing intensity, have exacerbated soil erosion, wetland loss and livestock productivity problems. The continued dam development and growing regional water scarcity have contributed to growing regional interest in the country's wetlands as a source of water and sustainable hydropower generation. In some districts, progress has been made towards increased awareness and management of the rangeland and wetlands resources, while in others, this has been impeded largely by conflicts between the traditional and state government structures. At the same time, the national government, backed by an international donor and the United Nations Framework Convention on Climate Change (UNFCCC), has launched two separate wetlands conservation pilot projects, bringing far more resources to the wetlands issue than was previously available.

Context

The narrative is situated in Lesotho, a small, land-locked country in southern Africa. Constituting 30 352 square kilometres, the 'Mountain Kingdom' with its lowest point at 1483 m ranging up to 3598 m above sea level, contains the sources of the Orange-Senqu River. The Orange-Senqu River Basin includes the Gauteng Province, South Africa's economic powerhouse. Though the region is characterised by relatively low rainfall, Lesotho itself receives significantly more precipitation than its neighbours. The sources of the Orange-Senqu river lie in Lesotho's highland region of the Maloti-Drakensburg range, where annual precipitation varies across districts from 900 to 1300 mm. With a population of around 2 million, an estimated 85 per cent of Basotho people live in rural areas where subsistence agricultural and livestock make up a significant part of their livelihoods. Since the coming on-stream of the Lesotho Highlands Water Project in the 1980s, the country has been able to generate hydropower and sell its water resource to South Africa, transferring it via tunnels to the Gauteng region. The construction of the third dam in this project will begin in 2009.

Environmental degradation: links to adaptation

Awareness of land degradation and soil erosion in the Lesotho highlands has existed amongst government ministries, international donors, and, to an extent, local actors for some time. The United States Agency for International Development (USAID) and the Canadian International Development Agency (CIDA) first launched range management development projects in the early 1980s, though these failed due to poor understanding of local politics and culture associated with land and livestock (Ferguson, 1990). The National Adaptation Plan of Action (NAPA) submitted in 2007 identified improved rangeland management and wetlands conservation as adaptation priorities. At the regional level, the Orange-Senqu River Basin Commission (ORASECOM) launched a wetlands monitoring project in 2007. Visible effects of soil loss and the increased frequency of droughts in recent years, especially in 1998, 2004 and 2008, have increased interest in range and wetland conservation. The extent of scientific data on rangeland conditions, carrying capacity and wetlands extent remains quite limited and has not yet informed adaptation policy to a large degree.

Benefits of adaptation: a question of scale

The increasing importance of the hydrological functions of wetlands to the regional and national economies, with projected increasing water scarcity and further dam construction, has affected the interests of actors in the Orange-Senqu river basin. This is reflected in funding from the Millennium Challenge Corporation for a wetlands conservation and rangeland management project. Similarly, ORASECOM is taking steps towards wetlands conservation. The NAPA process opened up another avenue for funding through the Adaptation Fund and Least Developed Countries Fund. At the local level, however, adaptation in grazing practices remains strongly influenced by networks related to traditional governance systems. Benefits for resource use are still largely framed in shorter, seasonal time horizons. In some districts, the Principal Chief has co-operated with initiatives to establish community based grazing associations, while in others the Principal Chief has refused to give up the traditional authority of granting and enforcing grazing rights, undermining efforts to adapt grazing practices.

Building trust: engaging communities

Rangeland and wetland conservation are affected directly by the legal and institutional context in which individual households make decisions; land acquisition rights, land use rights and traditional institutions (i.e. chieftan-ship) largely determine incentives on livestock-related decisions. Changes in livestock practices locally have been notoriously slow to take shape in Lesotho for a number of reasons, including the traditional institutions associated with livestock in Basotho culture (Showers, 2005). In some areas, grazing associations supported by the Rangelands Department have increased the involvement and capacity of farmers to manage the pasture and wetlands, limiting conflicts between villages. In other areas, past failures have made local leadership distrustful of new initiatives and changed grazing practices have been beset with conflict.

International institutions driving change

At the national level, institutional developments have been driven by international processes. The influence of international conventions and discourse in the form of, for example, the Millenium Development Goals, the Convention on Biological Diversity, the Ramsar Convention on Wetlands, the UNFCCC, and the UN Convention to Combat Desertification, have had a discernible effect on policy processes in Lesotho, sparking several developments aimed at adaptation. For example, as a Ramsar Convention signatory, Lesotho has developed a National Wetlands Programme to raise awareness and co-ordinate action on wetlands conservation amongst Lesotho's government departments and private actors, though, as mentioned, due to a lack of capacity, it has limited effectiveness. The involvement of new actors, more resources and technical support through the UNFCCC process has the potential to increase the capacity of national level institutions. However, while climate change awareness may help to increase institutional capacities at the national level, conflicting interests and responsibilities must still be addressed in order to implement adaptation in natural resource management at the local level.

Conclusions

We have seen that, in Lesotho, adaptation to climate change involves sectors that have previously been identified as requiring institutional and behavioural changes. The increased awareness of the climate change issue at the international level has brought increased resources for, and interest in, policies and projects which address climate risks through formal government structures. Scientific knowledge has played a significant role by increasing the awareness of national level actors and thus reinforcing identified priorities in natural resource management and conservation. As adaptation moves into implementation, the engagement of traditional local institutions and actors is key. Several pilot projects supported by international donors and centring on conservation have been undertaken since 2006; however, in order for these initiatives to be successful, a process of social learning must involve the traditional local authorities in order to address conflicting framings of benefits and time scales of adaptation.

5.6 Discussion

While the four adaptation situations presented above are diverse, they share some common factors. Not all enabling factors are, however, effective in each situation. This section discusses which factors prove to be relevant to each situation.

Experiencing shocks

The experience of climate-related extreme events can play a major role in stimulating climate change adaptation, in particular for raising public awareness about different climate hazards. Evidence shows that extreme events such as the heatwave of 2003, and the severe floods in 1998 and 2000, led to changes in both national planning as well as emergency responses in cities across Europe. However, whilst there is strong evidence that shocks can lead to policy changes (e.g. Johnson *et al.*, 2005), it also needs to be recognised that 'windows of opportunity' for change can be short-lived and that 'unlearning' can occur, with people often reverting back to their previous behaviour over time, having not learnt the lessons of weather-related events.

Shocks also play a role in enabling adaptation, because they can reveal a lack of individual and institutional capacities to respond. The heatwave, for example, showed many indirect effects that may not have been anticipated in the original plan such as the impact on critical infrastructure. Furthermore, in many instances, the shocks have spotlighted issues relating to the responsibilities of different actors.

An important factor that moderates whether shocks lead to adaptation is whether the shock is seen as part of a trend rather than a one-off 'freak event'. People will only start to contemplate the future implications of the event if they see it as part of a trend, because the following type of reasoning is triggered: 'The unanticipated cost of that heatwave for our business was a certain amount of money, and the projections are that these heatwaves will occur every seven years. What does that mean for our future business plan?'

The right information at the right time

Scientific information has played varying roles in each of the situations. The cases also illustrate that it is not just the type of information that matters, but the whole communication process. Several factors such as whether the information comes from a trusted authority, is formulated in the right language and fits the needs of the moment, all influence the usability of information (Haas, 2004). Interviews and workshops conducted in ADAM show that stakeholders often only need to receive headline messages, such as 'summers will be wetter, winters warmer and wind speeds higher'. On this basis, they can make decisions knowing that there is uncertainty and that, after a period of time, the decisions will need to be revisited.

The national adaptation policy case shows the importance of individuals in the process of information provision as well as the importance of 'economic information'. Information on the economic consequences of climate change is likely to be picked up by a different audience who perhaps do not hear the climate information or do not perceive it to be relevant to them. In the urban risk management case, information was only picked up after the experience of the heatwave of 2003. Scientific knowledge concerning the urban character and vulnerability of elements at risk is now being used to inform responses. Building on this awareness, more targeted and accessible information is being produced, for example in the form of guidelines and checklists, to further develop local capacity to respond.

In the cases of the Guadiana and Lesotho, there is little evidence that information on climate change had any effect on local action. In the Guadiana, information is only available in form of regional impact scenarios (MMARM, 2008; Junta de Andalucía, 2009); the local agents, however, expected information on specific adaptation options for their particular sectors. In Lesotho, local actors have generally not incorporated information on climate change into their decision making because other stresses such as climate variability and environmental degradation are considered more pressing.

Social networks and space for collaboration and learning

It is unlikely that experiencing shocks, or having access to information, on their own will automatically lead to adaptation. Rather, it is the availability of social networks and space for learning that increases capacities to respond and help transform past experiences and information into action. Indeed, the success of UKCIP in promoting adaptation is not so much attributable to mere information provision, but also to the creation of stakeholder networks. In the Guadiana case the formation of co-operatives helped individuals to increase income and to become more resilient against shocks. The heatwave example shows that those with limited access to social networks are the most vulnerable. Networks provide the necessary space to share experiences and information, as well as to discuss concerns and ideas with people in similar situations. They offer opportunities for collectively thinking through the implications of climate change and what adaptation might mean in practice, both individually and for the whole community.

Providing incentives through higher-level institutions

All of the four situations presented demonstrate the importance of higher-level institutions for enabling adaptation at lower levels, through the creation of social networks and the provision of information or other incentives. For example, UKCIP and the EU INTERREG fund have played important roles in supporting the creation of networks, which have led to an increase in local capacity to respond. The national adaptation indicator, developed in the context of the UK adaptation policy framework, acts as an incentive for local authorities because they now have to consider it in their reporting. Their funding and status will depend on how well they perform in implementing it. Providing the right incentives is, however, by no means trivial. The Lesotho case illustrates decades of failure to do so. In some areas, incentives have improved management, in others not. Furthermore, past failure has led to a distrustful attitude towards new incentives.

The capacity of institutions to provide incentives is often limited by a misfit in responsibility (see also Young, 2006). In the London case, for example, the ability of authorities to address heatwaves comprehensively is restricted due to a lack of direct influence on the form and design of the city. Furthermore, adaptation is often facilitated by institutions designed for other purposes than that of climate adaptation. At the national level, the Millennium Development Goals and the Ramsar Convention on Wetlands have been at least equally important as the UNFCCC for raising awareness of climate issues in the case of Lesotho. At local levels, the INTERREG funds enhanced climate adaptation in the Guadiana, and the hydropower and water export development increased interest in the wetlands issues in Lesotho. These arguments show the importance of 'mainstreaming' adaptation. Owing to the cross-cutting nature of adaptation, it needs to be considered by existing institutional mechanisms, policies and networks, rather than relying on the design and implementation of independent ones (see also Chapter 12, Gupta *et al.*).

Further factors

The list of factors given above is not meant to be comprehensive and indeed there are further factors that may enhance adaptation, such the leadership of influential key players and the clear definition of actors' roles and responsibilities. All in all, there is no one single factor that enables adaptation, rather a combination and a succession of factors, which are uniquely intertwined in each particular situation. Given local circumstances, it is therefore important to understand what the key factors are and how key barriers can be overcome.

5.7 Conclusions

We argued that, for many adaptation situations, the impact modelling and decision analytical framing of climate change adaptation is not appropriate, as adaptation is not a matter of a one-shot decision based on certainty about impacts and effective adaptation options. Rather, adaptation is a continuous learning process between actors and institutions at all levels of decision making. Owing to the complexity of the issue and the different high stakes involved, there is ambiguity of exactly what is the problem to be solved and what are the perceivable solutions. Therefore, adaptation research needs to shift from focusing on the reduction of uncertainties in impact modelling to exploring what enables and hinders adaptation and the associated learning process. To demonstrate this, we presented four cases of distinct adaptation situations and discussed barriers to implementation and enabling factors.

We find that at the national and international levels, scientific evidence plays an important role in raising awareness and promoting policy making and adaptation planning. At more local levels, however, scientific evidence was, in the cases considered here, less important. Often, headline messages such as 'winters are likely to be warmer' are sufficient. If adaptation is constrained by a lack of capacity or motivation to adapt, providing scientific information may not be relevant at all.

While the experience of weather-related shocks has been important for raising general awareness, it is the building of new social networks which plays the most central role in enhancing individual capacities for adequate responses. Institutions may foster the creation of social networks and provide other crucial incentives for promoting local action. Designing effective institutions, is difficult, however, and failure is more common than success. Generally, there is a danger that institutional design does not sufficiently take into account local views and culture. Relevant enabling institutions often stem from processes not related to climate change, such as economic development. A redistribution of institutional responsibilities is often necessary to remove barriers to adaptation.

We stressed the need to facilitate adaptation. More research and space for social learning are needed to find out what kind of facilitation is most effective for a given adaptation situation. Towards this end, higher level decision making should aim at understanding what hinders and promotes adaptation at lower levels. Particularly for the local scale, it can be said that enabling adaptation is less an exercise of information provision but more one of facilitating a social process, in which scientific information may or may not play a role.

In sum, we believe that there are no panaceas to climate change adaptation. While there is a lot that can be learned from success stories in climate change adaptation, 'good practice' may not be transferable from one situation to the next due to subtle differences in context. It is not one single factor that leads to adaptation; rather it is the combination and succession of many, often subtle factors. The general answer that can be given to climate change adaptation at this early stage, in our opinion, is to further develop a culture of social learning.

References

Adger, W. N., Agrawala, S., Mirza, M. M. Q. *et al.* (2007) Assessment of adaptation practices, options, constraints and capacity. In *Climate Change 2007: Impacts, Adaptation and Vulnerability. Contribution of Working Group II to the Fourth Assessment Report of the Intergovernmental Panel on Climate Change*, ed. Parry, M. L., Canziani, O. F., Palutikof, J. P., van der Linden, P. J. and Hanson, C. E. Cambridge, UK: Cambridge University Press, pp. 717–43.

Armitage, D., Marschke, M. and Plummer, R. (2008) Adaptive co-management and the paradox of learning. *Global Environmental Change*, **18**(1), 86–98.

Berkes, F. (2007). Community-based conservation in a globalised world. *Proceedings of the National Academy of Sciences*, **104**(39), 15188–93.

Carter, T. R., Parry, M. L., Harasawa, H. and Nishioka, S. (1994) *IPCC Technical Guidelines Assessing Climate Change Impacts and adaptations*. Department of Geography, University College London, UK and the Center for Global Environmental Research, National Institute for Environmental Studies, Japan.

Carter, T. R., Jones, R. N., Lu, X. *et al.* (2007) New Assessment Methods and the Characterisation of Future Conditions. In *Climate Change 2007: Impacts, Adaptation and Vulnerability. Contribution of Working Group II to the Fourth Assessment Report of the Intergovernmental Panel on Climate Change*, ed. Parry, M. L., Canziani, O. F., Palutikof, J. P., van der Linden, P. J. and Hanson, C. E. Cambridge, UK: Cambridge University Press, pp. 133–71.

Chapman, J. (2004) *System Failure: Why Governments Must Learn to Think Differently*. London, UK: Demos.

Cash, D. W., Clark, W. C., Alcock, F. *et al.* (2003) Knowledge systems for sustainable development. *Proceedings of the National Academy of Sciences*, **100**, 8086–91.

Cots, F., Tàbara, J. D., McEvoy, D., Werners, S. and Roca, E. (2009). Cross border organisation as an adaptive water management response to climate change. The case of the Guadiana river basin. Government and Policy. *Environment and Planning*. In press.

Darwin, J., Johnson, P. and McAuley, J. (2002) *Developing Strategies for Change*. London, UK: FT Prentice Hall.

Ferguson, J. (1990) *The Anti-Politics Machine: 'Development', Depoliticisation and Bureaucratic Power in Lesotho*. Cambridge, UK: Cambridge University Press.

Folke, C., Hahn, T., Olsson, P. and Norberg, J. (2005) Adaptive governance of social-ecological systems. *Annual Review of Environmental Resources*, **30**(8), 1–33.

Greater London Authority (2005) *Adapting to Climate Change: A Checklist for Development*. London, UK: Greater London Authority.

Greater London Authority (2008) *The London Climate Change Adaptation Strategy*. London, UK: Greater London Authority.

Haas, P. (2004) When does power listen to truth? A constructivist approach to the policy process. *Journal of European Public Policy*, **11**(4), 569–92.

Johnson, C. L., Tunstall, S. M., and Penning-Rowsell, E. C. (2005) Floods as catalysts for policy change: historical lessons from England and Wales. *International Journal of Water Resources Development*, **21**(4), 561–75.

Junta de Andalucía (2009) *Las Políticas de Cambio climático de la Junta de Andalucia*. Presentación en el proyecto ADAM Sevilla, 06.02.2009. Sevilla: Consejeria de Medio Ambiente.

Kovats, S. R. and Ebi, K. L. (2006) Heatwaves and public health in Europe. *The European Journal of Public Health*, **16**(6), 592–9.

Massey, E. (2007) *Framework Comparing Adaptation Initiatives within Countries*. EIONET workshop on impacts, vulnerability and adaptation. Copenhagen: European Environment Agency.

McEvoy D., Lonsdale K. and Matczak P. (2008a) *Adaptation and Mainstreaming of EU Climate Change Policy*. Brussels: CEPS policy briefing note for the European Commission.

McEvoy, D., Cots, F., Lonsdale, K. and Tàbara, J. D. (2008b) The role of institutional capacity in enabling climate change adaptation. The case of the Guadiana river basin. In *Transborder Environmental and Natural Resource Management*, ed. W. Long Kyoto, Japan: Center for Integrated Area Studies, Kyoto University.

MMARM (2008) *Plan Nacional de Adaptación al Cambio Climático. Primer informe de seguimiento del plan nacional de adaptación al cambio climático*. Madrid: Ministerio de Medio Ambiente Rural y Marino (MMARM).

Mostert, E., Pahl-Wostl, C., Rees, Y., Searle, B., Tàbara, J. D. and Tippett, J. (2007) Social learning in European river-basin management: barriers and fostering mechanisms from 10 river basins. *Ecology and Society*, **12**(1): 19, availble at: http://www.ecologyandsociety.org/viewissue.php?sf=28.

NHS (2008) *Heatwaves: Supporting Vulnerable People Before and During a Heatwave*, London: NHS.

Pahl-Wostl, C. (2007) Transition towards adaptive management of water facing climate and global change. *Water Resources Management*, **21**(1), 49–62.

Pasteur, K. and Scott-Villiers, S. (2004) If relationships matter, how can they be improved? Learning about relationships in development. *Lessons for Change in Policy and Organisations*, **9**, Brighton: Institute of Development Studies.

Pelling, M. (2006) Measuring vulnerability to urban natural disaster risk. *Open House International*, Special edition on managing urban disasters, **31**(1), 125–32.

Pelling, M. and High, C. (2005) Understanding adaptation: what can social capital offer assessments of adaptive capacity? *Global Environmental Change A*, **15** (4), 308–19.

Perkmann, M. (2007) Policy entrepreneurship and multilevel governance: a comparative study of European cross-border regions. *Environment and Planning C: Government and Policy*, **25**(6), 861–79.

Rittel, H. W. & Weber, M. M. (1973) Dilemmas in a general theory of planning. *Policy Sciences*, **4**, 155–169.

Senge, P. (1990) *The Fifth Discipline: The Art and Practice of the Learning Organisation*. London, UK: Century.

Showers, K. B. (2005) *Imperial Gullies: Soil Erosion and Conservation in Lesotho*. Ohio, USA: Ohio University Press.

Tàbara, J. D. (2005) *Sustainability Learning for River Basin Management and Planning in Europe*. HarmoniCOP integration report. Osnabrück, Germany: University of Osnabrück. Available at: http://www.harmonicop.info.

Tàbara, J. D., Cots, F., Dai, X. *et al.* (in press) Social learning on climate change among regional agents and institutions. Insights from China, Eastern Europe and Iberia. In Leal, W. (ed.), *Interdisciplinary Aspects of Climate Change*. Frankfurt / New Cork / Bern / Vienna: Peter Lang Scientific Publishers.

TCPA (2007) *Adaptation by Design: A Guide for Sustainable Communities*. London, UK: Town and Country Planning Association.

WHO Europe (2008) *Heat-health Action Plans*. Copenhagen, Denmark: WHO Europe.

Young, O. (2006) Vertical interplay among environmental and resource regimes. *Ecology and Society*, **11**(1), 27, available at: http://www.ecologyandsociety.org/vol11/iss1/art27/

Part II

Strategies within Europe

6

How do climate policies work? Confronting governance dilemmas in the European Union

Lead authors:

FRANS BERKHOUT[1], CONSTANZE HAUG, TIM RAYNER,
HARRO VAN ASSELT, ROGER HILDINGSSON, DAVE HUITEMA,
ANDREW JORDAN, SUVI MONNI, JOHANNES STRIPPLE

[1]Co-ordinating lead author

Summary

This chapter introduces the concept of *governance dilemmas* and discusses its meaning for European Union (EU) climate policy. We argue that climate policy is shaped by choices regarding six key issues: problem definition; the level of action (whether EU, national or local); the modes of governance (whether regulatory, market or network based); the timing of action; the distribution of costs and benefits; and the manner of enforcement. Dilemmas arise in making these choices in policy design and implementation. Legitimate and effective governance implies trade-offs, often under conditions of uncertainty about the effectiveness, impacts and costs of policies. We assess how these six governance dilemmas have been confronted over the past 15 years, basing our findings on a meta-analysis of 262 EU climate policy evaluation studies. We also analyse how climate governance dilemmas have influenced climate policy outcomes. We find that: (i) the climate problem is typically framed as a market failure; (ii) climate policy is deeply multi-level, but with a trend towards harmonisation at the EU level; (iii) in terms of scope, market instruments are now increasingly important in EU climate policy (particularly the emissions trading scheme), but regulatory instruments still play a significant role; (iv) striking a balance between policy predictability and flexibility remains contentious; (v) climate policies are often mildly regressive in their economic impacts; and (vi) monitoring and enforcement are generally very poor. We conclude that EU climate policies have tended to lead to slow and incremental emission reductions. Our attention is focused primarily on mitigation policy owing to the comparatively recent advent of policy attention for adaptation to climate change impacts.

Making Climate Change Work for Us: European Perspectives on Adaptation and Mitigation Strategies,
ed. Mike Hulme and Henry Neufeldt. Published by Cambridge University Press © Cambridge University
Press 2010.

6.1 Introduction

In the Kyoto Protocol the EU negotiated a target of an 8 per cent reduction in greenhouse gas emissions in 2008–2012 compared with 1990. Although ambitious, EU policy makers were confident that the target would be achieved. Over the past 15 years a wide range of new 'climate' policies has been introduced to reduce greenhouse gas emissions in Member States and at the EU level[1]. This chapter deals with the question of how these policies have evolved in Europe and why they have taken the form that we observe today. We are also interested in the question of whether the complex system of objectives, institutions and measures that are included under the banner of climate policy are effective in reducing greenhouse gas emissions. We address the question of how EU climate policies work in three different ways: (i) in practical terms, how does the policy system fit together? (ii) in political terms, how have climate policies come to be positioned within the broader architecture of policy commitments and traditions within the EU? and (iii) in policy terms, how far have they led to their expected outcomes?

Given the novelty of climate change as a domain of EU policy, and the high degree of experimentation which we observe in the policies and measures adopted, early confidence in the effectiveness of policy (based on a starting point for EU climate policy at about 1990) now seems remarkable. The study of public policy suggests that there is a complex chain of causality between the framing of a problem, its codification in a piece of policy and its outcomes in terms of changed behaviour amongst people and organisations. Often, policies come into being as compromises between conflicting objectives and, even more often, they have unexpected effects, in their own right or through their interaction with other policies or trends in society (the price of energy, for instance). For the policy analyst, the confidence of climate policy makers is not self-evident. Indeed, in this the analyst is not alone, because there exists what we might call 'climate policy schizophrenia' amongst policy makers. While the policy makers have a vested interest in the success of EU policy as a front-runner in the international climate regime, they are also confronted with a surprising degree of ignorance about how exactly climate policy works in the EU. Although effort has increased, particularly since 2002, monitoring, reporting and evaluation of policy is still far from comprehensive (Rayner *et al.*, 2009).

The question, 'how does EU climate policy work?' has an even more urgent significance as the first commitment period of the Kyoto Protocol comes to an end. Even if the EU does achieve its Kyoto target, it remains difficult to say how much of this will have been achieved as a result of its climate policy. While aggregate EU greenhouse gas

---

[1] Although policies and strategies to deal with the impacts of climate change (adaptation) have developed in the past 5 years, these will not be discussed in this chapter.

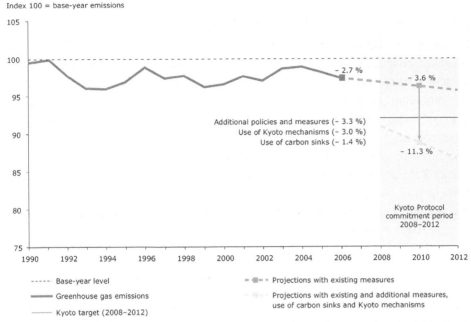

Figure 6.1. EU-15 greenhouse gas emissions and projections for the Kyoto commitment period 2008–2012 (EEA, 2008: 8).

emissions have shown a downward trend, it is also apparent that much still needs to happen before 2012 for the target to be reached (see Fig. 6.1). New climate policy will play a role, but other factors, such as growth in the EU economy, will also have a profound influence. If tougher emission reduction targets are agreed at the UN's 2009 Copenhagen conference or thereafter, then European climate policy will need to become more ambitious and effective than before. Understanding how emissions targets are translated into policy and how this influences changes in behaviour in economies and societies leading to a radically changed future carbon footprint for Europe therefore becomes even more important. Beyond this, with the advent of major new policy activity dedicated to reducing vulnerability and increasing resilience to climate change impacts in Europe, there is a need to understand how climate adaptation policies may work.

This chapter begins by setting out a 'governance dilemmas framework' for looking at EU and Member State climate policies in the round. Six dilemmas are briefly discussed (for a fuller treatment, see Jordan *et al.*, 2010). We then provide a discussion of the legal basis and evolution of climate policies in Europe, emphasising a growing convergence around common measures adopted at the EU level. On the basis of a meta-analysis of evaluation studies, we assess EU climate policies using the governance dilemmas framework. Finally, we conclude with some more prescriptive reflections on what has worked in EU climate policy and what has not.

6.2 An analytical framework: six governance dilemmas

In governing complex problems, policy makers and other actors (industry, civil society) who make up the governance system are typically required to make difficult choices between alternative policies. Such choices involve different ways of defining problems, calibrating policy instruments and dividing the costs of action amongst those involved. As these characteristics are seldom aligned with each other and may well be sharply conflicting, each set of choices raises awkward dilemmas for governors to deal with. That the various actors typically adhere to deeply held norms and values, including equity, effectiveness, transparency and participation, makes the task of governing climate change an immensely difficult one (Jordan *et al.*, 2010). Governance dilemmas are at their most acute when policy choices require the weighing up of multiple and conflicting values.

6.2.1 Prevailing governance frameworks

By speaking of *dilemmas* associated with particular instances of policy making, we are introducing a strong element of *agency* into our perspective. Indeed, we assume that the emergence of particular policies is the product of an ongoing political interaction between various groups of actors, with different sets of demands and expectations. Choices in relation to particular policies are taken within the context of much broader governance frameworks of organisations and policies. The development of long-term governance frameworks in the EU cannot be seen as being predictable, rational or centrally co-ordinated by one single 'governor'. It is more appropriate to refer to the emergence of rules, institutions and norms over time and as part of a policy system or regime, rather than assume that they are planned and implemented in a single step, with only one objective – namely, climate protection.

Here we are primarily concerned with the development of EU climate policies. These policies are shaped through interactions between Member States and the EU institutions, and relevant pressure groups from business and civil society. We are also seeking to understand the relationship between climate policies and prevailing governance frameworks. Put simply, how do governors create policies to handle climate change within the context of existing institutions, ideas and wider political relationships? One set of theories argues that individual policy choices are rationally designed by Member States to maximise their preferences (Moravscik, 1998); another argues that governing frameworks (themselves the product of earlier policy choices) gradually become institutionalised and in so doing have a constraining effect on future developments (Pierson, 1996).

6.2.2 Six governance dilemmas

Our analysis seeks to find a balance between rationalist and institutionalist approaches to policy analysis by identifying six policy dilemmas and using these as a series of perspectives through which to view policies. Each dilemma represents a trade-off that must be made in policy design and action. Different actors will have quite different attitudes to these trade-offs, and to the relationships between them. In other words, policy actors seek to secure their own interests but within a system of interconnected trade-offs. Moreover, given substantive and procedural uncertainties about the effects of policy action, discourses and perceptions of actors will also play a major role.

Dilemma 1: What is 'the problem'?

Problems are not simply 'out there' in the world waiting to be resolved by actors working in the policy system. On the contrary, policy and other actors often engage in struggles to present or 'frame' problems in ways that suit their political interests. The most important distinction here is between a problem framing which is stable and one which is open to change and contestable (due to actor learning, new scientific understandings, new perceptions of costs and benefits, evidence on effectiveness, etc.). This implies a dilemma in which governors must choose which aspect and/or framing of the problem to adopt.

Dilemma 2: At what level should policy makers act?

Here, the main choice is between acting at low levels of the policy system (according to the principle of subsidiarity, for instance), or working through action at higher levels, further away from the specificities of the particular situations in which most decisions are made. The implicit dilemma here is between choosing to emphasise one or more of the following: flexibility, accountability and transparency (likely better at more local levels), and co-ordination and collaboration (through higher level action). Greater local diversity may offer more opportunities for learning, but fewer opportunities for successfully implementing what is learned across the EU through harmonisation. In practice, the choice between the levels is seldom open – governors in the EU typically have a stronger legal competence to act at some levels (and via some modes of governance) than others.

Dilemma 3: When should policy makers act?

There are several dimensions to this dilemma: risk taking (moving first without knowing what others will do), profit making (the potential for 'first mover advantages'), and legitimacy (asking others to act without acting oneself may not be credible). Related to this, is the question of whether governors therefore make policy stable and predictable, clearly signalling the change envisaged to societal actors, or opt for flexibility, to leave open the possibility of updating or revising policy goals later. An important choice is whether to act early in a more assertive, preventative or precautionary manner, or to wait for others who do not share the same problem framing, or want more time to consider the evidence or evaluate alternative policy options.

Dilemma 4: How should policy makers act?

The literature usually distinguishes three modes of governance: markets, networks and hierarchies. Each of these modes exhibits certain theoretical strengths and weaknesses. The main dilemma confronting governors is how to put together a package of modes, and their associated policy instruments, in a way that max-imises their respective strengths and reduces their weaknesses (Goulder and Parry, 2007). These imply dilemmas related to the balance to be struck between the relevant degrees of authority, responsibility, participation and effectiveness (where state action is needed) and 'governance' (where non-state actors enjoy more influence and responsibility). Critically, the choice is between creating and imposing a set of enforceable social norms hierarchically (i.e. through regulation), allowing them to emerge and disseminate in natural or artificial markets, or relying on dynamic incentives associated with voluntarism in networks of social and economic actors.

Dilemma 5: Who wins and who loses?

While policy interventions may seek to generate net social benefits, they tend to generate losers as well as winners. Therefore, the most important dilemma con-fronting governors is what costs and benefits to take into account in decision making, what to leave out and whether (and if so, how?) losers will be compensated. Judgements on these issues are tied to considerations of policy effectiveness, fair-ness and legitimacy. Often, there is an asymmetry between perceptions of benefits among winners and the voice of perceived losers; costs are often over-stated by the most directly affected parties, who are usually more effective in making their case than potential winners. This is especially obvious in intergenerational issues like climate change. There are also potential spill over impacts on other dilemmas. For example, in terms of timing, moving quickly to achieve more ambitious policy targets may disrupt the status quo (and thus create more losers) to a greater extent than a series of incremental adjustments.

Dilemma 6: How are policy outcomes secured?

The key choice in terms of implementation and enforcement of policy is whether to use informal, or less hierarchical instruments without sanctions (e.g. reporting requirements, etc.) or more formal instruments with financial and legal sanctions to secure the preferred policy outcomes. In this, there are strong links to the modes and instruments dilemma (4). The main governance dilemma here is therefore between achieving co-ordination through social norms and peer pressure, or through hierarchical means, backed by sanctions for compliance. In practice, there is always a balance between the two, so that systems with clear penalties also develop normative elements, and vice versa.

Summarising, climate policy, like most other domains of public policy, faces a set of governance dilemmas leading to six overarching questions:

(i) *Problem perception and policy objectives*: How do policy makers frame the climate problem?

(ii) *Levels and scales of governance*: At which level (or scale) of governance do policy makers seek to act?

(iii) *Timing and sequencing*: How do policy makers strike the balance between the need for predictability of a long-term policy framework and the need to adapt it to new evidence and changing circumstances?

(iv) *Modes and instruments of governance*: Which modes and instruments of governance do policy makers adopt?

(v) *Costs and benefits*: How are the costs and benefits of climate policies distributed, and how are any losers compensated?

(vi) *Implementation and enforcement*: How are policy outcomes secured?

6.3 The EU climate policy landscape

Climate policies in the EU have accumulated as a complex texture of national-level policies[2] and EU-wide emission reduction 'common and coordinated policies and measures' (CCPMs). These common measures, embodied in the EU Climate Change Programme (ECCP), have developed over the period since 1992 and have been adopted to ensure cost-effectiveness and minimise competitive distortions. Since 1996, there has been political agreement in the EU that in order to prevent the most damaging effects of climate change, international policy should aim for a maximum global temperature increase of 2 °C over pre-industrial levels and a stabilisation of atmospheric carbon dioxide concentrations below 550 parts per million (CEC, 2005a)[3]. Only recently has the EU turned its attention to policies for climate adaptation (CEC, 2007c).

---

[2] Many policies now recognised as climate policies were originally implemented for other reasons, as energy or environmental measures, for instance.

[3] More recent research has shown that the likelihood of achieving the 2 °C target with an atmospheric concentration of 550 ppm CO_2e is small (den Elzen *et al.*, 2006) and attention has now turned to the potential for achieving 'very low' stabilisation (400–450 ppm) (see Chapter 3, van Vuuren *et al.*).

6.3.1 Rationale and legal basis for EU climate policy

As we show below, the most common function of climate policy in the EU is to deal with the problem of a failure of the energy markets to address the environmental damage costs of greenhouse gas emissions. Climate change represents a global externality however, with the EU contributing only a relatively small share of global emissions (about 11 per cent in 2006). A unilateral EU climate policy would therefore be ineffective in preventing dangerous climate change. Additional arguments, relating to the historical responsibility of wealthy industrialised countries for causing climate change and to the moral responsibility of richer countries to be 'role models' in emissions reduction have played a significant role (Dellink *et al.*, 2008). In the international context the costs of unilateral action have been much debated, especially with regard to European manufacturers of traded goods whose competitiveness may be affected by the higher costs that could be generated by an ambitious EU climate policy.

The founding Treaties do not contain a single legal basis for Community Action in the area of environment, sustainable development or climate change. Depending on the nature of the climate policy measures adopted, they have, to date, been based on a range of provisions, including Articles 71 (Transport), 95 (Approximation of Laws), 133 (Common Commercial Policy), and Article 175 (Environmental Policy) (Lacasta *et al.*, 2002). This fragmented legal basis also reflects the inherently cross-cutting or boundary-spanning nature of climate change policy. A major constraint on its development has been the limited EU competence in the energy sector (Collier, 1997; Dahl, 2000). In the absence of any formal Community competence, energy-related legislation has been based on the EU's exclusive competence on internal market issues, and on environmental powers. However, by invoking the subsidiarity principle, Member States have frequently managed to retain their sovereignty in this domain and prevent the EU from developing a co-ordinated 'energy policy' (Collier 1997).

6.3.2 The evolution of EU climate change governance

Climate policy has featured on the EU policy agenda since the late 1980s, with an early landmark being the first Joint Energy–Environment Council in 1990 agreeing to the goal of stabilising carbon dioxide emissions by 2000. The demands of negotiating the 1992 United Nations Framework Convention on Climate Change (UNFCCC) later further catalysed the EU's response. The EU's initial climate change programme (CEC, 1992) involved four components: (i) a framework directive on energy efficiency, in the context of the existing 'Specific Actions for Vigorous Energy Efficiency' (SAVE) programme; (ii) promotion of renewable energies through the ALTENER programme (an EU programme aimed at promoting the use of renewable energy sources); (iii) a monitoring mechanism; and (iv) a proposed combined carbon/energy tax. The initial objectives of SAVE and ALTENER fell victim to subsidiarity concerns, reporting duties were cut, the targets made non-mandatory and the necessary finance not granted. The third component, the so-called monitoring mechanism, required the Member States to '...devise, publish and implement national programmes for limiting anthropogenic emissions of carbon dioxide', in order to fulfil UNFCCC commitments and contribute to the achievement of the EU's collective target. Through this mechanism the Commission took on the role of ensuring that national programmes added up to the initial emissions reduction target and of raising the alarm if they did not. An EU carbon tax would have constituted a fourth component of the Programme and failure to agree it (following opposition from the UK and powerful industrial lobbies) constituted a major early setback for EU efforts in climate policy.

During preparations for the Kyoto Conference of the Parties (COP) to the United Nations Framework Convention on Climate Change (UNFCCC) in 1996–7, there were attempts to define climate change policies and measures (PAMs) that could be adopted internationally by all Organisation for Economic Co-operation and Development (OECD) and industrialised countries. However, non-EU OECD nations led by the USA rejected the proposal, insisting on complete autonomy to define policy (Long *et al.*, 2002). Despite its inability to agree on a carbon tax, the EU did agree on differentiated targets for each Member State. According to this 'burden sharing agreement' (BSA), quite substantial reductions by the more progressive Member States gave less developed Member States 'headroom' to grow and increase their emissions. On this basis, the EU brought a 15 per cent reduction target to the negotiating table (even though the individual targets added up to only 9.2 per cent). Ultimately, the outcome of Kyoto was an 8 per cent reduction target, (see Table 1) which the EU was allowed to meet through internal burden sharing. The BSA adopted in 1998 has been acclaimed as one of the most successful aspects of the early stages of the EU climate change governance regime (Wettestad, 2000).

In 2000 (and in response to the perceived danger of the EU failing to meet its overall Kyoto target), the Commission launched the European Climate Change Programme (ECCP). Its aim was to identify, develop and introduce additional 'common and coordinated measures' in the most cost-effective, least market-distorting way. The most important and innovative initiative, the EU Emissions Trading Scheme (EU ETS), was implemented in 2005. The ETS, which constitutes a significant break from the norm of EU environmental policy as a whole (Sbragia, 2000), covers all the big industrial emitters across the Member States, which together account for about half of the EU's total carbon dioxide emissions. A so-called 'Linking Directive' allows credits from emissions reductions from Joint Implementation (JI) and Clean Development Mechanism (CDM) projects, as introduced by the Kyoto Protocol, to count towards

Table 6.1. *Key EU Greenhouse gas emission reduction targets (Kyoto and 2008 Energy and Climate Package)*

| Commitment | Coverage | | Emissions reduction target | Dates |
	Member States	Other		
Kyoto Protocol	EU-15		− 8%	1990–2010 (average over period 2008–2010)
EU Council 2008	EU-27		− 20%	1990–2020
			− 30%*	1990–2020
		EU-ETS	− 21%	2005–2020
		Non-ETS	− 10%	2005–2020

* If a new global climate agreement is negotiated.

installations' targets under the EU ETS. The programme also incorporated voluntary agreements with European and Japanese and Korean car manufacturers reached in 1998 and 1999. The European Automobile Manufacturers Association (ACEA) undertook to reach average emissions of 140 grams of carbon dioxide per kilometre for new cars by 2008; the Japanese Automobile Manufacturers Association (JAMA) and Korea Automobile Manufacturers Association (KAMA) agreed to meet the same target by 2009. The only sanction was the threat of future legislation.

The 2008 Climate and Energy Package (CEC, 2008), agreed in the run-up to the 2009 Copenhagen COP, represents another significant step in the evolution of EU climate policy. Central to the package were a new set of unilateral targets: (i) to cut EU greenhouse gas emissions by 20 per cent (relative to 1990); (ii) to establish a 20 per cent share for renewable energy; and (iii) to improve energy efficiency by 20 per cent. The climate package divided the EU economy into two parts, depending on whether they were inside or outside the EU ETS. While emission allowances are allocated in a centralised way under the ETS from 2012, the 'effort-sharing' decision assigned national targets for the sectors not covered by the trading scheme. Emissions reductions here are, in line with the principle of subsidiarity, to be achieved through domestic measures and through a variety of EU Common and Co-ordinated Policies and Measures (CCPMs). These include a broad mix of traditional command and control mechanisms, market-based instruments, informational measures, funding for technology and innovation, voluntary agreements and networking.

6.4 Assessing EU policy against the governance dilemmas

Our empirical analysis of EU climate policies against the six governance dilemmas discussed in Section 6.2 is based on a meta-analysis of climate policy evaluations conducted between 1998 and 2007 (Haug *et al.*, 2008). These evaluations reflect the climate policy discourse as it has evolved over the past decade and also track the opinions of policy communities making choices, designing policies and observing their impacts once implemented. The evaluations were collected through journal database searches, the Internet and contacts with policy makers and others in the policy community. All are effectively in the public domain. Our search resulted in an extensive (although not exhaustive) list of evaluations conducted since 1998 in six EU Member States – Germany, UK, Italy, Finland, Portugal and Poland – and at the EU level. From this list, we identified studies that offered a systematic assessment of policies already in place (*ex post* evaluations) and excluded those that were either not sufficiently systematic (such as position papers by lobby groups) or that were wholly *ex ante*. Policies were classified as 'climate change' policies if they were reported as such in respective National Communications to the Secretariat of the UNFCCC. This selection process resulted in a dataset of 262 evaluations.[4]

6.4.1 Problem perceptions and framing

Problem perceptions and framing are fundamental to the handling of a policy problem (Majone, 1989; Schön and Rein, 1994). How a problem is perceived by governments and the public at large shapes decisions over whether policy action should be taken, when it should be taken, what instruments will be employed, at what administrative level that action should be focused and what resources will be devoted to it. Drawing from our evaluations dataset, this section reflects on how the problem of climate change is constructed in EU and Member State policy discourse.

Overall, almost all studies we reviewed emphasise the urgency with which measures to cut greenhouse gas emissions need to be taken. A number of evaluations support a higher degree of ambition with regard to the emission reduction targets, including a stronger link to the EU's 2 °C target (e.g. FoEE, 2005; Lechtenböhmer *et al.*, 2005). Two themes emerge from the evaluations with regard to problem perception: the framing of climate change as a market failure and/or state failure; and the degree to which climate policy has been designed to deliver 'co-benefits' (beyond climate protection, such as employment and health benefits that result from lower emissions, see e.g. van Vuuren *et al.*, 2006). In the evaluations reviewed, the underlying diagnosis of the climate change problem is that it derives from a *combination* of market failure (the inadequate allocation of property rights and/or prevalence of unpriced external costs), and state failure (the lack of effort made to integrate environmental considerations into the day-to-day working of government). Hence, the strong implication of most evaluations is that climate

---

[4] Available at www.adamproject.eu/.

change could be dealt with adequately provided appropriate measures are taken to correct these failures (e.g. SRU, 2004; Lechtenböhmer *et al.*, 2005; Egenhofer *et al.*, 2006). This is despite the dissonance between the measures needed to internalise external costs (e.g. through a carbon tax) and the measures needed for climate policy integration (e.g. by carrying out a 'climate assessment' of all new policy measures).

Framing the problem as a market failure leads to the promotion of market-based instruments to incorporate the wider 'social cost of carbon' into the price of goods and services (e.g. OECD, 2002b). The removal of perverse incentives in the form of subsidies to polluting activities, such as coal mining, is also prominent in many evaluations (e.g. SRU, 2000; OECD, 2002a; Michaelowa, 2003). Special provisions (derogations) for 'losers' have been a stock-in-trade for EU climate policy. Evaluations also note that several long-standing measures reported as climate policies were initially designed as responses to other problems. Evaluation studies vary in the extent to which they seek to quantify co-benefits of climate policies, but their existence is believed to have helped policy makers in securing support for key policies (e.g. Szarka and Blühdorn, 2006).

We conclude that climate policy evaluations confirm dominant framings in terms of market and state failure. Only a small number of studies argue that more radical problem framings are needed, an example being a 'transitions' perspective, suggesting the transformation of energy systems towards renewables (Prins and Rayner, 2007). The conservatism of most evaluations serves to further embed the idea of reducing emissions through pricing climate damages and by implementing policy integration (Kuik *et al.*, 2008). However, evaluations also acknowledge that there are frequently political obstacles to policy formation and show that co-benefits are often invoked to justify action. A range of different objectives related to energy security, fuel poverty, industrial innovation, health and employment are proposed. Stressing co-benefits has been a way of expanding the political viability of a measure. Rarely has the 'best' policy (such as a carbon tax) been adopted. In the context of second-best policies and multiple objectives, effectiveness may be hard to measure and the goal of emission reduction may even be compromised. The link between policy objectives and specific outcomes of policies therefore remains ambiguous.

6.4.2 Levels and scales of governance

In the EU, the question of which level policy makers should act at has been prominent since subsidiarity entered into policy discourse in the early 1990s (Jordan, 2002). The essential governance dilemma here is that, acting autonomously, Member States may not make sufficient headway to ensure that the EU meets its collective commitments; however, too much central steering might entail unacceptably high regulatory costs and could meet strong political resistance from states concerned about economic competitiveness and losing their policy making competence to the EU. The dilemma about levels and scales also relates to the question of who should be regulated, specifically which sources of greenhouse gas emissions should be addressed and how.

Questions regarding the relationship between the European dimension of policy making and national action were addressed in about half of the reports reviewed. Interestingly, none of the studies refers to negative interactions between climate policy activity at the EU level and in Member States, although Agnolucci (2006), for example, notes how uncertainty over a draft EU directive temporarily stifled renewables growth in Germany. On the contrary, there appears widespread support for greater harmonisation at the EU level in policy evaluations, especially with regard to the ETS and carbon taxation, mainly on the grounds that this would lead to greater effectiveness. In practice, harmonisation has occurred in a variety of ways; sometimes leading to common rules (as with minimum performance standards for consumer goods and energy labels) and sometimes leading to agreed national targets, with Member States retaining the freedom to determine how to implement these.

The initial design of the EU ETS left many important decisions to the discretion of Member States. These decisions specified the total number of allowances to be awarded, and the distribution between sectors and installations (Grubb *et al.*, 2005). These issues were taken up in the 2008 revision of the ETS under which a single EU cap for installations covered by the ETS was imposed and auctioning of emissions permits will be introduced from 2013. This major step in European harmonisation gave rise to relatively little debate.

Another area considered a candidate for increased harmonisation efforts by a number of evaluations was energy policy, more specifically carbon and energy taxation, and support for renewables. Although efforts to introduce a European-level carbon tax failed in the early 1990s, calls for harmonisation are still a recurrent theme in policy evaluations (Prime Minister's Office and Economic Council, 2000; Interwies *et al.*, 2002). For the most part, these remain at a fairly general level, emphasising the economic superiority of a common EU tax over a patchwork of national schemes, whilst acknowledging the political difficulties. Concerning renewables, proponents of policy harmonisation stress increased competitiveness and reduced costs of technologies that might result from a more harmonised approach. While agreement on EU renewables targets was achieved in 2007, harmonisation of policy measures remains '...difficult to achieve in the short term' (CEC, 2005b: 11) because of disagreement over which of the two main policy instruments used for renewables promotion in Europe (i.e. feed-in tariffs and tradable quota systems) better encourages growth in a liberalised electricity market.

Perhaps most interesting are the findings regarding the relationship between the CCPMs and Member State action on climate change. At issue here is the question of how a balance is struck between the need for co-ordination and the desire for flexibility at the national and local level. Relatively few reports (seven in total) considered the relation between CCPMs and national policies. Almost all these reports were either commissioned or written by the European Commission or the European Environmental Agency. One of the reasons for this relative scarcity may be that Member State evaluation studies do not refer to CCPMs directly, even though national policy is often an outcome of EU legislation. Yet the available studies suggest that CCPMs often work as drivers for national mitigation action, especially in cases where a Member State has not been active in the field before (e.g. Golder Europe EEIG, 2005; EEA, 2006; CEC, 2007a). The quantitative targets of many CCPMs support their effectiveness, even though not all Member States comply with the targets set. There is also little sign that 'Europeanising' policies has hindered or reduced the effectiveness of the national efforts of 'leaders' like Germany, Sweden and the UK.

In summary, evaluation studies in our sample indicate the benefits of harmonisation of climate policies in Europe. Aspects of energy policies, especially taxation and renewables promotion, have been the focus of calls for greater harmonisation. Recent developments have shown a progressive harmonisation, although this has not reduced the diversity of measures implemented at the Member State level, where there continues to be policy experimentation. The evaluations also demonstrate the benefits of CCPMs as an effective means to support national climate policies.

6.4.3 *Timing and sequencing*

This dilemma encompasses questions such as whether to act early or to postpone action, the timescale over which policy should be introduced, and the dangers of becoming 'locked in' to inappropriate policy pathways (Pierson, 1996). In an area like climate policy that is characterised by long time horizons and great uncertainty over the potential costs and benefits of different courses of action, there is an important dilemma between sending clear signals to the market, through regulatory stability, and leaving sufficient flexibility to respond to new information. Some evaluations refer to the need to see climate change as a long-term problem, emphasising the need for ambitious long-term emission reduction targets (e.g. EAC, 2007; FoEE, 2005). This corresponds with the target-setting approach that has been so dominant in EU climate policy making. Evaluations also stress the significance of intermediate targets as points along a desired trajectory (EAC, 2007).

Attention to the timing and sequencing dilemma was most explicit in the evaluations addressing two policy domains: the EU ETS and renewables support. Concerning the EU ETS, the dilemma is manifest with regard to the length of trading periods. Several evaluations considered their relatively short duration (five years) a significant drawback (Egenhofer *et al.*, 2006). Member States governments' ability to go 'back to the drawing board' prior to each new allocation period means that certainty only exists for up to five years ahead. Longer trading periods would provide greater predictability, potentially reducing investment risks. At the same time, they decrease the flexibility to commit to new targets in the light of new science or changed political conditions. The revised EU ETS directive has taken a middle course by extending trading periods beyond 2012 to seven years. However, concerns about the time limit on trading periods signal a deeper problem about the credibility of long-term political commitment to emissions reductions in the EU.

With respect to renewable energy, our dataset identified continuous political support and policy stability as key conditions for enabling its successful expansion (CEC, 2005b). Moreover, the specific design of the mechanism chosen for the promotion of renewables, and especially the ability of an instrument to reduce the risk for renewable energy investors, emerged as crucial (Jacobsson and Lauber, 2006). Feed-in laws appear to stimulate renewable energy sources (used for example in Germany and Portugal) and reduce commercial risks to investors more effectively than certificate schemes (e.g. Mitchell *et al.*, 2006; Szarka and Blühdorn, 2006). Feed-in tariffs guarantee prices to renewables generators, whereas certificate schemes operate by setting a fixed quota of renewable electricity consumption or production. Under a feed-in scheme the price paid for renewable energy is stable and producers bear no 'volume risk' because network operators are required to accept all generated renewable energy. By contrast, under a certificate scheme such as the Renewables Obligation in the UK, the value of the certificates depends on supply and demand in the certificate market, which has proven unpredictable (Mitchell *et al.*, 2006). The available evaluation studies also suggest that there need not be a trade-off between predictability and flexibility in renewables policy, provided that the support mechanism in place is well designed. The German system of feed-in tariffs, for instance, while creating a more stable investment climate, also has substantial flexibility, due to a system of regular review (Szarka and Blühdorn, 2006).

To conclude, policy evaluations tend to endorse the 'early mitigation' strategy, which at least rhetorically, has been central to the EU's political position on climate change since 1992. Temporality has influenced policy design especially in debates about the EU ETS and renewables policy, both of which have sought to balance predictability against flexibility in policy design. For renewables support, the German experience suggests that an emphasis on predictability need not compromise the capacity to review and adjust policies. For the ETS, the main concern has been with the length of allocation periods, although this may have become less pressing as the general political commitment to climate policy has grown within the EU since 2005, thereby reassuring would-be investors of the economic wisdom of investing into emissions reduction.

6.4.4 Modes and instruments

A recurrent dilemma in environmental policy making is whether to rely on hierarchical, market or network-based modes and instruments of governance (Jordan et al., 2003). The main challenge confronting policy makers is how to choose governance modes (and their associated policy instruments) in a way that maximises their strengths and reduces their weaknesses.

Looking at the evaluations that have been conducted, fiscal measures appear to be the most evaluated type of instruments. This is not surprising, given their dominant role in policy discourses and the controversy they generate amongst target groups. Market-based instruments (and especially the EU ETS) receive their fair share of attention – taking into account the limited number of such measures across our sample countries (Jordan et al., 2003) and the fact that the EU ETS was launched only in 2005. Traditional regulation, on the other hand, is relatively under-evaluated. This resonates with another finding from the meta-analysis, namely the small number of regulatory measures which include a requirement to evaluate performance. Amongst the 262 evaluations reviewed, only 19 stated that they were produced to fulfil a specific legal reporting requirement.

Almost half of the evaluation effort in our sample is devoted to network-based instruments, such as voluntary/negotiated agreements (VAs) and information-based measures. In this context, the findings regarding the (in)effectiveness of voluntary approaches in climate policy are particularly striking. Voluntary approaches are applied in all six countries covered by our meta-analysis. Among the evaluations we reviewed, approximately 60 cover voluntary approaches in more detail. The evaluation community has paid special attention to four climate-related VAs: the Declaration of German Industry on Global Warming (1996); the Finnish Energy Conservation Agreement (1997); UK Climate Change Agreements (2001); and the ACEA-JAMA-KAMA covenants (1998/1999). These were examined in roughly 30 studies in our sample. Evaluation reports tend to agree that the Declaration of German Industry on Global Warming Prevention, an example of a unilateral industrial commitment at least in its initial versions, has largely failed in stimulating significant emission cuts. Many of the targets were set close to business-as-usual projections, implying that most reductions would have taken place irrespective of the declaration (Krarup and Ramesohl, 2000; SRU, 2000; RWI, 2000).

Similar findings apply for the UK Climate Change Agreements (CCAs) from 2001. Despite showing a high degree of participation, the effectiveness of the agreements is contested by a number of studies (e.g. EAC, 2004; de Muizonand and Glachant, 2007). Most sectors exceeded their quantitative targets for 2010 well in advance of the official deadline, without requiring special effort (FES, 2005). These targets, projected to reduce greenhouse gas emissions by nine per cent, were initially perceived as ambitious. There is now a widespread recognition that *ex ante* targets tend to be highly conservative, primarily due to strategic behaviour by the industries involved in negotiating them (FES, 2005; Glachant and de Muizon, 2007).

More positive conclusions on the effectiveness of voluntary agreements can be drawn from the Finnish experience. The Finnish Energy Conservation Agreement scheme (ECAs) appears to demonstrate that public voluntary programmes can be successful in stimulating participation and energy-saving measures, provided they are accompanied by proper incentives (in this case government subsidies for energy auditing and energy saving investments). The Finnish scheme has precipitated a high degree of participation in the industrial and public sectors, but to a lesser extent in the private services sector (Heikkilä *et al.*, 2005; Pöllänen and Kalenoja, 2005). The energy and climate objectives of the programme are projected to be exceeded (Motiva, 2006). The scheme, however, seems mainly to collect 'low hanging fruits'; most measures have relatively short payback time (zero to two years), whereas measures with longer payback time are rarely implemented (Khan, 2006).

These findings resonate with a common argument in the literature that VAs lead to incrementalism rather than radical change (Jordan *et al.*, 2003). This also applies to the negotiated agreements between the European Commission and European car manufacturers. The ACEA-JAMA-KAMA covenants (1998/1999) are judged to have induced mainly gradual improvements in engine technology, with most progress attributed to the 'dieselisation' of the European car fleet, rather than innovative power trains for low-emitting cars (e.g. Kågesson, 2005; Ricardo Consulting Engineers, 2005). Overall, despite some progress, projections that the industry would miss its 2008/2009 target (CEC, 2006a) prompted the Commission to propose a legislative framework including mandatory fuel efficiency standards for new passenger cars (CEC, 2007d).

In summary, besides fiscal measures, network measures were the most evaluated instruments across the reports we reviewed. In terms of evaluation results, the meta-analysis confirmed the widely held view that voluntary agreements are unable to bring about far-reaching changes in behaviour by firms and sectors. Most reports explain the disappointing performance of voluntary agreements by citing unambitious targets and the lack of incentives for participation and compliance mechanisms. The Finnish experience shows, however, that if applied in the right setting and with adequate (financial) incentives, voluntary agreements can play a useful role in a policy mix.

6.4.5 Costs and benefits

When deciding the timing, level and means by which to act, policy makers have to weigh the associated costs and benefits. Although policies may seek to generate net social benefits, they inevitably generate winners and losers (Bennett, 1991). The most important governance dilemma in this context is what costs and benefits to take into account when decision making, and whether losers will be compensated. Judgements on these issues are tied to considerations of policy effectiveness, fairness, legitimacy and the relative power of affected parties. One trade-off that policy makers may be faced with is between equity and cost-effectiveness: should carbon savings be made at least cost to the economy as a whole, even if that means an economic loss for certain groups?

Sharing the costs of abatement effort between EU Member States ('burden sharing' pre-2008, 'effort sharing' since 2008) has long been recognised as being crucial to EU climate policy (Skjaerseth, 1994). The EU was able to overcome the 'asymmetrical interests' regularly identified as key obstacles to agreement on EU climate change response strategies using the so-called Triptych Approach (Blok *et al.*, 1997). The most significant feature of the approach was its *sectoral* focus. The Triptych framework divided domestic economies into three broad sectors. While identical targets were imposed on the energy-intensive industries to alleviate competitiveness concerns, a more differentiated approach was used for the other sectors. This approach was abandoned in 2008, when GDP per capita in the 27 EU Member States became the sole indicator for Member State national targets in the non-trading sectors after 2012 (Haug and Jordan, 2010).

The issue of distributive equity in climate policy is not widely covered in the evaluation studies we reviewed. Only a small number used fairness (which in our assessment includes equity as well as competitiveness analyses) as an evaluation criterion (see Figure 6.2). This lack of attention may, if Weidner (2005) is correct, reflect a degree of political expediency, whereby governments prefer not to highlight the regressive nature of their policies.

Amongst our sample of six states, official UK evaluations did the most to record the distribution of costs and benefits across three groups: the Treasury, i.e. taxpayers, firms and consumers. Yet judgements about where the burden ultimately falls are complicated by uncertainty over the extent to which additional costs incurred by firms are passed on to consumers. Economic models suggest that this is likely to be the case for between 50 per cent and 100 per cent of any cost increases, depending on the competitiveness constraints faced by businesses (Defra, 2006).

Along with competitiveness concerns, the distributional implications of carbon taxes are a major issue in determining their political acceptability. Evidence from the evaluations suggests that carbon taxes may be mildly regressive (i.e. lower income

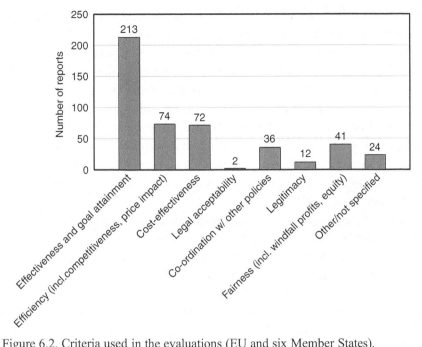

Figure 6.2. Criteria used in the evaluations (EU and six Member States).

groups pay proportionately more in taxation than wealthier groups), but this often depends on the modelling framework used (see e.g. Tiezzi, 2005). Significantly, however, this regressive effect can be greatly attenuated, or even reversed, where the tax is applied in a revenue-neutral context. Effects vary across Member States, with Symons *et al.* (2002) finding that, while energy or carbon dioxide taxation is regressive in Finland (where the heaviest burden falls on low-income households), Germany, France and, to a lesser degree, in Spain, it is progressive in the UK (except for the highest income group) and neutral in Italy. Moreover, regional distribution is uneven, with households in rural areas disproportionately negatively affected (Mustonen and Sinko, 2000).

In handling revenues from carbon taxation, by contrast, a key trade-off emerges between efficiency and equity. Interestingly, this dilemma is treated differently across Member States. Where a part of the revenue is used to compensate poorer households suffering adversely from the tax, the scope for revenues to be used to maximise efficiency gains from reductions in other existing distortionary taxes, such as value added tax (VAT), will be lost. In Germany, for example, ecological tax reform recycled tax revenues by cutting both employers' and employees' social security contributions, showing less concern '…with fiscal orthodoxy and more with political appeal' (Andersen, 2005: 523).

Debates over efficiency (the functioning of the market) and equity (redistribution of wealth) have also been prominent in the context of the EU ETS (Endres and Ohl, 2005). The generous allocation of allowances to installations covered by the scheme during the initial period shifted the burden for achieving Europe's Kyoto targets to the 'non-traded' sectors (Rogge *et al.*, 2006). Moreover, the ETS led to an increase in electricity prices (Sijm *et al.*, 2006). Given the state of the electricity market in Europe, power producers were able to pass through the extra costs of the ETS to their customers, leading to windfall profits. The market power of the electricity generators may have adverse consequences for industries falling both within and outside the scope of the ETS (see e.g. Egenhofer *et al.*, 2005; Sijm *et al.*, 2006; Neuhoff *et al.*, 2006).

In summary, the degree to which explicit attention is given to the costs and benefits of climate policy (here narrowly interpreted as distributive equity) is most evident with regard to the EU ETS and carbon taxation, both highly visible and relatively controversial policy instruments. The meta-analysis indicates that effective climate policies involving carbon taxation tend to be slightly regressive, but the different experiences of countries suggest this is not unavoidable. In the case of the ETS, certain powerful industrial sectors appear to be able to pass the costs of emission reductions to household consumers.

6.4.6 Implementation and enforcement

In implementation the chosen instruments are applied to a target group, the response monitored, and adjustments made for deviations from expected outcomes. The key dilemma here relates to whether provisions should exist for sanctions against non-compliance, and what level of monitoring and enforcement is optimal. The idea of an 'implementation gap' in EU environmental policy (Jordan, 2002), long prominent in European policy circles, is also reflected in the reports we reviewed. A number of EU-level evaluations express concern at the incomplete implementation of the measures outlined in the ECCP. For EU policy this involves a long chain of procedures including political adoption of the proposed policies by EU institutions, transposition into Member State law, enforcement by national authorities, and recourse to European-level sanctions in cases of non-compliance (e.g. CEC, 2006b; CEC, 2006c; CEC, 2007b).

About two-thirds of the evaluations we reviewed devoted attention to the implementation and enforcement of climate policies at a general level. Three more specific themes emerged from the meta-analysis: loopholes in the design of some instruments; poor provisions for monitoring; and the weakness or lack of enforcement mechanisms for many policies. Starting with the first of these, many evaluations demonstrate that policy instruments, initially designed with specific ambitions, were watered down in the final political stage before implementation, with very negative implications for their effectiveness. The reason for these shortcomings is often the same; there is a need to secure support from influential lobbies that would otherwise block the policy. The many special rules and exemptions from energy taxation, for example in Italy and in Germany, are one example of this (OECD, 2002a). Allocations of emission allowances under the EU ETS are another notorious example. In the case of Germany for instance, the leading environmental policy watchdog deplored that for the first commitment period '...[a]lmost half the allocated emissions allowances were issued based on special circumstances which exempt privileged installations from reduction obligations' (SRU, 2006, p. 12).

Moving on to monitoring, *ex ante* and *ex post* measurements of the performance of policy instruments are widely seen as a precondition for establishing their effectiveness and for learning from implementation experience (Ecofys, 2007). One particularly striking theme to emerge from the meta-analysis was the widespread criticism of the poor provision for monitoring of effectiveness that characterises many policy instruments (e.g. RWI, 2000; EAC, 2005; Maslin *et al.*, 2007). In their attempts to evaluate the effectiveness of 20 energy efficiency policies, Ecofys (2007) were hampered because most instruments lacked a comprehensive monitoring system. Even though methodologies are available, both the availability and quality of monitoring data turned out to be inadequate. Lack of reliable *ex ante* data at the installation level also caused difficulties in determining the cap and distribution of allowances in the first phase of the EU ETS (Ellerman and Buchner, 2006).

A broader institutional issue that is raised in some evaluations is the lack of early-warning mechanisms to inform policy makers whether their intended results are being delivered. On this question, it is instructive to note that, even in a country with a relatively advanced climate policy such as the UK, a leading watchdog has criticised the time taken by the government to acknowledge that the policy was '...so far off course' (EAC, 2005: 40) from its 2010 targets. This has been attributed to inadequate monitoring. On top of this, policy makers are faced with the possibility that apparent successes may in fact be less significant than supposed, owing to 'rebound effects' (Defra, 2006; Maslin *et al.*, 2007). Evaluations in our sample often acknowledged the need for more sophisticated monitoring of such effects and treatment in calculations of emission reductions obtained (e.g. FES and PSI, 2005).

Finally, another frequently highlighted concern is the absence of sanctions for non-compliance (e.g. SRU, 2005; IPA Energy and Water Consulting, 2007; Maslin *et al.*, 2007). This is particularly salient in the case of voluntary agreements (RWI, 2000), but also, for instance, in renewables certificate trading schemes (CEC,

2007b). Furthermore, non-compliance is a cause of concern in the building sector, where energy standards for new buildings are often poorly enforced, apparently due to concerns over the impact on property developers (see e.g. Oxera, 2005; Maslin *et al.*, 2007).

In conclusion, the most striking theme to emerge from the meta-analysis of EU climate policy evaluations with regard to the implementation and enforcement of climate policies is the generally poor provision made for monitoring which characterises many policy instruments. The meta-analysis further points to the need for stronger political will to design and implement policies without loopholes and to put in place adequate enforcement mechanisms.

6.5 Conclusions

This chapter has reviewed the development and impacts of EU climate policy over the past 15 years or so, framing the analysis through six policy dilemmas. A complex multi-level texture of climate policies has emerged in Europe in the period since 1990. Much of this policy is made and implemented by the EU Member States, but 'common and co-ordinated policies and measures' have become more important, especially with the advent of the Emissions Trading Scheme in 2005. An initial period of high policy experimentation may be coming to an end, and there appears now to be a trend towards greater harmonisation at the EU level. The 'Europeanisation' of climate policy (Jordan and Liefferink, 2004) is explained by the widely recognised need to co-ordinate emissions reduction at the European level, both to allocate effort within the EU and to present a more powerful European voice in international negotiations (Jordan *et al.*, 2010). Gradually, climate change has also become an important *raison d'etre* for the European project, at a time when European citizens and their governments are becoming more sceptical about deeper European integration. The EU continues to play a leading role in international climate politics and political leaders have found this an important vehicle for projecting the EU's 'soft' power globally.

Through an analysis of some 262 evaluations of EU climate policies we have been able to draw some conclusions about the nature and impact of these policies. Broadly we conclude that:

(i) Co-benefits are often used to justify emission reduction policies, but reliance on such a strategy may compromise climate effectiveness.

(ii) Emission reduction goals of climate policies are often compromised by provisions to deal with distributional issues (whether socially or in industrial sectors).

(iii) EU-level policies support national policy efforts amongst laggard countries and do not compromise policy efforts by leading countries (like Germany and the UK).

(iv) Voluntary action generally tends to be ineffective in climate policy.

(v) Policy offering a long-term perspective to energy producers and users is more effective than policy which retains a large measure of short-term flexibility.

(vi) Successful implementation of climate policy instruments depends to a large extent on sound monitoring, but the provisions made for this are generally poor.

However, we still know surprisingly little about the effectiveness of EU climate policies. Although goal attainment has been an important criterion for many policy evaluations, very few evaluation studies make quantitative estimates of the emission reduction impacts of particular policies, or of portfolios of policies. The value of EU climate policies in terms of their primary goal – emission reductions – is therefore still an open question. The span of opinion on the effectiveness of EU climate policy ranges from Eichhammer *et al.* (2001: 1), who assert that, in the UK and Germany, '…a mix of policies at the national and regional levels added considerably to the reduction of greenhouse gases', to Kerr (2007), who argues that emission reductions were purely serendipitous and that there is no statistically significant data to show that climate policies had indeed produced an effect. One of the greatest failings is the lack of adequate *ex ante* baselines from which it would be possible to estimate the impacts of policies, against those of other contingent factors, or of counterfactuals (i.e. what would have happened without them).

Looking ahead, EU policy needs to confront a range of possible international outcomes, both politically and in terms of physical climate. Politically, current EU strategy is to try to construct a more ambitious international mitigation regime after 2012 by offering tougher emission reduction targets on the condition that other countries (especially the US, China, India and Brazil) also work towards binding targets. However, international climate regimes may emerge that are very different from the one the EU is trying to build, and which require a much greatrer focus on adaptation to the impacts of climate change (see Jordan *et al.*, 2010 for a discussion of possible scenarios). Alternatively, evidence that climate change impacts are greater than predicted, and recognition that much lower atmospheric concentrations are required to avert dangerous climate change, may lead to the development of more transformative, radical mitigation policies. Emerging understanding of the climate system and climate change impacts suggests that more extreme physical outcomes are a distinct possibility.

References

Agnolucci, P. (2006) Use of economic instruments in the German renewable electricity policy. *Energy Policy*, **18**, 3538–48.

Andersen, M. S. (2005). Regulation or coordination: European climate policy between Scylla and Charybdis. In *Emissions Trading for Climate Policy*, ed. Hansjurgens, B. Cambridge, UK: Cambridge University Press.

Bennett, G. (1991) *Dilemmas: Coping with Environmental Problems*. London: Earthscan.

Blok, K., Phylipsen, G. J. M., and Bode, J. W. (1997) *The Triptych Approach, Burden Sharing Differentiation of CO₂ Emissions Reduction among EU Member States.* Discussion paper for the informal workshop for the European Union Ad Hoc Group on Climate, Zeist, the Netherlands, January 16–17, 1997. Utrecht: Utrecht University.

CEC (1992) *Commission of the European Communities. A Community Programme of policy and action in relation to the environment and sustainable development – Towards Sustainability – Fifth Environmental Action Programme,* COM (92) 23.

CEC (2005a) *Winning the battle against global climate change.* COM(2005) **35** final.

CEC (2005b) *Communication from the Commission: The Support of Electricity from Renewable Energy Sources.* Brussels: CEC.

CEC (2006a). *Implementing the Community Strategy to Reduce CO₂ Emissions from Cars: Sixth annual Communication on the effectiveness of the strategy.* Brussels: CEC.

CEC (2006b) *Report from the Commission to the Council and the European Parliament on Implementation of Community Waste Legislation.* Directive 75/442/EEC on waste, Directive 91/689/EEC on hazardous waste, Directive 75/439/EEC on waste oils, Directive 86/278/EEC on sewage sludge, Directive 94/62/EC on packaging and packaging waste and Directive 1999/31/EC on the landfill of waste for the Period 2001–2003. Brussels: CEC.

CEC (2006c) *The Second European Climate Change Programme. Working Group ECCP Review – Topic Group Energy Supply. Final Report.* Brussels: CEC.

CEC (2007a) *Communication from the Commission to the Council and the European Parliament: Biofuels Progress Report. Report on the progress made in the use of biofuels and other renewable fuels in the Member States of the European Union.* Brussels: CEC.

CEC (2007b) *Green Paper follow-up action: Report on progress in renewable electricity. Communication from the Commission to the Council and the European Parliament.* Brussels: CEC.

CEC (2007c) *Adapting to Climate Change in Europe – Options for EU Action. Green Paper from the Commission to the Council, the European Parliament, the European Economic and Social Committee and the Committee of the Regions.* Brussels: CEC.

CEC (2007d) *Setting emission performance standards for new passenger cars as part of the Community's integrated approach to reduce CO₂ emissions from light-duty vehicles.* COM(2007) **856** final

CEC (2008) *20 20 by 2020: Europe's climate change opportunity.* COM 2008 **30** final. Brussels: CEC.

Collier, U. (1997) The EU and climate change policy: the struggle over policy competences. In *Cases in Climate Change Policy: Political Reality in the European Union.* ed. Collier, U. and Lofstedt, R. London: Earthscan, pp. 43–64.

Dahl, A. (2000) Competence and subsidiarity. In Gupta, J. and Grubb, M. (eds.) *Climate Change and European Leadership: A Sustainable Role for Europe?* Dordrecht: Kluwer Academic Publishers, pp. 109–34.

Defra (2006) *Greenhouse Gas Policy Evaluation and Appraisal in Government Departments.* London: Department for Environment, Food and Rural Affairs (Defra).

Dellink, R., Dekker, T., den Elzen, M. *et al.* (2008) *Sharing the Burden of Adaptation Financing.* IVM Report 08/05. Amsterdam: Institute for Environmental Studies (IVM).

den Elzen, M., Meinshausen, M. and van Vuuren, D. (2006) Multi-gas emission envelopes to meet greenhouse gas concentration targets: cost versus certainty of limiting temperature increase. *Global Environmental Change,* **17**(2), 260–80.

EAC (2004) *Environmental Audit Committee. Budget 2004 and Energy. Tenth Report of Session 2003–4, House of Commons.* London: Stationery Office.

EAC (2005) *Pre-budget 2004 and Budget 2005. Tax, Appraisal and the Environment.* House of Commons. London: Stationery Office.

EAC (2007) *The EU Emissions Trading Scheme: Lessons for the Future, 2nd report of session 2006–07.* London: Stationery Office.

Ecofys (2007) *From Theory Based Policy Evaluation to SMART Policy Design – Summary report of the AID-EE project.* Politecnico di Milano: Ecofys, Lund University, Wuppertal Institute for Climate, Environment and Energy.

EEA (2006) *Energy and environment in the European Union: Tracking progress towards integration.* EEA Report No 8. Copenhagen: European Environment Agency.

EEA (2008) *Greenhouse Gas Emission Trends and Projections in Europe 2008.* EEA Report No 5/2008. Copenhagen: European Environment Agency.

Egenhofer, C., Fujiwara, N. and Gialoglou, K. (2005) *Business consequences of the EU Emissions trading scheme.* Brussels: Centre for European Policy Studies (CEPS).

Egenhofer, C., Jansen, J. C., Bakker, S. J. A. and Jussila Hammes, J. (2006) *Revisiting EU Policy Options for Tackling Climate Change: A social cost-benefit analysis of GHG emissions reductions strategies.* Brussels: CEPS.

Eichhammer, W., Schleich, J., Boede, W. U. *et al.* (2001) Greenhouse gas reductions in Germany and the UK – Coincidence or policy induced? *An Analysis for International Climate Policy.* Karlsruhe, Brighton, Berlin: Fraunhofer Institute for Systems and Innovation Research, SPRU, DIW.

Ellerman, D., and Buchner, B. (2006) *Over-Allocation or Abatement? A Preliminary Analysis of the EU ETS Based on the 2005 Emissions Data.* Venice: Fondazione Eni Enrico Mattei.

Endres, A., and Ohl, C. (2005) Kyoto, Europe? An economic evaluation of the European Emission Trading Directive, *European Journal of Law and Economics*, **1**, 17–39.

FES (2005) *Future Energy Solutions. Climate Change Agreements – Results of the Second Target Period Assessment.* Didcot: AEA Technology/Defra.

FES and PSI (2005) *Future Energy Solutions and the Policy Studies Institute. Evaluation of the Government's Energy Efficiency Policies and Programmes, A report for DEFRA as part of the Climate Change Programme Review.*

FoEE (2005) *Friends of the Earth Europe. How the European Union responds to the global threat of climate change. An assessment by Friends of the Earth Europe.* Brussels: FoE Europe.

de Muizon, G. and Glachant, M. (2007) Climate change agreements in the UK: a successful policy experience? In *Reality Check: The Nature and Performance of Voluntary Environmental Programs in the United States, Europe and Japan.* ed. Morgenstern, R. D. and Pizer, A. Washington, DC: 64–85.

Golder Europe EEIG (2005) *Report on implementation of the landfill Directive in the 15 Member States of the European Union.* Berkshire: Golder Europe EEIG.

Goulder, L. H. and Parry, I. W. H. (2007) *Instrument Choice in Environmental Policy.* RFF DP 08–07, Washington DC: Resources for the Future.

Grubb, M., Azar, C., Persson, M. (2005) Allowance allocation in the European emissions trading system: a commentary. *Climate Policy*, **5**, 127–36.

Haug, C. and Jordan, A. (2010) Burden sharing: distributing burdens or sharing efforts? In *Climate Change Policy in the European Union: Confronting the Dilemmas of Mitigation and Adaptation.* ed. Jordan, A. *et al.* Cambridge, UK: Cambridge University Press.

Haug, C., Rayner, T., Huitema, D. *et al.* (2008) *Navigating the Dilemmas of European Climate Policy. A Meta-analysis of Recent Policy Evaluations*, Climatic Change *[under review].*

Heikkilä, I., Pekkonen, J., Reinikainen, E., Halme, K. and Lemola, T. (2005). *Energiansäästösopimusten kokonaisarviointi*. Helsinki: Finnish Ministry of Trade and Industry.

Interwies, E., Blobel, D. and ten Brink, R. (2002). *Ökosteuer – Stand der Diskussion und der Gesetzgebung in Deutschland, auf der EU-Ebene und in den anderen europäischen Staaten*. Berlin: Ecologic.

IPA Energy and Water Consulting (2007). *UK Power Sector Emissions – Targets or Reality. Final Report to the WWF*. Edinburgh: IPA Energy and Water Consulting.

Jacobsson, S. and Lauber, V. (2006). The politics and policy of energy system transformation. *Energy Policy*, **34**, 256–76.

Jordan, A. (2002). The implementation of EU environmental policy: a policy problem without a political solution? In *Environmental Policy in the European Union* ed. Jordan, A. London: Earthscan.

Jordan, A., Wurzel, R. K. W. and Zito, A. R. (2003). *'New' Instruments of Environmental Governance? National Experiences and Prospects*. London: Frank Cass & Co.

Jordan, A. and Liefferink, D. (eds.) (2004) *Environmental Policy in Europe: the Europeanisation of National Policy*. London: Routledge.

Jordan, A., Huitema, D., van Asselt, H., Berkhout, F. and Rayner, T. eds. (2010) *Climate Change Policy in the European Union: Confronting the Dilemmas of Mitigation and Adaptation?* Cambridge, UK: Cambridge University Press.

Kågesson, P. (2005). *Reducing CO₂ Emissions from New Cars: A Progress Report on the Car Industry's Voluntary Agreement and an Assessment of the Need for Policy Instruments*. Brussels: European Federation for Transport and Environment.

Kerr, A. (2007). Serendipity is not a strategy: the impact of national climate programmes on greenhouse-gas emissions. *Area*, **39**, 418–30.

Khan, J. (2006). *Evaluation of the Energy Audit Programme in Finland within the framework of the AID-EE project*. Lund: Lund University.

Krarup, S. and Ramesohl, S., (2000). *Voluntary Agreements in Energy Policy – Implementation and Efficiency. Summary report of EU VAIE research project*. Wuppertal: Wuppertal Institute for Climate, Environment and Energy.

Kuik, O., Aerts, J., Berkhout, F., *et al.* (2008). Post-2012 climate policy dilemmas: a review of proposals. *Climate Policy*, **8**, 317–36.

Lacasta, N. S., Dessai, S. and Powroslo, E. (2002). Consensus among many voices: articulating the European Union position on climate change, *Golden Gate University Law Review*, **32** (4), 351–414.

Lechtenböhmer, S., Grimm, V., Mitze, D., Thomas, S. and Wissner, M. (2005). *Target 2020: Policies and Measures to Reduce Greenhouse Gas Emissions in the EU*. Wuppertal: Wuppertal Institute and Wissenschaftszentrum Nordrhein-Westfalen.

Long, T., Salter, L. and Singer, S. (2002). WWF: European and global climate policy. In Pedler, R. (ed) *European Union Lobbying*. Basingstoke: Palgrave.

Majone, G. (1989). *Evidence, Argument and Persuasion in the Policy Process*. New Haven, CT: Yale University Press.

Maslin, M., Austin, P., Dickson, A., Murlis, J., Owen, M. and Panizzo, V. (2007). *Audit of UK Greenhouse Gas Emissions to 2020: will current Government policies achieve significant reductions?* London: UCL Environment Institute.

Michaelowa, A. (2003). Germany. A pioneer on earthen feet? *Climate Policy*, **1**, 31–43.

Mitchell, C., Bauknecht, D. and Connor, P. M. (2006). Effectiveness through risk reduction: a comparison of the renewable obligation in England and Wales and the feed-in system in Germany. *Energy Policy*, **3**, 297–305.

Moravcsik, A. (1998) *The Choice For Europe*. Ithaca, New York: Cornell University Press.

Motiva (2006). *Energy Conservation Agreements – Progress Review 2005.* Helsinki: Motiva Oy.

Mustonen, E. and Sinko, P. (2000). *Hiilidioksidiveron Vaikutus Kotitalouksien Tulonjakoon,* Helsinki: Government Institute for Economic Research.

Neuhoff, K., Åhman, M., Betz, R. *et al.* (2006). Implications of announced phase II national allocation plans for the EU ETS. *Climate Policy,* **6**, 411–22.

OECD (2002a). *Organisation for Economic Cooperation and Development. Environmental Performance Reviews: Italy.* Paris: OECD.

OECD (2002b). *Environmental Performance Review: United Kingdom.* Paris: OECD.

Oxera (2005). *Policies for Energy efficiency in the UK Household Sector, Report prepared for DEFRA.* London: Department for Environment, Food and Rural Affairs (Defra).

Pierson, P. (1996). The path to European integration. *Comparative Politics,* **29** (2), 123–63.

Pöllänen, M. and Kalenoja, H. (2005). *Linja-autoalan Energiansäästösopimuksen Arviointi.* Helsinki: Finnish Ministry of Trade and Industry.

Prime Minister's Office and Economic Council (2000). *Environmental and Energy Taxation in Finland – Preparing for the Kyoto Challenge. Summary of the working group report.* Helsinki: Prime Minister's Office.

Prins, G. and Rayner, S. (2007). Time to ditch Kyoto. *Nature,* **449**, 973–5.

Rayner, T., Huitema, D., Massey, E. *et al.* (2009) Climate policy evaluation across Europe. Tyndall Warhung Paper. Norwich: Tyndall Centre for Climate Change Research.

Ricardo Consulting Engineers (2005). *Study into Passenger Car CO_2 Reduction for the Department of Transport.* Wetherby: DfT publications.

Rogge, K., Schleich, J., Betz, R. and Cozijnsen, J. (2006). EU emission trading: an early analysis of National Allocation Plans for 2008–2012. *Climate Policy,* **6**(4), 361–94.

RWI (2000). *Rheinisch-Westfälisches Institut für Wirtschaftsforschung. Die Klimaschutzerklärung der Deutschen Industrie unter neuen Rahmenbedingungen.* Monitoringbericht 1999. Essen: RWI.

Sbragia, A. (2000). Environmental policy. In Wallace, H. and Wallace, W. *Policy Making in the European Union. 4th edn.* Oxford, UK: Oxford University Press.

Schon, D. A. and Rein, M. (1994). *Frame Reflection: Towards the Resolution of Intractable Policy Controversies.* New York: Basic Books.

Shepsle, K. A. (1989). Studying institutions: some lessons from the rational choice approach. *Journal of Theoretical Politics,* **1**, 131–47.

Sijm, J., Neuhoff, K. and Chen, Y. (2006). CO_2 cost pass-through and windfall profits in the power sector. *Climate Policy,* **6**, 49–72.

Skjaerseth, J. (1994). The climate policy of the EC: too hot to handle?, *Journal of Common Market Studies,* **32**(1), 25–45.

SRU (2000). *Sachverständigenrat für Umweltfragen. Umweltgutachten 2000. Schritte ins nächste Jahrtausend.* Berlin: SRU.

SRU (2004). *Umweltgutachten 2004. Umweltpolitische Handlungsfähigkeit sichern.* Berlin: Sachverständigenrat für Umweltfragen.

SRU (2005). *Umwelt und Strassenverkehr. Hohe Mobilität – Umweltverträglicher Verkehr.* Berlin: Sachverständigenrat für Umweltfragen.

SRU (2006). *Die nationale Umsetzung des europäischen Emissionshandels: Marktwirtschaftlicher Klimaschutz oder Fortsetzung der energiepolitischen Subventionspolitik mit anderen Mitteln?* Berlin: Sachverständigenrat für Umweltfragen.

Symons, E., Proops, J. and Speck, S. (2002). The distributional effects of carbon and energy taxes: the cases of France, Spain, Italy, Germany and UK. *European Environment,* **12**, 203–12.

Szarka, J. and Blühdorn, I. (2006). *Wind Power in Britain and Germany. Explaining Contrasting Development Paths*. London: Anglo-German Foundation for the Study of Industrial Society.

Tiezzi, S. (2005). The welfare effects and the distributive impact of carbon taxation on Italian households, *Energy Policy*, **12**, 1597–1612.

van Vuuren D. P., Cofala J., Eerens, H. E. *et al.* (2006). Exploring the ancillary benefits of the Kyoto Protocol for air pollution in Europe. *Energy Policy*, **34**(4), 444–460.

Weidner H. (2005). *Global Equity Versus Public Interest?*. Berlin: Wissenschaftszentrum Berlin.

Wettestad, J. (2000). The complicated development of EU climate policy. In Grubb, M. and Gupta, J. (eds) *Climate Change and European Leadership*. Dordrecht: Kluwer Academic Publishers.

7

Transforming the European energy system

Lead authors:
GUNNAR S. ESKELAND[1], PATRICK CRIQUI,
EBERHARD JOCHEM, HENRY NEUFELDT
[1]Co-ordinating lead author

Contributing authors:
GIACOMO CATENAZZI, WOLFGANG EICHHAMMER,
ANNE HELD, MARTIN JAKOB, KRISTIN LINNERUD,
TORBEN K. MIDEKSA, SILVANA MIMA, THURE TRABER,
WOLFGANG SCHADE, ULRICH REITER, NATHAN RIVE,
HAL TURTON

Summary

This chapter looks at ways in which the European energy system can be transformed to meet the 2 °C target. For this, we draw on ADAM research that combines results from detailed bottom-up technology and economy-wide top-down modelling approaches to draw a broad-brush picture on Europe's future energy system for a high ('adaptation') and low ('mitigation') temperature increase world. Based on common assumptions of income growth and population development in Europe, we provide an integrated view of the energy system that includes all energy-using and primary-energy-converting sectors. The differentiation between adaptation and mitigation scenarios allows us to specify the costs and technological transitions needed to attain the 2 °C target, as well as the avoided adaptation costs.

In the 'adaptation scenario', climate impacts in the energy system become more important towards the middle of the century. In the 'mitigation scenario', costs of achieving the target in the first decades will be modest through combining energy efficiency improvements, increasing use of renewables and gas, and a decline in energy intensive manufacturing. However, more challenging efforts towards far-reaching technological change will be necessary to reach not only the European target of 2020, but also the overall goal of limiting global temperature increase to 2 °C by the end of this century. Benefits of early emphasis on technological change include: (i) preparing for further emission reductions throughout the middle of the century both in Europe and in the rest of the world, and (ii) increased energy supply security for Europe.

Making Climate Change Work for Us: European Perspectives on Adaptation and Mitigation Strategies, ed. Mike Hulme and Henry Neufeldt. Published by Cambridge University Press © Cambridge University Press 2010.

7.1 Introduction

The energy system comprises use and production of the major energy carriers such as fuels and electricity and currently emits around 80 per cent of Europe's 5.2 billion tonnes carbon dioxide equivalents. Electricity generation alone produces some 27 per cent of this output. The energy system is therefore likely to play a pivotal role in the European Union's efforts to achieve greenhouse gas reductions of at least 20 per cent by 2020, compared to 1990 levels (European Commission, 2008). It is also likely to play an important role in the medium to long-term goals of decarbonising Europe's economy by 50 to 80 per cent by 2050.

This chapter analyses the energy system with the help of a set of interconnected bottom-up technology models (Jochem *et al.*, 2009a), an oligopolistic partial equilibrium model (Traber and Kemfert, 2009) and economy-wide, top-down computable general equilibrium models (Rive and Mideksa, 2008, Eskeland *et al.*, 2009a; Traber and Kemfert, 2009). Greatest attention is generally paid to electricity generation and use, but, importantly, electricity can be substituted by other energy carriers (in space heating or industrial processes, for instance). Thus, this chapter provides a broader analysis, including the employment of general equilibrium analysis. The aim is to assess the technical transformation, economic costs and political instruments needed to achieve Europe's goals and maintain opportunities to reach the overall target of preventing 'dangerous anthropogenic interference of the climate system' (UNFCCC, 1992). Due to the focus on technical feasibility and the policy instruments to exploit this potential, some attention is paid to present policies and proposals (e.g. in the policy review section of Section 7.5), but a major review of proposals and recent policy developments is not proposed (see, for instance, EEA 2008, and associated documents on the CARE package).

The model system is described in Section 7.2 and uses two different scenarios to envisage the transformation and its costs. The adaptation scenario assumes a continuation of current policies (but with little or no mitigation beyond 2012) and maintains present consumption and investment patterns, eventually reaching global average temperatures around 4 °C above pre-industrial levels. The second scenario, or mitigation scenario, requires strong mitigation consistent with stabilising the global climate at temperatures 2 °C above pre-industrial levels. See Chapter 3, van Vuuren *et al.*, for a detailed discussion of the different scenarios used in ADAM research.

Section 7.3 discusses the technical potentials for improved efficiencies and energy substitution to achieve the 2 °C target in Europe, together with the adaptation of the energy system and the investments needed for a 4 °C scenario by 2050. Important technical potentials for reducing emissions exist on both the demand side and the supply side of the energy system. These include more efficient use of materials and energy on the demand side and fast diffusion of renewable energy sources and low emitting technology (including carbon capture and storage) on the supply side. These options are explored using bottom-up models that allow for a high degree of technical detail and include all end-energy-using sectors (the demand side), namely residential, services, agriculture, transport and industry, together with the primary-energy-transforming sectors (the supply side).

In Section 7.4 attention is narrowed, to analyse options for the transition of electricity generation towards a carbon-lean system, building on the development of available technologies. How the necessary technological change is organised will affect the costs of energy.

Section 7.5 appraises a set of different available policies from the perspective of cost optimisation within a global context. The challenge of adaptation and mitigation will influence consumption patterns and production systems, affecting the generation of goods and services in Europe. This analysis represents a shift in analytical perspective towards general equilibrium, examining issues such as trade and intersectoral links. It is shown that a stringent climate change policy in Europe may have the potential to guide and accelerate the world's technological shift towards sustainability. Whether an early emphasis on a least-cost path will be combined with other economic developments and goals such as structural change, energy security, technological change and carbon leakage is questioned.

This chapter ends in Section 7.6 with some general conclusions based on the previous analysis, focusing in particular on research and development in the context of long-term transformation of the energy system.

7.2 The model framework

Simulating the options of adaptation and mitigation policies in Europe faces two challenges:

(i) As fossil fuels make up 80 per cent of the energy demand in Europe, there are thousands of technical and organisational possibilities to reduce energy demand in all sectors of the economy. These have to be considered at an adequate level of detail in order to identify related costs, reductions of emissions and polices needed.

(ii) The large variety of possibilities has to be appropriately aggregated to allow analysis at the economy wide level of energy in order to appraise climate change strategies at the national and EU level.

These two demands for detailed technological and sectoral analysis, on the one hand, and macroeconomic analysis, on the other, led to the development of the hybrid energy model system used to assess the European energy system within the ADAM project (see Figure 7.1). In this system all the bottom-up models and the macroeconomic models are combined for maximum consistency. Several process-orientated

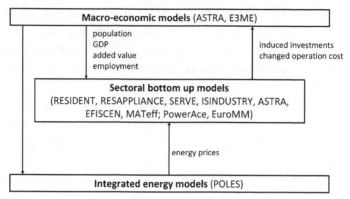

Figure 7.1. Overview of the hybrid energy model system. (Source: Jochem *et al.*, 2009a.)

bottom-up models allow detailed analysis for the residential sector (RESIDENT for buildings and homes and RESAPPLIANCE for electrical appliances), the commercial sector (SERVE), industry (ISINDUSTRY), the transport sector (ASTRA-Transport), the renewable energies (PowerACE ResInvest) as well as for the energy conversion sector, particularly for the thermal power plants and large co-generation (EuroMM). The macro-economic drivers and impacts are simulated by two macro-economic models, E3ME and ASTRA (see Jochem *et al.*, 2009a). An integrated energy model (POLES) was used to maintain consistency at the global level. Finally, a European forestry model (EFISCEN) and a model simulating potentials of efficiency of energy-intensive materials (MATEFF) were used to complete the hybrid energy model system. The results of this model system are presented and discussed in the following Sections 7.3. and 7.4 (see also Jochem *et al.*, 2009b, and Schade *et al.*, 2009). All models used are detailed in the Model Appendix.

Many issues are covered by this set of models, however, at different levels of detail. Energy efficiency improvements, for instance, can be represented as discrete investments in building technology and insulation in one model (bottom-up), but as a simple energy demand curve with substitution between energy and capital in another (generally highly aggregated in macroeconomic models). Emissions reductions stemming from substitution between energy technologies in electricity generation are treated similarly, but the consequences of changing electricity costs through the competitiveness in sectors using electricity need to be traced out in a macroeconomic model. In the case of the hybrid model system (see Figure 7.1) this was achieved through direct model interaction, while with the following two general equilibrium models the links are less explicitly developed.

For the economic analysis of alternative policies and costs of mitigation, two economy-wide general equilibrium models were developed. Firstly, the

GRACE-EL model (Aaheim and Rive, 2005, and Rive and Mideksa, 2009) represents how all sectors interact through trade, and has detailed technology representation in electricity generation. Developments include the role of longlived assets in the electricity sector. Secondly, the EMELIE model uses detail on power generating technology and ownership, allowing analysis of how enterprises are affected, and of market power. The results of these two models are discussed in Section 7.5.

The welfare costs of mitigation are driven by the 'direct' cost of abatement, being the marginal cost of emission incorporating the permit price and resulting economy-wide effects. These include the substitution by consumers away from transport (with limited fuel options and high abatement costs) towards electricity (with many fuel substitutability options) and services (with low exposure to the cost of carbon). Meeting a climate policy target requires numerous adjustments in outputs, inputs, and trade patterns which are all interlinked. This is why it is important to go the extra step from analysing abatement costs by sector to applying measures of welfare costs drawn from a general equilibrium model.

7.3 The future European energy system under adaptation and mitigation

This section describes the main differences in energy demand for two contrasting scenarios. Firstly, the adaptation scenario is based on a business-as-usual pathway and a temperature increase of around 4 °C above pre-industrial levels by the end of the century (and continuing temperature rise thereafter). Secondly, the mitigation scenario is based on strong and early mitigation efforts that lead to a temperature stabilisation at 2 °C above pre-industrial levels (see Chapter 3, van Vuuren *et al.*, and Chapter 11, Knopf *et al.* for a description of the ADAM scenarios). For an overview, we first provide an outline of how the energy system as a whole, i.e. all sectors that use energy (residential, services, transport and industry) and the primary energy converting sector, changes under the two scenarios between now and the middle of the century. Then we give a detailed description of the transformation of the energy system for each sector. For the mitigation scenario, these subsections also detail the possible policies that would lead to the required transformation. For the adaptation scenario, the expected impacts to the sectors are highlighted.

7.3.1 Overview of the whole energy system

The energy system incorporates the four final energy using sectors: (i) residential; (ii) service; (iii) transport; and (iv) industry, with a non-energy sector that uses fossil fuel based products for the production of plastics, fibres, etc. (further disaggregated, inter alia in Section 7.5). It also includes the primary energy conversion sector that converts primary energies (e.g. crude oil, nuclear energy, hydropower, wind,

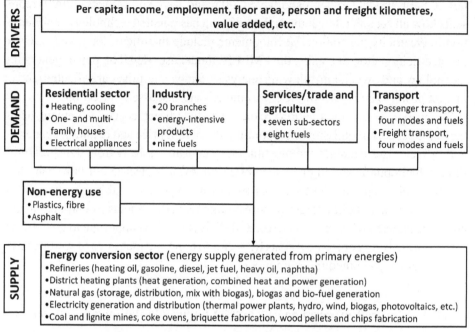

Figure 7.2. Structure of the European energy system in the final energy sectors and the conversion sub-sectors, as represented by the bottom up models and POLES

wood, solar radiation) into final energies such as electricity (*electricity generation*), petroleum products, coke, wood pellets or chips, and district heat (see Figure 7.2).

In each sector there are huge efficiency losses during the conversion from primary energy to end energy. The energy losses in the conversion sector are almost 30 per cent, mainly due to low efficiencies of thermal power plants. Twenty-six per cent of total final energy delivered by the conversion sector in 2005 (almost 50 EJ) was used in the residential sector, 14 per cent in the commercial/agricultural sector, 31 per cent in industry (including construction), and 29 per cent in the transport sector (excluding international air and ship transport). The conversion of final energies into useful energy (e.g. heat, power, light, electronic communication and calculations) incurs losses of about 40 per cent. Well-insulated houses and buildings, highly efficient lighting, and use of waste heat in industry can substantially reduce the demand for energy. Reduced primary and final energy demand and, hence, reduced fossil fuel use would cut related greenhouse gas emissions. These would also be cut through an increasing use of renewables.

Compared with the current situation, energy-related carbon dioxide emissions in the EU27+2 countries (EU27 plus Switzerland and Norway) remain at a high level of approx. 4.7 billion tonnes per annum until 2050 for the adaptation scenario. Emissions rise slightly until 2030 due to more intensive coal use in electricity production and

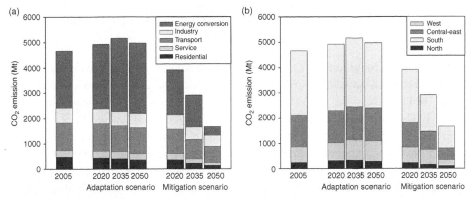

Figure 7.3. Changing energy-related CO_2 emissions (a) in energy using and converting sectors and (b) in different regions of Europe (EU27+2) for the adaptation and the mitigation scenarios until 2050 (Source: EuroMM).

decline slightly thereafter due to an increase in renewable energies (see Figure 7.3, left bars). On the other hand, for the mitigation scenario, energy-related carbon dioxide emissions first decrease slightly until 2020 and then fall rapidly until 2050 to achieve 1.67 billion tonnes per annum, being a reduction of 65 per cent compared to 2005.

The emissions reductions are largest for the energy conversion sectors, which fall from 48 per cent currently to 36 per cent in 2050, due to the increase of renewable energy sources. Contributions from transport and industry, on the other hand, rise from 24 and 12 per cent to 30 and 19 per cent, respectively, because of growing demand and smaller reduction potentials. The contributions of the residential and service sectors remain rather stable at around nine and six per cent, respectively. Interestingly, for the adaptation scenario, the proportion of emissions from the transport and industrial sectors decreases slightly, while the proportion from the energy conversion sector rises because of the increased use of coal.

Figure 7.3 also shows the distribution of energy-related carbon dioxide emissions in four major regions of Europe[1]. The distribution between the regions is quite stable over time but demonstrates the predominance of emissions from the West with about 53 per cent followed by the South with 26 per cent. The Central-East and North contribute approximately 15 per cent and six per cent emissions, respectively. This distribution does not change very much for the mitigation scenario, suggesting that all regions contribute similarly to the emissions reductions. A more detailed analysis of the energy system and its transformation under the adaptation and the mitigation scenario is provided in the following subsections.

---

[1] *North*: Denmark, Finland, Norway, Sweden; *West*: Austria, Benelux, France, Germany, Ireland, Netherlands, Switzerland, United Kingdom; *Central/East*: Baltic States, Czech Republic, Hungary, Poland, Slovakia, Slovenia; *South*: Bulgaria, Greece, Italy, Malta, Cyprus, Portugal, Romania, Spain.

Box 7.1. Boundary conditions of the scenarios, EU27 plus Switzerland and Norway

The boundary conditions of the macroeconomic models, such as the development of the European population or the temperature increase due to climate change, were adopted from the ADAM work on scenarios (see Chapter 3, van Vuuren *et al.*). The European population is stagnating around 2020 and declines by four per cent in 2050 compared to the year 2000. Economic development in Europe based on GDP continues to grow, but at a declining rate from 2.2 per cent annually for the first 15 years to 1.3 per cent per annum for 2030 to 2050. The share of the manufacturing sector in total GDP is almost stable, while that of the service sector continues to increase.

Global oil prices were projected by the POLES model and converted to nominal prices assuming a two per cent annual global inflation rate. The nominal price of world crude oil rises from $70 to $120 per barrel in 2030. In real terms, however, the oil price remains almost constant over the period until 2030, then increases to 100 $_{2005}$ per barrel in 2050. This trajectory assumes that the constraint of the peak oil (see IEA 2009) can be compatible with a relatively smooth pathway for oil production and prices. Details of boundary conditions are found in Table 7.1.

Table 7.1. *Boundary conditions of the energy system and electricity sector studies.*

Sector	2000	2015	2030	2050	2015/2000 % p.a.	2030/2015 % p.a.	2050/2030 % p.a.
Total GDP	9542	13 252	17 277	22 370	2.2	1.8	1.3
-agriculture	297	315	354	469	0.4	0.8	1.4
-manufacturing	1691	2316	3108	4253	2.15	1.9	1.58
-services	6702	9733	12 641	17 649	2.19	1.62	1.68
Population	496	506	502	477	0.13	−0.05	−0.25

Gross Domestic Product (GDP), value-added of European economic sectors in billion €$_{2000}$, and development of population in million inhabitants, 2000 to 2050. (Source: Results from E3ME for 2000 to 2030 and from ASTRA for 2030 to 2050.)

7.3.2 The residential sector

Despite an increase in the number of dwellings by 25 per cent from 2005 to 2050, total final energy demand of the residential sector rises slightly before it stabilises (see Figure 7.4, left) in the adaptation scenario. This decoupling is due to more efficient buildings and the warmer climate, which reduces heating demand by about 15 per cent in 2050 and final energy demand of the residential sector by 10 per cent (relative

Table 7.2. *Shares of efficiency classes of residential buildings, EU27+2*

	Adaptation			Mitigation		
	Standard	Low energy	Very low energy	Standard	Low energy	Very low energy
Year		%			%	
2000	100.0	0.0	0.0	100.0	0.0	0.0
2010	99.7	0.3	0.0	99.6	0.4	0.0
2020	99.3	0.7	0.0	96.8	3.0	0.2
2030	98.8	1.2	0.0	85.6	12.1	2.3
2040	98.1	1.8	0.1	69.2	24.9	5.8
2050	97.2	2.7	0.1	54.3	36.2	9.5

Adaptation and mitigation scenarios, 2000 to 2050.
(Source: POLES Adaptation and Mitigation Scenario.)

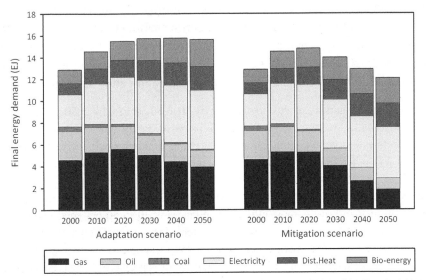

Figure 7.4. Final energy demand (in EJ) of the residential sector of EU27+2, 2005 to 2050, in the adaptation and mitigation scenarios. (Source: POLES Adaptation and Mitigation Scenario.)

to the baseline[2]). The impact is more important for the European countries north of the Alps, which can expect a net decline in final energy demand due to warmer winters and little impact on air conditioning in summers (Table 7.2). Residential electricity demand in Southern Europe increases due to additional air conditioning, while a

[2] The baseline is a fictitious scenario, which assumes no adaptation to changing temperatures. (See Chapter 3, van Vuuren *et al.* for details of the scenarios.)

country like France is fairly balanced by the adaptation scenario's increase in cooling demand and reduction in heating demand (Eskeland and Mideksa, 2008).

While oil and gas for heating are decreasing with time, electricity is expected to rise by about 30 per cent until 2050 in the adaptation scenario. The increase is driven by further purchases of 'white goods', particularly dishwashers, and of new electronic goods by private households. This suggests that the increase is not driven by climate change effects, but rather by general market developments, increased comfort expectations and higher income that also stimulates the usage of cooling appliances. Although air conditioning affects the increase in electricity demand, its contribution within the residential sector remains limited (five per cent of the sectoral demand in 2050). It is only in Southern Europe that there is a significant increase in electricity usage due to additional space cooling.

In the adaptation scenario, energy price increases induce small improvements in the efficiency of buildings. However, the impact is too limited to overcome non-economic obstacles and trigger a significant development of low and very low energy buildings[3]. By 2050, such buildings are projected to represent only a few percent of the building stock. In the mitigation scenario, policies have a significant impact on the market penetration of low energy buildings, representing more than 40 per cent by 2050 (see Table 7.2). As a result, heating and electricity demand of residential buildings is substantially reduced in the mitigation scenario.

- Heating fuel demand is expected to decrease by 50 to 60 per cent by 2050 in the mitigation scenario. This is mainly due to a rapid diffusion of highly efficient new buildings; in addition to extensive retrofitting rates in the existing building stock, increased use of heat pumps and other renewable energies, e.g. modern forms of fuel wood, solar thermal systems. The latter substantially contribute to the carbon dioxide reductions in this sector.
- Substantial efficiency gains are achieved for electric appliances. Even today, specific electricity demand of the most efficient appliances is 30 to 50 per cent lower than the standard versions; most of these efficiency potentials are cost-effective from a life cycle cost point of view. Assuming accelerated market diffusion of the high-efficient appliances, the electricity demand of the residential sector is projected to be reduced by 30 to 40 per cent compared with the adaptation scenario.

In order to achieve the envisaged efficiency gains and reduction of energy demand, early and strict regulation in building codes, appliance standards and mandatory labelling, together with development of policies in professional training and information campaigns will be needed. Building codes and labels will have to be harmonised within Europe to ambitious levels (e.g. passive houses). Minimal

---

[3] The two bottom-up models used here include the diffusion of new low energy or very low energy buildings (LE-VLE or passive buildings), using one-half or one-quarter of the average demand of existing buildings in each region. The VLE building concept reflects current efforts in many countries to develop zero or even negative emission houses, and, when associated with integrated solar PV panels, positive energy buildings.

energy performance standards are an effective policy instrument with respect to heating equipment, electric appliances and stand-by electricity demand of electrical appliances.

A specific bundle of policy measures is also needed to meet the major challenge of retrofitting the existing building stock. Although cost-effective measures are potentially significant, there are various barriers to be overcome. An effective policy bundle may include economic, financial, informational, educational and organisational measures. A carbon tax, the revenues of which could be partly used in promotional programmes, energy-efficiency labels, energy audits and other information programmes will be needed to realise the greenhouse gas reductions required in a 2 °C target policy.

In the mitigation scenario, total annual investments amount to almost €70 billion in order to increase energy-efficiency to the projected level in the residential sector and to substitute the use of fossil fuels by renewables. These investments enable a reduction of consumers' annual energy bill by €15 billion by 2050. By contrast, in the adaptation scenario, additional investments in air conditioning would amount to around €0.5 billion, mostly invested in the Mediterranean countries.

7.3.3 The service sector

Although the floor area of commercial and public buildings is projected to increase by about 50 per cent until 2050, fuel demand of the tertiary sector (mostly for heating and warm water) is expected to increase by only 10 per cent by 2050 in the adaptation scenario. This result incorporates the increase in average temperatures due to climate change. Regarding electricity demand, the service sector, which includes subsectors such as public and private offices, retail, restaurants and hotels, health and other services, is one of the most dynamic. Electricity demand of this sector is projected to double. In Southern Europe, electricity increases are greater due to additional space ventilation, air conditioning and cooling demand in trade and the food supply chain. In Northern Europe the increase in energy demand drops due to warmer winters without increasing the need for air conditioning in summer.

In the mitigation scenario, carbon dioxide reduction in the service sector is equivalent to that of the residential sector due to similar investments in buildings, heat generation and electrical appliances:

(i) Important reductions of specific heating and cooling demand can be achieved by insulation, triple glazing and improved control techniques and building management in low energy or passive buildings. Heat recovery from many processes (e.g. bakeries, laundries, hotels and hospitals) will reduce the demand for process heat.

(ii) Electricity use by appliances in buildings can be reduced by 40 per cent to 80 per cent, through modern systems of ventilation, pumps, elevators or lighting. Electricity efficiency measures include adequate powering, demand-oriented controls and partial use of power electronics. Large electrical efficiencies are also feasible for information technology equipment.

Policies for this sector are similar to those in the residential sector, including extensive retrofit rates in the building stock. Total annual investments in the mitigation scenario in the order of €45 billion are needed to achieve the efficiency gains and to substitute the use of fossil fuels. This leads to a net reduction of energy costs of about €10 billion per annum. In the adaptation scenario, investments in air conditioning reach about €2 billion per annum in 2050.

7.3.4 The transport sector

Energy demand in the transport sector needs to be differentiated by transport category in order to devise appropriate policies and measures. There are also contrasts between the EU15 and the Central European regions.

(i) Passenger surface transport is expected to grow moderately by 25 per cent until 2050. This is due to the stabilisation and subsequent decline of population growth, combined with a tendency for increased urbanisation. On the other hand, in Central Europe, the increase reaches over 50 per cent in 2035 but then stagnates until 2050;

(ii) Growth of intercontinental air transport is expected to continue at a rate between two and four per cent annually;

(iii) Freight transport is growing faster than passenger transport and rises by 150 per cent between 2005 and 2050. The increase for Central Europe is even more pronounced at over 270 per cent until 2050, due to higher GDP growth and increasing international trade. Trucks maintain a high proportion of freight transport.

The next 40 years will be characterised by high prices for motor fuels, particularly after 2030. Alternative propulsion technologies and lighter vehicles will play an important role for improving efficiencies and reducing emissions. In the medium term, these include compressed natural gas and bio-ethanol, hybrid or full battery electric vehicles for use in cities. In the long term, there may also be hydrogen fuel-cell vehicles and an extended use of battery electric vehicles, even for vans (see Figure 7.5, left). Stock effects are less important than in the building sector due to the shorter lifetime of vehicles. Electricity and hydrogen thus penetrate quite rapidly in the mitigation scenario (see Figure 7.5, right).

In the adaptation scenario, the per capita fuel use for transport in EU15 countries stabilises and then falls back to the present levels or below, reflecting the large potential for improved fuel efficiency in vehicles. The existing high level of

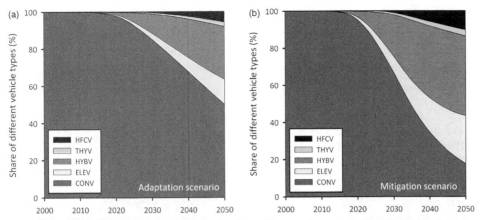

Figure 7.5. Share of different types of light and low emission vehicles, EU27+2, (a) adaptation scenario and (b) mitigation scenario, 2000 to 2050. (Source: POLES Adaptation and Mitigation scenario.)

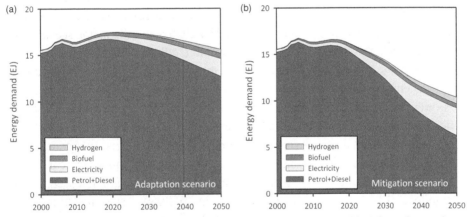

Figure 7.6. Energy demand of the transport sector, EU27+2, (a) Adaptation and (b) mitigation scenarios, 2000 to 2050. (Source: POLES adaptation and mitigation Scenario.)

ownership leaves little room for an increase in the stock or utilisation of vehicles. The situation is different for central Europe, where the vehicle stock and utilisation are still rising.

In the adaptation scenario, total final energy demand of the transport sector follows a plateau between 2015 and 2035 and then drops to current levels with around 90 per cent being provided by fossil fuels. Conversely, in the mitigation scenario the demand decreases considerably to 10 EJ after 2025. Electricity becomes more important over this period, but fossil fuels still dominate (see Figure 7.6).

Table 7.3. *Fossil-based electricity generation, EU-27+2, adaptation and mitigation scenarios, 2000 to 2050*

Year	Adaptation		Mitigation	
	Without CCS	With CCS	Without CCS	With CCS
	(TWh)		(TWh)	
2000	1620	0	1620	0
2010	2150	0	2150	0
2020	2810	0	2140	4210
2030	3310	0	1490	1470
2040	3810	0	1060	2250
2050	4210	0	8120	2630

(Source: POLES.)

7.3.5 The industrial sector

Although industrial production, measured in value added (i.e. as in GDP accounting), more than doubles between 2005 and 2050, industrial energy demand increases only by about 30 to 40 per cent to about 18 to 21 EJ by 2050 in the adaptation scenario (see Table 7.2). This decoupling is possible due to more efficient technologies in electric motors, motor systems (compressed air systems, pumps, etc.), efficient lighting and steam generation. New industrial production processes also contribute to carbon dioxide emission reductions. In the adaptation scenario no carbon capture and storage options are considered for industrial processes. The impact of climate change on energy demand is small, as the share of heating in overall industrial energy demand is below 10 per cent, and for cooling processes the share is only four per cent.

Electricity demand rises by 50 per cent until 2050 in the adaptation scenario, but total industrial fuel demand falls by six per cent by 2050. Associated changes are technology shifts (e.g. electric arc steel instead of oxygen converter steel), further automation and fuel substitution.

Mitigation options are numerous in all industries. In the basic products industries, material efficiency and substitution still offer substantial opportunities for reduction in energy demand and related greenhouse gas emissions. Depending on the model, final energy demand is reduced by 30 to 40 per cent by 2050 compared with the adaptation scenario by realising these potentials (see Table 7.3). The decline is mainly due to a rapid diffusion of highly efficient cross-cutting technologies and new industrial processes that are presently at the market entry or envisaged by intensive research and development (R&D) during the next few decades. Among all final energy forms, electricity continues to maintain its leading role, but its share decreases from 45 per cent in 2050 in the adaptation scenario to 40 per cent in the mitigation scenario. The role of biomass doubles in the mitigation scenario to around 20 per cent of total demand in 2050. The demand for coal, oil and gas decline by 78, 55 and 24 per cent, respectively.

One important policy instrument for the basic products industries is the EU Emission Trading Scheme (EU-ETS). Continuous tightening of the emission cap and the phasing out of free allocation of allowances over time will intensify R&D effort and innovation in the industrial sector (see also the CARE package, EEA, 2008). In addition, specific bundles of policy measures are needed. Although the potential for energy-efficiency improvement is in most cases large and cost-effective from a company perspective, various barriers still impede tapping into this potential. An effective policy may therefore develop along policy portfolios combining mandatory minimum standards (e.g. labels for industrial technologies), energy efficiency funds to promote energy management schemes and investment in energy-efficient technologies. Energy audits and other information programmes would make corporate energy use transparent and raise pressure to invest in energy-saving programmes.

Total annual investments would amount to around €10 billion in the mitigation scenario in order to achieve the projected energy efficiency levels and the substitution of fossil fuels and electricity by renewables. These investments would enable a (net) reduction of the annual energy bill of industrial consumers by about €2.5 billion per annum by 2050.

7.3.6 Primary energy converting sector: refineries, district heat, co-generation

At present, refineries have adapted their output mix of fuels to the current demand levels within each European country. These country-level demands depend much on differential taxation and environmental regulations. In addition, there is some trade in refinery products due to refinery process constraints and different domestic demand for oil products. In the future, under either an adaptation scenario or a mitigation scenario, the mix of oil products on the demand side is expected to change.

In the adaptation scenario, relatively modest changes are expected, leading to a slight increase in demand for transportation fuels and lower demand for oil heating. In contrast, the mitigation scenario will lead to a substantial shift in demand to alternative fuels and severe losses in heating oil demand, i.e. to a reduction in refining capacity (see Figure 7.7).

The demand for district heat will fall in Europe by 2050 (see Figure 7.7), when Nordic countries in particular will benefit from rising air temperatures due to climate change. Therefore less investment in district heating systems will be required and the need for combined heat and power generation plants (cogeneration) will be reduced. Due to efficiency measures, heat-only plants will be replaced by co-generation plants.

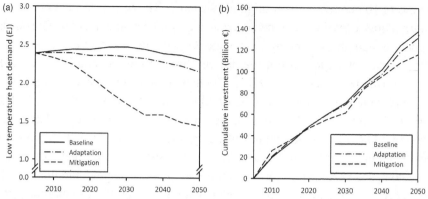

Figure 7.7. (a) Low temperature heat demand in EU27+2, and (b) cumulative investment in co-generation and district heat plants, lower lines: Adaptation scenario; higher lines: Mitigation scenario, 2005 to 2050. (Source: EuroMM.)

7.4 Future European electricity use and conversion in the adaptation and mitigation scenarios

The previous section looked at the energy system as a whole, including energy using and primary energy transforming sectors. This section focuses more narrowly on the important electricity subsystem (see also Reiter and Held, Climate change adaptation and mitigation in the European electricity sector, unpublished). The structure of electricity generation may change substantially during the next few decades, in part through an increasing use of renewables (see Section 7.4.2). The level of electricity demand in the adaptation or the mitigation scenarios also affects the utlisation of thermal power plants, nuclear energy, fossil fuels and carbon capture and storage technologies.

7.4.1 Challenges for the electricity sector in the adaptation scenario

As climate change affects European regions differently, the necessary adaptation policies will vary across Europe (see also Chapter 4; Aaheim *et al.*). In broad terms, Southern European regions will need to undertake more costly adaptation measures in their electricity use and generation, whereas the energy sector in Nordic countries will benefit from climate change. In the latter case, climate change will reduce electricity demand for heating in winter and lower demands for district heat. Additionally, the hydropower potential in Scandinavia may increase due to more abundant precipitation.

In contrast, Southern European regions will face major challenges to adapt to climate change because of two simultaneous and conflicting effects. Firstly, electricity demand will rise during summer periods (see Section 7.2) to fulfil an increasing demand for space cooling. At the same time, the electricity generation from thermal power plants is likely to be limited, since the power output is highly dependent on the availability and temperature of cooling water. With rising air temperatures from climate change, the efficiency of conventional thermal power plants is reduced by approximately 0.1 per cent for each additional degree Celsius of the cooling agent. In addition, water availability for cooling purposes is likely to decrease due to changing runoff patterns, changes in precipitation and evaporation and more frequent heatwaves.

Generally, regulations on the use and release of water into rivers stipulate a threshold discharge temperature of around 25 °C; if this temperature is reached in rivers used for cooling, utility companies will be unable to release cooling water and hence will have to decrease their power output. In Southern Europe and under extreme weather conditions, some power plants may have to shut down completely. Avoiding such events will require additional investment to install advanced cooling systems, such as dry cooling towers. However, these technologies are more expensive and need more electricity than conventional cooling systems.

From 2020, the monthly mean temperatures for rivers in the months of July and August are likely to reach values higher than 25 °C in Southern European regions. To avoid reduced electricity output from thermal power plants and to cover increased electricity demands in summer, increased generation capacities (in aggregate) will be needed together with installation of advanced cooling systems. As a result, a significant share of all thermal electricity generation in 2050 will be from power plants with advanced cooling systems.

More frequent heatwaves in Central European countries will also induce shortages of electricity production in summer, as was observed in August of 2003, when German and French utility companies were forced to reduce power output from thermal power plants due to reduced cooling water availability. Additional investments will also be necessary in these regions to ensure stable electricity grid operation with sufficient reserve capacity. These extreme events, however, are not considered in this analysis, due to a lack of information concerning their frequency and duration.

Two hydrological effects from climate change will also influence the power output of hydropower plants, forcing further adaptations. Firstly, the annual water balance will change depending on the region, with increases in runoff of water up to 25 per cent in Nordic countries and decreases of up to 25 per cent in Southern European regions. Additional investment in new dams and changed reservoir management will offer Nordic countries a chance to increase their output of hydropower by approximately 13 per cent until 2050 in the adaptation scenario. In Southern Europe, however, it may be necessary to increase reservoir volume to overcome the losses from reduced river runoff; however, this may be only partially effective since it leads to larger reservoir surface areas and hence greater evaporation. In the adaptation scenario, Southern Europe may face a reduction of approximately 15 per cent in hydropower generation.

Secondly, the seasonal pattern of precipitation and snow/ice melting will also change. Two main precipitation patterns exist with direct influence on hydropower production: one pattern is dominated by rainfall with a river runoff peaking; the other one is dominated by snowfall and snowmelt, with a spring maximum. The impact of climate change is expected to shift the pattern of snow-dominated regimes towards the rainfall-dominated regime, leading to earlier (winter) peak runoffs.

In addition, increasing ambient air temperatures are expected to lead to higher efficiency losses in electricity transmission and distribution, as higher ambient air temperatures increase the electricial resistance of conductors. These losses may add up to one per cent of total European electricity demand in 2050 in the adaptation scenario, thus requiring additional generation.

In the adaptation scenario, electricity generation increases in part with a higher share of coal-based power plants in the next decades. Low international coal prices, abundant coal reserves in some EU-member countries (e.g. Poland and Germany) and availability of advanced coal-fired power generation technologies will foster the competitiveness of coal-based generation. In comparison, the share of nuclear power in total electricity production remains roughly constant in the adaptation scenario. After a period of decline until 2020, resulting mainly from the impact of nuclear phase-out policies in some EU-member countries, total investment begins to rise again as countries with a relatively high acceptance of nuclear power replace existing capacities and install additional power plants.

Unlike coal and nuclear power, natural gas plays only a minor role for electricity generation in Europe in the adaptation scenario. The depletion of natural gas reserves together with high and volatile gas prices (compared with coal market prices) are unfavourable to new investments in electricity generation in the adapation scenario.

Renewables, on the other hand, make an increasingly important contribution to the European electricity system, even in the absence of additional mitigation policies. They increase their share in electricity generation to 23 per cent in 2050 (see Figure 7.8) as discussed below in Section 7.4.2. The effects of climate change on renewable generation other than hydropower (e.g. wind power), however, are not considered in this study, mainly due to a lack of reliable information on extreme events.

In the adaptation scenario, significant changes in the capacity mix are needed in the European electricity sector in order to adapt to the impacts of climate change. In particular, climate change poses major challenges for adaptation in Southern Europe. Utility companies and regulatory authorities need to plan appropriate investment and policy measures to guarantee secure power generation under the conditions of a changing climate. It has to be stressed that some of the potential impacts could not be considered here, such as the higher incidence of extreme

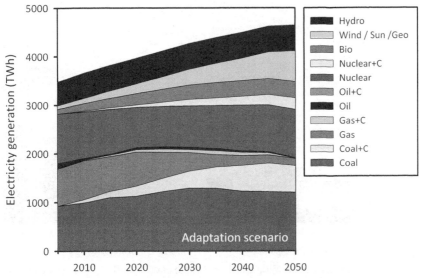

Figure 7.8. Electricity generation and related primary energy shares, EU-27+2, adaptation scenario, 2005 to 2050. (Source: EuroMM.)

heatwaves, or additional electricity needs for irrigation and desalination; they could pose further adaptation needs to electricity generation.

7.4.2 The rise of renewable electricity in Europe

The boost in the use of renewable energy technologies (RET) in the electricity sector plays a decisive role in the adaptation scenario and even more so in the mitigation scenario for achieving ambitious emission reduction targets. At current prices, with limited internalisation of external damages, RET is not in all cases cost-competitive. It is, however, supported by means of market stimulation policies, such as feed-in tariffs and quota obligations, given their low external cost and high cost reduction potentials. The main determinants of future RET development consist of the available technical potential, the economic performance, the expected cost reductions, and the policy framework for RET support.

According to the modelling results, the share of RET in electricity generation is expected to increase from 15 per cent in 2007 to about 23 per cent in 2050, even in the adaptation scenario where electricity demand increases by almost 30 per cent to 4700 TWh until 2050 (see Figure 7.9). A central influencing factor behind this development is the assumption of a continuation of currently implemented RET support policies on a national level, even if it is assumed that government support is reduced progressively.

Next to the positive contribution of RET to mitigation policies, some technologies are also vulnerable to climate changes such as modifications in temperature and

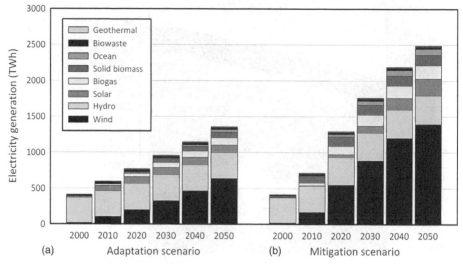

Figure 7.9. Electricity generation from renewable energies, EU-27+2, (a) adaptation and (b) mitigation scenario, 2000 to 2050. (Source: PowerACE ResInvest.)

precipitation patterns, evaporation, wind speeds and cloudiness. However, climate change impacts on the use of RET appear moderate, as far as they can be quantified by means of existing climate models. This analysis takes into account climate change induced impacts on the use of hydropower, the most affected renewable energy source, and on wood fuel availability. Results for a modified hydropower production are described in more detail in Section 7.3.1, whereas the consequences of climate change on the production of electricity from woody biomass remain marginal for Europe.

A substantially greater increase in the use of RET in the electricity sector is essential for the achievement of climate policy targets under the mitigation scenario. The use of RET is projected to grow at 3.8 per cent p.a. on average to 2600 TWh by 2050. This corresponds to more than 50 per cent of total electricity generation, which is itself limited by a lower overall demand. Wind energy is projected to overtake the currently dominating role of hydropower from 2015 onwards. But the full range of RET has to be deployed in the mitigation scenario, including less mature technologies, such as solar PV or ocean energy, which are expected to undergo further cost reductions. In particular, biomass technologies and solar power are expected to grow considerably (see Figure 7.9, right).

A high proportion of fluctuating RET supply in the electricity mix, as experienced by wind power, solar PV, concentrating solar power, ocean energy and hydropower does pose a challenge for grid integration and stability. It will require a new management of the electricity system with significant additional system integration costs. However, these challenges can be managed in the long term by reinforcing the grid infrastructures, including international transmission capacities and by using back-up capacities, smart grids and energy storage technologies.

The competitiveness of RET improves in particular in the mitigation scenario as a result of rising carbon prices, technological learning and economies of scale. Additional investments for the accelerated diffusion of the RET for electricity generation are about €18 billion p.a. on average. Increasing the reserve coefficient in a cost-effective way may require the application of well-designed policy measures adapted to the specific requirements of RET. This might include, for example, feed-in tariffs, and the use of general climate policies, such as cap and trade policies.

7.4.3 The electricity generation in Europe and related policies to 2025 and 2050

All analyses of future energy demand conclude that electricity is very likely to increase; the growth in demand depends on the scenario (see Section 7.3). New generating capacities are required in any scenario in order to restore adequate reserve margins and to replace existing power plants, many of which will reach the end of their lifetime in the next two decades.

In the adaptation scenario, electricity demand in Europe is projected to nearly double (POLES, see Figure 7.12) or to increase by 50 per cent (bottom-up models) from today's level by 2050. This difference in the anticipated electricity demand is due to two effects: (i) a higher growth of industrial production in the POLES model, (ii) a higher level of efficiency improvement and intra-industrial structural change to less energy-intensive products in the bottom-up projections. Fossil fuels provide nearly 55 per cent of total electricity production, increasing significantly by the middle of the period in both modelling systems. Due to this predominance of fossil fuels in the generation mix, there are large negative environmental externalities caused by electricity generation.

Electricity is a major contributor to greenhouse gas emissions currently representing 39 per cent of European carbon emissions. By 2050, the figure rises to about 52 per cent in the adaptation scenario, as energy-related carbon dioxide emissions for electricity generation are projected to increase by two per cent p.a. (see Figure 7.12 left) and by 0.5 per cent p.a. according to the bottom-up models. This is due to the assumption of continued existing policies on renewables between 2005 and 2050 (which is not assumed in the POLES adaptation scenario). In spite of the improvements in the thermal efficiency of fossil fuel-based power stations, the carbon dioxide emissions from power generation grow faster than from any other sector, due to the increasing electricity demand.

In the mitigation scenario, the European electricity generation continues to increase during the whole period, but it does so more slowly than in the adaptation scenario in the POLES model (see Figure 7.10). Electricity generation stagnates in

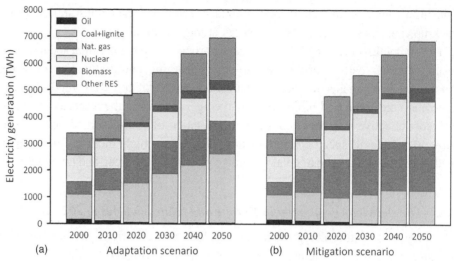

Figure 7.10. Electricity generation, EU27+2, (a) adaptation and (b) mitigation scenario, 2000 to 2050. (Source: POLES.)

the mitigation scenario of the bottom-up models until 2050. Furthermore, as a result of the shift in the mix of fuels used to generate electricity, the fossil fuel share falls to less than 45 per cent by the end of the period.

In the mitigation scenario, Europe is expected to develop a robust portfolio of new electric power technologies for generation, transport and end-use. No major technological option is neglected for reducing total European greenhouse gas emissions to a level compatible with the Europe's climate objective of 2 °C. By 2050, the most strongly impacted technologies in the POLES projections are nuclear and carbon capture and storage (CCS) technologies: in 2050, nuclear power generation maintains its share of total power generation at 26 per cent, while 41 per cent of the European electricity production is provided by technologies with CCS (see Table 7.3).

Major opportunities for reducing GHG emissions in electricity generation

Six major options can be identified to achieve the carbon emissions reduction target in the mitigation scenario: (i) energy-efficiency and very low emission buildings or vehicles, (ii) changes in the energy mix at the level of final energy demand, (iii) changes in the energy mix in the electricity generation towards natural gas, (iv) renewable energies, (v) nuclear energy, and finally (vi) carbon capture and storage. Demand-side options, and in particular energy efficiency, play a major role in the short to medium term, while carbon capture and storage and changes in the energy mix make an increasingly large contribution to the reduction effort in the long term (see Figure 7.11).

To conclude on the major assumptions and results, the differences in the contributions of technologies and sectors according to the modelling framework

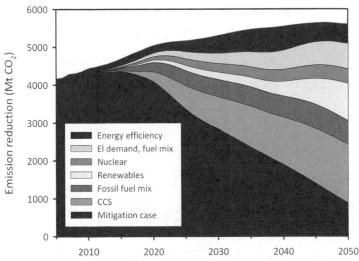

Figure 7.11. Carbon dioxide emission reductions of EU-27+2 from Adaptation to Mitigation scenario by several major technologies, 2000 to 2050. (Source: POLES.)

can be analysed. Obstacles to efficiency in the final energy sectors are more pronounced in the analysis by POLES which leads to higher electricity and fuel demand than in the bottom-up models. Carbon dioxide emissions, however, are reduced by higher shares of nuclear energy and carbon capture and storage technologies. Further, material efficiency improvements in the bottom-up models lead to reduced demand of energy-intensive basic products and a 10 per cent reduction in industrial energy demand.

7.5 Economic analysis of the near term European mitigation policy

In this section, the economic dimension of the mitigation scenario is addressed with the emphasis on policy instruments and consequences. The focus is on the economic implications of the mitigation scenario where Europe meets its goal of reducing its emissions by 20 per cent in a cost-effective way (European Commission, 2008). To obtain estimates of overall costs of this mitigation strategy, general equilibrium models have been developed and used[4]. This allows, *inter alia*, the examination of impacts in sectors using inputs from emission intensive sectors, and to analyse whether trade in energy-intensive goods, such as metals and other basic materials, is an important issue.

[4] Most results reported here are from GRACE-EL (Rive and Mideksa, 2009), while those emphasising the longer run, and market power, are from EMELIE (Traber, 2008a). See annexes.

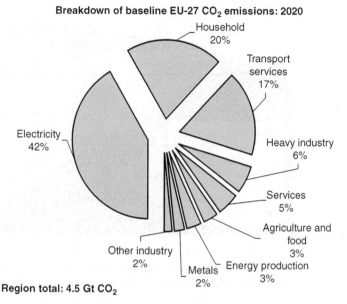

Breakdown of baseline EU-27 CO$_2$ emissions: 2020

Household 20%

Transport services 17%

Electricity 42%

Heavy industry 6%

Services 5%

Agriculture and food 3%

Other industry 2%

Metals 2%

Energy production 3%

Region total: 4.5 Gt CO$_2$

Figure 7.12. Breakdown of carbon dioxide emissions by sector in a 2020 adaptation scenario. (Source: Eskeland *et al.*, 2009a.)

7.5.1 The economic impacts of 20 per cent emission reductions by 2020

For the case of 20 per cent abatement by 2020 – a variant of the mitigation scenario – we first establish an adaptation scenario, assuming that neither Europe's present Emission Trading Scheme (ETS) nor any other mitigation policy is implemented from 2012 onwards. This would lead to sectoral breakdown of carbon dioxide emissions by 2020 as depicted in Figure 7.12.

Electricity generation represents 42 per cent of European carbon dioxide emissions in the adaptation scenario, followed by transportation, services and households (including their cars), with about 20 per cent each. These three major emitting sectors alone account for 80 per cent of the total emissions until 2020. The picture indicates that emission reductions in Europe essentially require transformation of these three largest emitting sectors.

At the time of writing, the implementation details of European climate policy to 2020 were not clear. As such, our scenarios were designed as simply and cost effectively as possible: climate policy is modelled as a uniform cap on carbon dioxide emissions across all sectors, with no imports of permits (e.g. from CDM projects, nor 'hot air' from Eastern Europe)[5].

---

[5] Other variants of the mitigation scenario have been run, though not reported on here (adjustment costs, renewables' support, border tax adjustments, fragmented quota markets). We believe the most important observations are conveyed in the present discussion. CDM (clean development mechanism) denotes emission reduction credits from developing countries, and the term 'hot air' is used for quotas from countries that might not be effectively bound by their quotas, because of restructuring for other reasons.

The medium-term goal of Europe is modelled here as reducing carbon dioxide emissions by 20 per cent by 2020 (or 30 per cent, if there was international co-operation). The costs of achieving this goal can be expressed in a number of ways. Three measures are as follows: the price of emission quotas (or permit price), equivalent variation as a measure of welfare costs and changes in the electricity price due to the policy.

Firstly, the permit price under carbon trading captures the marginal abatement cost of carbon. This is the price each agent would prefer to pay rather than undertake further abatement. These prices are €64 and €108 per tonne of carbon dioxide in the 20 and 30 per cent mitigation scenarios, respectively. This is about three to five times higher than that seen in the first ETS period (2005–2007), or 2008. The import of CDM permits could reduce the emission prices, as would support for renewables and for energy efficiency (analysis not shown here). Conversely, if a sectorally segregated ETS system had been assumed, prices in some sectors would be higher, in others lower, with higher aggregate costs.

Secondly, estimates of the economic impact of climate policy in terms of welfare loss compared to the adaptation scenario can be presented[6]. In the model, economic welfare is measured in terms of aggregate consumption and reported relative to the adaptation scenario (with no mitigation policies). The EU-wide welfare cost of mitigation scenario variants with 20 and 30 per cent emission reduction by 2020 are estimated to be 0.7 and 1.1 per cent of consumption, respectively[7]. It is important to note that this notion of welfare cost excludes any benefits from avoided climate damage and avoided adaptation cost (see Chapter 4, Aaheim *et al.* for details on residual damages, the benefits of avoided damages and adaptation costs for different mitigation targets). These avoided costs must hence be taken into account in a full benefit–cost assessment (not considered feasible here).

Thirdly, the cost of such a policy is reflected in the final price of electricity. Table 7.4 shows percentage changes in electricity prices by 2020 under two mitigation scenario variants. Electricity prices (or tariffs) rise in the mitigation scenarios by a third to two-thirds relative to the adaptation scenario.

In the mitigation scenarios, the contribution to the total electricity price from the cost of carbon permits is also of interest. In countries with relatively carbon-lean generation portfolios (Norway, France), carbon permits directly contribute relatively little to the cost of electricity, but nevertheless the increases in electricity prices are similar to those in their neighbouring regions (Germany, UK). This is the result of

---

[6] The analysis shares with many that the costs of mitigation are uncertain but seem to be moderate. Two important major references in terms of reviews are IPCC (2007) and Stern (2006). Recently, model results reflecting lower stabilisation goals of 400 ppm (see Chapter 11, Knopf *et al.*) conclude that these can also be achieved at moderate costs.

[7] The measure of welfare cost is equivalent variation as percent of consumption, allowing consumers income for consumption that compensates for mitigation so as to keep utility constant. The measure will be approximately comparable to percentage of income, or GDP, used by others (such as in Stern, 2006).

Table 7.4. *Percentage changes in electricity price by 2020 relative to 2000 under the 20% and 30% variants of the mitigation scenarios*

EU-Regions	20% reduction		30% reduction	
	Rise in electricity price due to mitigation	Contribution of permit price to the increase price	Rise in electricity price due to mitigation	Contribution of permit price to the increased price
Nordics	32%	14%	41%	19%
UK and Ireland	41%	50%	64%	42%
France and Switzerland	29%	9%	44%	8%
Germany and Austria	41%	37%	61%	31%
BeNeLux	31%	32%	49%	28%
Baltics and Poland	70%	45%	111%	25%
Iberian Peninsula	32%	35%	43%	36%
Italy	34%	8%	43%	7%
Greece	71%	49%	107%	37%
Rest of Eastern Europe	59%	25%	85%	18%

(Source: Eskeland *et al.*, 2009a.)

cost-effective implementation in a supposedly liberalised Europe: electricity prices will be the same (or similar) across all countries. Electricity markets are not fully integrated due to transmission constraints, but there is only one permit market. In countries with predominantly fossil-based electricity production capacity, the price increase will in part be used for the necessary transition towards low-carbon technologies, while in countries with larger non-fossil assets, more will result in a windfall to owners of existing plants.

The permit cost contribution to this tariff increase, however, is on average about 20 to 30 per cent, assuming permits are fully auctioned. The difference between the increase due to permit price and the full price increase reflects the higher costs of expansion for low-carbon technologies, and thus higher returns to all established capacity (average costs increase much less than marginal costs). In France, for instance, not much transformation of the industry is expected, but electricity prices still increase by about 25 per cent. The carbon price contribution is limited to three per cent, even in the 30 per cent mitigation scenario. It has been observed also in other settings that climate and environmental policies raise average costs less than marginal costs, so that sectoral profits rise or are unchanged, an observation that served to limit political resistance from the industry (see Buchanan and Tullock, 1975, and Burtraw *et al.*, 2005).

Table 7.5. *Sectoral distribution of abatement in 2020 under 20% and 30% mitigation scenario variants*

Sector	Share of adaptation secenario emissions (2020)	Share of EU-wide mitigation contributed by sector (2020)	
		2020–20%	2020–30%
Transport Services	17%	5%	6%
Heavy industry	6%	7%	6%
Agriculture and food	3%	3%	3%
Metals	2%	3%	3%
Services	5%	4%	4%
Energy prod	3%	2%	2%
Electricity	42%	70%	68%
Household	20%	6%	7%
Other industry	2%	1%	1%
Total	100%	100%	100%

(Source: Eskeland *et al.*, 2009a.)

In summary, evaluated from the perspective of the welfare cost and changes in electricity price due to the climate policy, the cost of achieving the 20 per cent goal by 2020 appears to be moderate. Despite the carbon price increasing three to five times compared to 2010, the welfare cost is less than 1 per cent of GDP. Furthermore, abatement costs rise only slightly, reflecting the proportional cost rises when the emission reduction target is raised from 20 to 30 per cent.

When comparing the goals of 20 and 30 per cent emission reductions by 2020, Table 7.5 suggests that in both cases the largest proportion of abatement will come from the electricity sector. This reduction is proportionally higher than the original share of emissions for both scenarios. Since the emission reductions are achieved through a simulated cap and trade system covering all sectors, the role of electricty generation reflects comparatively low-cost opportunities available in this sector. Alternative policy simulations with higher adjustment costs retain the important role of the electricity sector as providing more than its proportional share of emission reductions (not shown).

The sectoral distribution of abatement in both mitigation scenario variants indicates that the bulk of the abatement stems from the electricity sector. This is driven by the homogeneity of the electricity output, which results in a high degree of substitutability between generation technologies. As we find below, the broad pattern is a shift away from coal towards low-carbon technologies. Emissions from transport and house-holds (much of the latter is from household-operated vehicles) fall proportionally less. This lower contribution to emission reductions in part reflects the fewer fuel substitu-tion options available in transport. A long history of air quality standards and fuel

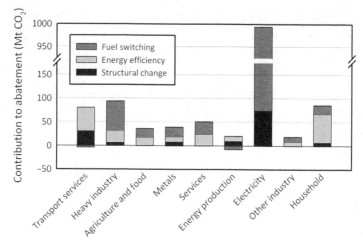

Figure 7.13. Decomposition of EU abatement by sector in 2020; fuel switching (top), structural change (bottom) under 20% mitigation. (Source : Eskeland *et al.* 2009a.)

taxes for European vehicles means that a significant part of the potential for leaner and more efficient vehicles and fuels is already exploited.

The contributions to abatement by each sector can be analysed further in terms of fuel switching, energy efficiency improvements, and structural change (mainly output reduction in emission intensive industries). This is seen in Figure 7.13 for the 20 per cent target. The electricity generation technologies are described with fixed inputs for each primary energy source, a simplification that rules out efficiency improvements within electricity generation compared to the adaptation scenario. Figure 7.13 shows that, outside of the electricity sector, energy efficiency improvements are the largest contributor to abatement, followed by fuel switching and reduced output. Fuel switching plays an important role towards the goals for 2020 in the three sectors that contribute higher emission reductions than their share in emissions: electricity generation, heavy industries and metals.

In the electricity generating sector substantial issues are the availability and cost of natural gas, and the role Europe allows gas to play in its energy balance. For electricity generation, natural gas is attractive in the mitigation scenarios, but expansion is constrained by assumptions regarding resource availability, European energy security concerns, and adjustment costs (see Eskeland *et al.*, 2009a). The impact on the electricity sector is reflected by substitution in the generation portfolio (mix of fuels, wind, nuclear, etc.), illustrated in Figure 7.14.

In electricity generation, carbon reductions are primarily a result of a shift towards gas-based generation, a reduction of coal-based electricity, and a reduction in generation that corresponds to lower demand. It is worth noting that, in general equilibrium modelling, changes in sectoral composition of national output combine with pure within-sector abilities to substitute away from electricity. Thus, part of the

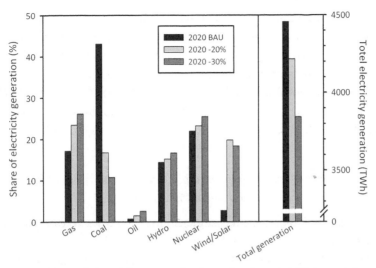

Figure 7.14. Share of electricity generation by technology in EU, in an Adaptation scenario and (left) and variants of mitigation scenarios: 20% (middle, black) and 30% (right, white). (Source : Eskeland, *et al.* 2009a)

electricity output reduction seen here is due to output reductions in electricity intensive manufacturing, part of which is compensated for by increased imports from less regulated industries in other regions (thus leading to 'carbon leakage'[8]). Renewable generation (wind/solar/biomass) also increases substantially.

The results are reported in Figure 7.14. The simplistic assumption of cost-effective implementation here (as distinct from more realistic formulations in the longer term simulations in the bottom-up models, reflected, for instance, in Figure 7.12) is important, as are exogenously imposed constraints. Expansions of natural gas and nuclear power, for instance, are limited by assumptions about resource availability, adjustment costs, and political feasibility. Generation technologies are still dispatched based on least cost, but only to the extent that short term capital constraints and these exogenous limits allow. The long-term nature of capital works in a similar way to limit the responsiveness in the real economy to policy and price changes. This treatment of present assets is central to the stakeholder focus in the electricity sector case study[9].

---

[8] Carbon leakage refers to the phenomenon that emission reductions in one part of the world can lead to increased production of emission intensive products (metals, cement), so that part of the emission reductions are canceled by increases elsewhere. See, for instance, Eskeland and Harisson (2003), Eskeland *et al.* (2009b).

[9] In equilibrium models, the assumption of constant returns to scale – sensible *ex ante* – risks giving too high flexibility downwards when serious emission reduction policies are introduced. GRACE-EL in part follows Sue Wing (2006) in using vintaging of capital to represent the sluggishness with which for instance coal-fired power plants will be phased out. This important role of sluggish present assets motivates the stakeholder focus in Skodvin *et al.*, in press, and in Eskeland, 2009. Also, adjustment costs limit expansion in nuclear- and gas-based generation.

A preliminary conclusion may be that the EU can meet its 2020 goals at moderate costs, mainly through (i) substitution within fossil-based generating technologies towards natural gas; (ii) substitution towards renewable energy; (iii) reduced electricity consumption due to energy efficiency improvements; and (iv) reduced output in energy intensive manufacturing (Eskeland *et al.*, 2009a). A concern regarding this cost-effective implementation is that it entails little or no far-reaching technological change, a concern strengthened by the review, below, of present policy mixes[10].

7.5.2 *Policy instruments and political feasibility: descriptive and normative issues.*

Although the cap and trade system is sufficient to provide the necessary incentives for emission reduction and shift to cleaner energy sources, in almost all European countries it is supplemented by a number of different policies. Eskeland and Linnerud (2009), draw on a broader policy assessment, which evaluates four types of instruments (rows, in Table 7.6, below). The policies – all employed by all European countries examined – are evaluated according to the following three dimensions:

(i) the full polluter pays principle (FPPP), i.e. whether or not the instrument confronts the polluter with the full marginal cost of external damages on the full quantity of emissions;
(ii) the neutrality principle, or whether the instrument conveys the same stimulus to emission reductions irrespective of sector, technology, etc;
(iii) power of incentives to far-reaching innovation, or whether the instrument gives sufficient incentives to research and development (R&D) that render emission reductions in the future, relative to those for present emission reductions.

The reasons for highlighting these dimensions are rather obvious when referring to the textbook recommendations of a pollution tax or tradable quotas. The polluter pays principle, adopted by EU and OECD in principle, is important on justice as well as on efficiency grounds, and absolutely compelling on efficiency grounds in the long run (Harstad and Eskeland, 2008). The textbook recommends a quota market (or emission tax) because it conveys the same incentive for emission reductions to all abatement options. In practice, neutrality is violated, both in quota systems and in price-based systems. ETS applies to some sectors, not others, and support for solar photovoltaics is higher than for hydropower[11]. Table 7.6, below, evaluates the instruments according to the three above-mentioned dimensions. The R&D dimension is important for obvious reasons.

---

[10] The question of technological change has been subjected to additional analysis, for instance by examining whether less-than-perfect competition in the energy sector would deter or accelerate technological change. Market power would raise the electricity price but reduce the emissions price, thus slowing technological change (Traber, 2008a).
[11] One can in principle find normative support for such a differentiation in the literature by assuming higher potential for 'learning by doing' in one area than another (e.g. Kverndokk and Rosendahl, 2007).

Table 7.6. *Policy appraisal of the policy mix*

	Full polluter pays' principle (FPPP)	Technology neutral-ity dimension	Innovation incentive dimension
EU ETS	Future ETS will better meet FPPP by tightening the cap and fewer gratis quotas	ETS is quite technology neutral for the sectors included	By failing to set (tight) future quotas, insufficient innovation incentives likely to result
Cross-subsidies of RES-E	Both the FIT and the TGC represent subsidies relative to FPPP	TGC typically is quite neutral, FIT less so	Assist deployment, but not far-reaching R&D
Energy efficiency support	Standards and labels make polluters pay for mitigation when investing, but are no substitute for full pricing of emissions or energy	Standards and labels are often not consistent with neutrality	Standards may assist in technological change, or prevent it
R&D support	Public expenditure on R&D is often not consistent with full polluter pays principle	R&D support is rarely technology neutral	R&D support may address insufficient incentives in expected future emission costs. Presently likely too weak

(Source: adapted from Eskeland and Linnerud, 2009.)

An interpretation of why, in almost all European countries, the cap and trade policy is augmented by a number of different policies has to do with both history and political feasibility. The main insights from the assessment of the existing policies are firstly that when a neutral cap-and-trade system (ETS) is supplemented with support for renewables and energy efficiency, it reduces prices for emissions and electricity. These price moderations have desirable but also undesirable effects. Among the undesirable ones are that low current prices for electricity and emissions may result in low expected prices for electricity and emissions. If so, investments in R&D and investments in energy efficiency will be too low.

Secondly, support for renewables and energy efficiency assists the transition, but rarely in a neutral fashion. A typical bias is towards options that are immediately market-ready and owned by named stakeholders, so that potential solutions, relying on knowledge and innovation that is not close to market-ready, will be given too little stimulus. Seen jointly, the bias towards immediately available solutions and the likely low prices for emissions and electricity mean that far-reaching technological change has unduly low incentives and support.

7.6 Conclusions: the challenges for EU climate and energy policy in a broader context

The analysis of two contrasted scenarios, adaptation and mitigation, allows identification of key challenges for European policy makers in the next decades. The adaptation scenario results in high-energy demand and slightly increasing carbon dioxide emissions, because of the low priority attached to energy and material efficiency as well as renewable energies. In the building sector, due to rising temperatures, heating demand may be significantly reduced in Northern Europe. But warmer summers, particularly in southern Europe, will raise electricity demand for cooling purposes. Also, thermal power generation would be considerably hindered by cooling difficulties. This could be solved by investments in new cooling systems or by raising temperature limits of rivers and lakes. The limits are, however, guided by ecological concerns, so this would require selective and careful assessment and implementation. In southern Europe, changes in precipitation and hydrology would also significantly limit hydro-power production and necessitate increased generation capacity. This scenario will thus result in additional investments in the demand and supply sides of the energy sector at around €10 billion in Europe in 2050.

Conversely, the mitigation scenario results in a lower energy demand in the building sector, in transport and in industry due to additional investments in energy and material efficiency. The result is 30 per cent below present energy demand levels and 65 per cent below present carbon dioxide emissions. Less investment will have to be dedicated to energy supply, and more attention and resources are needed for more efficient energy uses in the various sectors, along with changes in behaviours and consuming habits. The scenario thus requires deployment of several policy instruments, appropriate for sectors/ technologies addressed. Economic instruments will play a great role reflecting the external cost of fossil fuel use, for example the European Emission Trading Scheme, already partially implemented. Carbon taxes or climate and energy taxes will probably have to be introduced, particularly for transport and building sectors. Feed-in tariffs and other incentives for renewable energy may also be needed to generate the strong increase in renewable energy production required by the mitigation scenario. Economic instruments are essential components of comprehensive climate and energy strategies at European level. They will not suffice alone however, because high transaction costs, a lack of information, inadequate decision routines or preferences and social prestige are also important factors. These obstacles suggest a mix of policy instruments is required.

Early and strict regulation will be needed in case of mass-produced products to mobilise the cost-effective options of energy-efficient solutions and decentralised use of renewables. Ambitious building codes and labeling should be harmonised, though not made uniform, within Europe. Energy performance standards will be applied in the case of heating equipment, electric drives, including pumps, ventilators, and compressors, electric appliances and stand-by electricity use of electronic appliances. Policy measures are also needed to meet the major challenge of retrofitting the existing building stock. The use of efficiency standards with a planned tightening will be an essential complement for accelerating the adaptation of the vehicle stock, even in the presence of carbon taxes or similar instruments.

Putting the climate and energy policy in a broader international context also obliges us to address the issue of the impact of the ETS on energy-intensive industries in Europe. Reducing output in such industries may be considered a low-cost mitigation option, but this might not prove true if it contributes to keeping down both emission prices and electricity prices, thus limiting technological change (Alfsen *et al.*, 2010). It may even be particularly worrisome if energy-intensive manufacturing migrates to non-abating countries. In such a case of carbon leakage, European emission reductions would not contribute to global emission reductions.

A concern with the least-cost way of achieving the 2020 targets is its reliance on natural gas. In the analysis, natural gas use is limited by assumptions that can be regarded as supply constraints or as politically invoked constraints on Europe's dependence on imported natural gas.

In summary, there are two important reasons to question what results as a least-cost solution for meeting the 2020 goals. Firstly is the issue of the disincentives to technological change caused by low prices for electricity and emissions. These low prices, in turn, result in part from low-cost abatement options such as natural gas, in part from carbon leakage, but also from subsidies through support for renewables and freely distributed quotas. Minimal immediate emphasis on far-reaching technological change would make the longer term 2 °C target of the EU extremely difficult to achieve, since little has been prepared both in terms of further European reductions and in terms of broader global participation. Secondly are the concerns about supply security for Europe through the expanded use of natural gas.

Both of these concerns are strengthened if strategies are also considered with an eye to the longer-term global emission reductions. Longer-term mitigation goals for Europe may necessitate lower natural gas use and higher user costs of energy in the intermediate term. Longer-term mitigation goals demand far-reaching technological change and lower energy demand in general. An immediate focus on the longer term will thus reduce Europe's dependence not only on gas but also on oil and probably coal.

Limited attention to far-reaching technological change is a concern as regards European climate policy, but also as regards the negotiation of an international climate regime. A treaty to boost R&D on climate friendly and low-cost technologies would be supportive of a cap-and-trade system such as the Kyoto protocol (and ETS). It would fix certain political weaknesses that otherwise result in mitigation efforts that are inadequate in the long term. Successful R&D results in low and zero emission technologies, may shift politics in the future towards tighter goals. R&D support, for instance in an international treaty, would significantly strengthen the results of the collective effort to combat climate change.

Acknowledgement

The researchers gratefully acknowledge research funding from the CELECT and ICEPS projects of the NORKLIMA and RENERGI programs of the Norwegian Research Council, in addition to that from the ADAM project.

References

Aaheim, H. A. and Rive, N. (2005). *A model for Global Responses to Anthropogenic Changes in the Environment (GRACE)*. Report 2005:05. CICERO, Oslo, Norway.

Alfsen, K., Eskeland, G. and Linnerud, K. (2010). Technological change and the role of nonstate actors. In *Global Climate Governance Beyond 2012: Architecture, Agency and Adaptation*, ed. F. Biermann, P. Pattberg and F. Zelli. Cambridge, UK: Cambridge University Press.

Buchanan, J. and Tullock, G. (1975). Polluters' profit and political response: direct control versus taxes. *American Economic Review*, **65**(1), 139–147.

Burtraw, D., Evans, D., Krupnick, A., Palmer, K. and Toth, R. (2005). Economics of pollution trading for SO_2 and NO_x. *Annual Review of Environment and Resources*, **30** (July), 253–289.

EEA (2008). Report No 6: *Energy and Environment Report 2008*. European Environment Agency.

Eskeland, G. and Harrisson, A. E. (2003). Moving to greener pastures: multinationals and the pollution haven hypothesis. *Journal of Development Economics*, **70**, 1–23.

Eskeland, G. and Mideksa, T. (2008). *Climate change and adaptation through electricity demand*. In ADAM deliverable D-P3c.3b+4b "Policy appraisal for the Electricity sector: Impacts, mitigation, adaptation, and long term investments for technological change".

Eskeland, G., Rive, N. and Mideksa, T. (2009a). *European Climate Goals for 2020 and the role of the Electricity Sector*. In ADAM deliverable D-P3c.3b+4b "Policy appraisal for the Electricity sector: Impacts, mitigation, adaptation, and long term investments for technological change".

Eskeland, G., Rive, N. and Mideksa, T. (2009b). *Long term investments and technological change*. In ADAM deliverable D-P3c.3b+4b "Policy appraisal for the Electricity sector: Impacts, mitigation, adaptation, and long term investments for technological change".

Eskeland, G. and Linnerud, K. (2009). *A review of European climate policy instruments in the electricity sector.* In ADAM deliverable D-P3c.3b+4b "Policy appraisal for the Electricity sector: Impacts, mitigation, adaptation, and long term investments for technological change".

Eskeland, G. (2009). *The consumer pays principle: towards a positive environmental economics.* In ADAM deliverable D-P3c.3b+4b "Policy appraisal for the Electricity sector: Impacts, mitigation, adaptation, and long term investments for technological change".

European Commission, (2008) *Eurostat Energy Statistics.* Available online: www.epp. eurostat.ec.europa.eu/pls/portal/url/page/SHARED/PER_ENVENE. 2008.

Harstad, B. and Eskeland, G. (2008). *Trading for the future: signaling in permit markets.* CMS-EMS Discussion Paper 1429.

IEA (International Energy Agency) (2009) *Oil Market Report.* Paris: OECD.

IPCC (2007) *Climate Change 2007: Synthesis Report, Contribution of Working Groups I, II and III to the Fourth Assessment Report of the Intergovernmental Panel on Climate Change*, Core Writing Team, ed. Pachauri, R. K. and Reisinger, A., Geneva, Switzerland.

Jochem, E., Barker, T., Scrieciu, S. *et al.* (2009a). *EU-Project ADAM: Adaptation and Mitigation Strategies: description of models used in Work Package M1.* Annex to Deliverable M.1.1. Karlsruhe: Fraunhofer ISI.

Jochem, E., Barker, T., Scrieciu, S. *et al.* (2009b). *EU-Project ADAM: Adaptation and Mitigation Strategies: supporting European climate policy.* Deliverable M1.2: Report of the Reference (4°C) Scenario for Europe. Karlsruhe: Fraunhofer ISI.

Kverndokk, S. and Rosendahl, K. E. (2007): Climate policies and learning by doing: Impacts and timing of technology subsidies. *Resource and Energy Economics*, **29**, 58–82.

Rive, N. and Mideksa, T. K. (2009) *Disaggregating the Electricity Sector in the GRACE Model.* Report 2009:02. CICERO, Oslo, Norway.

Schade, W., Barker, T., Criqui, P. *et al.* (2009). *EU-Project ADAM: Adaptation and Mitigation Strategies: Supporting European Climate Policy.* Deliverable M1.3: Final Report Europe. Karlsruhe: Fraunhofer ISI, in press.

Skodvin, T., Aakre, S. and Guldberg, A. T. *et al.* (in press) Target-group influence as a determinant of political feasibility: The case of climate policy design in Europe. *Journal of European Public Policy.*

Stern, N. (2006). *The Economics of Climate Change.* Cambridge, UK: Cambridge University Press.

Traber, T. (2008a), Impact of market power on price effects of the German feed-in tariff under emission trading, *5th International Conference on the European Electricity Market EEM 2008*, IEEE Explore, ISBN: 978–1–4244–1743–8, Lisbon, 4.

Traber, T. (2008b). Is oligopoly a good friend of the environment under emission trading? *The impact of market power on the European electricity market on the diffusion of low carbon technologies*, Conference paper presented at the IAEE international conference, Istanbul.

Traber, T. and Kemfert, C. (2009), Impacts of the German support for renewable energy on electricity prices, emissions, and firms. *The Energy Journal* **30**(3).

UNFCCC (1992). United Nations Framework Convention on Climate Change, http://www. unfccc.int/resources.

8

A risk management approach for assessing adaptation to changing flood and drought risks in Europe

Lead authors:
REINHARD MECHLER[1], STEFAN HOCHRAINER,
ASBJØRN AAHEIM, ZBIGNIEW K. KUNDZEWICZ,
NICOLA LUGERI, MARCO MORIONDO
[1]Co-ordinating lead author

Contributing authors:
ILONA BANASZAK, MARCO BINDI, ADAM CHORYŃSKI,
ELISABETTA GENOVESE, HARVIR KALIRAI, CARLO LAVALLE,
JOANNE LINNEROOTH-BAYER, PIOTR MATCZAK,
DARRYN MCEVOY, MACIEJ RADZIEJEWSKI, DIRK RÜBBELKE,
HÅKON SÆLEN, MART-JAN SCHELHAAS, MALGORZATA SZWED,
ANITA WREFORD

Summary

Adaptation to climate change is an emerging theme of the climate policy agenda in Europe. Changing weather risks have become a major political concern, and climate change research and policy is increasingly seeking to join forces with the field of disaster risk management. Building on estimates of current vulnerabilities and risks, this chapter introduces a risk management framework for assessing adaptation to major weather hazards in Europe, drought and heatwave stress to agriculture, and flood risk.

Our study confirms the view that regions in Eastern Europe represent disaster hotspots for flood risk, and areas in Southern Europe for drought and heat stress. Flood hazards are likely to worsen over much of Europe, yet due to a lack of localised projections from climate models, we consider robust risk projections into the future not possible. In contrast, we feel more confident in projecting drought and heatwave risk as a function of changes in broader-scale average climates. Although these risks are likely to worsen across Southern Europe, effective adaptation interventions seem at least partially possible. Regional heterogeneity in risk and response will continue, leading to climate change 'winners' and 'losers'.

Irrespective of future changes, weather-related disasters already today pose substantial burdens for households, businesses and governments. Risk-based adaptation

Making Climate Change Work for Us: European Perspectives on Adaptation and Mitigation Strategies,
ed. Mike Hulme and Henry Neufeldt. Published by Cambridge University Press © Cambridge University Press 2010.

planning seems important. Focusing on governments, we identify large weather-related contingent disaster liabilities. Faced with budgetary pressures, many European governments have been turning to the EU's Solidarity Fund for assistance, which provides financial aid to Member States for post-disaster emergency measures. Given the increasing demand for support from this Fund for uninsurable risks, it may be necessary to increase or better leverage future funding. Our assessment reveals high levels of uncertainty remain in risk-related adaptation planning, not least due to uncertain changes in future climate risks. More research is required to quantify and manage these uncertainties.

8.1 The adaptation context: increasing disaster losses and a need for consistent and robust information

Losses from weather extremes such as floods, droughts and other climate-related events in Europe (and elsewhere) have escalated in recent decades. Annual monetary losses from large-scale events have increased globally within four decades by an order of magnitude in inflation-adjusted monetary units. The increase has been more rapid than population or economic growth can account for fully (Mills, 2005). According to the Fourth Assessment Report of the Intergovernmental Panel on Climate Change (IPCC) anthropogenic climate change is *likely* to *very likely* to lead to increases in intensity and frequency of weather extremes (Parry *et al.*, 2007).

As a response, attention has recently turned towards climate change adaptation in Europe, with the EU Adaptation Green Paper (GP) initiating the onset of a consolidated EU climate adaptation strategy further elaborated in a White Paper (EC, 2007a; EC, 2009). Managing disaster risk is now considered a priority area for action on adaptation and features prominently in the report. The GP asserts that early policy response contributes to the prevention of significant future costs associated with reactive, unplanned adaptation. Where current knowledge is sufficient, adaptation strategies should be developed in order to identify optimal resource allocation and efficient resource use for guiding EU action through sectoral and other policies. Where significant knowledge gaps exist, however, community research, exchange of information, and preparatory action may further contribute to expanding the knowledge base and managing the uncertainty. Our analysis can be understood as a contribution to the emerging adaptation discussion and an exploration of the scale and scope of these knowledge gaps.

We identify information gaps and examine options for improving the methodologies used to assess extreme weather events, as potentially affected by climate change, together with society's capacity to absorb them. We identify and deal with four main challenges: (i) a lack of open-source, European-wide and spatially explicit extreme weather risk information; (ii) extreme impacts are underestimated due to the aggregation process and often represented by expected values only; (iii) typically, the focus of research on extremes has been on direct impacts (losses) of disasters and less work exists on indirect effects and assessments of policy responses; and (iv) a lack of direct links from climate modelling activities to evaluations of current hazards, vulnerabilities, risks, as well as to implementable adaptation practices.

Our research suggests that pursuing a risk management approach may significantly inform the identification of risks and key adaptation policies across Europe. We examine suitable risk analytical approaches and illustrate the challenges encountered. We focus on riverine flood risk, and drought and heatwave stress to agriculture, which are among the most challenging extreme weather events requiring adaptation cited in the GP.

Already today, there are climate-related extreme weather events that limit Europe's willingness or ability to adapt. On the other hand, market forces help to mediate impacts, such as in agriculture where a crop loss need not necessarily lead to an economic loss. Adaptation may also be important in helping reduce adverse impacts, particularly in agriculture. Finally, a set of risk management options – financial risk sharing tools – mediate impacts while being flexible enough to include additional information over time.

The discussion in this chapter is organised specifically around the theory of disaster risk analysis and risk management, with a focus on policy-relevant insights for Europe. We consider a risk analytical approach important for focusing attention on current challenges to adapt to adverse weather. Starting from such a baseline, and cognizant of key uncertainties regarding future impacts and potential responses, we emphasise the role of incremental learning in adaptation in line with Chapter 2, Russel *et al.* Our discussion also refers to other chapters in this volume, such as Chapter 3, van Vuuren *et al.*, Chapter 4, Aaheim *et al.*, Chapter 9, Werners *et al.* and Chapter 12, Linnerooth-Bayer *et al.*

Following a review of the literature and key terms in Section 8.2, we present the risk-based framework guiding our analysis in Section 8.3. This is followed by a discussion of the flood, drought and heatwave risk mapping exercises conducted on a pan-European scale for today and the future. This information is subsequently used to inform the assessment of the monetary and economic risks of Section 8.4. Section 8.5 examines key risk management and adaptation responses with a special focus on risk financing and sharing mechanisms. We illustrate our approach by means of two relevant decisions: (i) fiscal planning for disaster risks; and (ii) using the EU Solidarity Fund (EUSF) to share disaster risks across Europe. Section 8.6 concludes with general insights and policy-related findings.

8.2 Review of the state of the art on assessing extreme event impacts, risk management and adaptation

In this section, we discuss the relevant scientific literature on managing disaster risks and its relation to climate adaptation. Important terms are defined and explained in Box 8.1.

Box 8.1. Defining and using vulnerability and risk

Vulnerability and *risk* are concepts with multiple and ambiguous meanings. As an analytical term, vulnerability has been used in a broad range of disciplinary contexts, including geography, anthropology, engineering sciences, ecology and economics. Vulnerability as commonly defined within the context of climate change (e.g. IPCC, 2001) is a function of both potential climate change impacts and society's capacity to adapt to these impacts. Turner *et al.* (2003) define vulnerability as the degree to which a system or subsystem is likely to experience harm due to exposure to a hazard, either as a perturbation or stressor. Risk, alternatively, is a function of the *hazard* (likelihood and severity), exposure of people and assets, and *vulnerability* (see UNISDR, 2008), but usually stops short of considering the coping capacity or resilience of the exposed system. *Resilience* is one of the key concepts in vulnerability research. It refers to the capacity of the system to absorb disturbances and reorganise, while undergoing changes to retain essentially the same function, structure and identity (Walker *et al.*, 2002).

Vulnerability, as a research-organising concept, is more complex than risk, which is both its strength and its weakness. Its strength lies in its integration across multiple stresses, recipients (human and environmental), and temporal/spatial scales. Its weakness lies in the difficulty of carrying out empirical research at this level of complexity, and vulnerability assessments have generally lacked the empirical rigour of risk assessments (see Smit *et al.*, 1999; Kelly and Adger, 2000). A second drawback is the mismatch between the vulnerability concept and the policy process (Kasperson and Kasperson, 2001), since few policy makers have a mandate broad enough to address issues of this scope (there are no 'vulnerability managers'). In contrast, risk (with vulnerability as an important input) is better governed and managed.

Risk management is considered here to overlap significantly with climate adaptation. It can be defined as:

... the systematic approach and practice of managing uncertainty and potential losses, involving risk assessment and analysis and the development of strategies and specific actions to control and reduce risks and losses. Risk ... can be addressed by preventative measures, such as avoiding settlement in floodplains and building strong buildings; monitoring, early warning and response measures to manage extreme events; and risk transfer, including insurance, to cope with unavoidable impacts (UNISDR, 2008).

The challenge is thus to build on the rigour of (more narrowly focussed) risk assessments and contribute to the complex scientific, institutional and policy processes necessary for effectively assessing and reducing vulnerability to climate change.

8.2.1 Information on observed and modelled climate-related risks

In Europe, far-reaching impacts of changes in climate extremes have been documented for sudden-onset events such as floods and storms, as well as slower-onset disasters such as droughts and heat waves. Among others, the 2002 large-scale flooding over central Europe and the 2003 heatwave of unprecedented magnitude resulting in 70 000 deaths, placed risk management and adaptation at the top of the agenda (EC, 2007b). Agricultural practices are climate (especially heat and rainfall) dependent and the agricultural sector has been particularly exposed to changes in climatic mean values and interannual variability with about three-quaters of drought and heatwave losses reported in Europe over the last 30 years accruing in agriculture (EC, 2007b). Increases in temperature, as already observed and simulated by climate models, are expected to have a great impact on agriculture. The summer heat wave of 2003 accompanied by precipitation deficits (Schär *et al.*, 2004) led to agricultural losses exceeding €13 billion and a 30 per cent reduction in gross primary production of terrestrial ecosystems (Parry *et al.*, 2007).

There is, with important exceptions, a continuity and consistency between the observed changes and those projected for the future. Negative impacts will include increased risk of flash floods, more frequent coastal flooding and increased erosion due to storminess and sea level rise. Furthermore, regional climatic differences are expected to become more pronounced. In Northern Europe, reduced demand for heating, fewer winter deaths, increased crop yields, longer vegetation seasons, expansion of agricultural land areas, increased forest growth and an expanding use of water power are projected. In Southern Europe, a region already vulnerable to climate variability, climate change is projected to exacerbate heat and drought problems and to reduce water availability, hydropower potential and crop productivity (Parry *et al.*, 2007).

On the other hand, adaptation is an important component of any policy response to climate change in this sector (Hulme *et al.*, 1999; Mizina *et al.*, 1999). Studies show that, without adaptation, climate change may create considerable problems in agricultural production, and agricultural economies and communities in many areas; but with adaptation, vulnerability can be reduced and there are numerous opportunities to be realised. For example, changes in growing season and carbon dioxide fertilisation may actually have a positive impact on agriculture in some regions (Rosenzweig and Parry, 1994; Wall and Smit, 2005).

Knowledge of disaster impacts and risks is limited and heterogeneous. The IPCC Fourth Assessment Report and the EU's Green Paper on Adaptation advocate further refinement to risk management methods and tools (see Carter *et al.*, 2007). A reasonable amount of knowledge exists on the direct risks (exposure, vulnerability and impacts) from sudden and slow-onset disasters linked to extreme weather. The re-insurance industry, consulting firms and multi-lateral financial institutions have worked together with the academic research community to estimate risks from such extreme events across the globe. A number of EU research projects (such as PESETA) have focused on weather-related hazards and risks in Europe. There are national-level assessments of current and future weather risks, mostly on flood risk in England and Wales (DEFRA, 2001), Germany (Apel *et al.*, 2004, and Merz and Thieken, 2004) and the United States (Scawthorn *et al.*, 2006a, 2006b). Hall *et al.* (2005) projected risks up to 2100 for England and Wales.

A very recent national-level study conducted by Feyen *et al.* (2009) for the PESETA project fills an important gap by conducting a first European-wide assessment of current and future flood risks in 2100. It is important to note that this study is similarly organised around a risk analytical approach. Yet, as discussed in the following section, and in contrast to the PESETA project, primarily due to short-comings in climate modelling, we decided not to project flood risks into the future. An extensive literature on the effects of climate change on agriculture has been published (see Parry *et al.*, 2007), yet no pan-European assessment of monetary losses in agriculture for current and future climates has been conducted.

8.2.2 Projecting climate and weather patterns into the future

Climate projections using multi-model approaches indicate increases in globally averaged mean water vapour and precipitation over the twenty-first century. Yet, precipitation scenarios show strong seasonal and regional differences in Europe. There is a marked contrast between future winter and summer precipitation change. Wetter winters are predicted for the entire continent; in many regions there will be less snow and much more rain. In summer, an apparent difference in precipitation change between wetter conditions in Northern Europe and drier conditions in Southern Europe is predicted. Generally, the behaviour of changing precipitation extremes is projected to be notably different from changes in mean precipitation over much of Europe. The highest quartiles of daily precipitation amounts and annual maximum daily precipitation are anticipated to increase over many areas, including some areas where the mean precipitation is projected to decrease. However, climate models remain limited in their reproduction of local weather extremes due to, *inter alia*, inadequate (coarse) resolution (Christensen and Christensen, 2003; Kundzewicz *et al.*, 2006). Projections of changes in future extreme weather events remain highly uncertain and hinder us from robustly pre-dicting future flood risk. On the other hand, drought and heatwave stress operate as slower onset phenomena and are more strongly characterised by mean climate conditions for which there is greater confidence in model projections.

8.2.3 Integrative risk and adaptation assessments

The presence of extreme weather event risk and adaptation within integrative modelling approaches is emerging (see also Chapter 3; van Vuuren *et al.*), but there is room for improvement in terms of more effectively applying modelling capacity and available data. Based on work by Nordhaus and Boyer (2000), extreme event risks in adaptation studies and modelling have usually been represented in a rather *ad hoc* manner, using add-on damage functions that are based on averages of past impacts and contingent on gradual temperature increase. The impact of climate change on socio-economic drivers of emissions, such as production, consumption and demographics, are sometimes taken into consideration. Such approaches, however, implicitly assume that climate impacts are marginal and do not strongly affect the social system. Finally, there is often little explicit reference made in integrated assessment to implementable adaptation practices (e.g. in Stern, 2006).

It seems that the climate change modelling community is gradually embracing a more risk-based approach (see, for example, Jones, 2004; Carter *et al.*, 2007). Regional climate modelling and statistical downscaling methods, as well as climate and socio-economic downscaling techniques, are increasingly being applied, yet with remaining deficiencies (Goodess *et al.*, 2003). Moreover, assessments of climate change impacts and vulnerability have changed in focus from an initial analysis of the problem to the assessment of potential impacts and, recently, to the consideration of specific risk management methods (Carter *et al.*, 2007). This stronger focus on a *risk-based* assessment of adaptation is particularly important for *fat-tailed* (i.e. non-normally distributed) catastrophic impacts that are potentially very large, uncertain, unevenly distributed, and may occur in a distant future.

8.3 Methodological approach

We attempt to partially close the gap related to the lack of information on risks by a consistent, probabilistic European-wide mapping of disaster risks on different spatial aggregation levels. Modelling extremes in line with a risk-based and more geographically explicit approach can result in a better and more consistent assessment of disaster risk as a function of a hydrometeorological signal, of socio-economic drivers and of vulnerability. Together, these three factors account for the variability of natural hazards. We highlight three key analytical aspects of the methodological approach taken here: (i) integration; (ii) probabilistic analysis; and (iii) spatial explicitness. We discuss each of these in more detail in the following.

8.3.1 Integration

The methodology builds on three key steps. The first step involves arriving at an understanding of losses, or *direct risks*, in terms of monetary impacts on assets and crops. Hazard analysis entails determining the types of hazards affecting a given region, including their intensity and recurrence, as well as possible changes therein due to climate change. Assessing exposure involves analysing population[1] and assets exposed to hazards in a particular region. Vulnerability (see Box 8.1) is a multidimensional concept comprising a multitude of factors, and our study focuses on physical and financial vulnerabilities. Figure 8.1 illustrates our flood risk analysis which integrates relevant information on the hazard (flood depth and extent, as well as probability in terms of return period and recurrence), exposure (land use categories and demography) and vulnerability (considered with regard to physical susceptibility of exposed land use classes to flood depth and hazard). The end product is a spatially explicit probabilistic estimate of losses in monetary terms, summarised by means of risk maps at different aggregation levels.

In a subsequent step, we assess the consequences of those direct risks and determine whether and how they can be absorbed economically. This involves understanding the system's underlying financial vulnerability as a convolution of direct risk and resilience. Financial vulnerability, for example, addresses the question: 'To what extent can losses be replaced and at what costs, given a certain event with discrete probability?' When all events and probabilities are considered, economic risk can be estimated. In the final step, based on an understanding of direct and indirect economic risk, risk management and adaptation options can be assessed in terms of viability, costs, and benefits, together with other criteria.

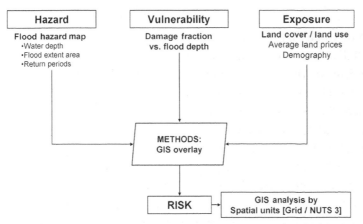

Figure 8.1. Operational framework for modelling flood risk as a function of hazard, exposure and vulnerability. (Source: Lugeri *et al.*, 2007.)

---

[1] We do not examine health aspects and loss of life.

The analysis becomes dynamic when climate and global change are explicitly introduced, since both have an effect on risk. Climate change, for example increasing temperatures, may affect hazards in the future and will consequently transform risk either by increasing or decreasing the frequency and/or intensity of events. On the other hand, economic development and population change, may also affect exposure and physical vulnerability of assets and lead to modified economic vulnerability.

8.3.2 Probabilistic analysis

The analysis of weather extremes and adaptation to their impacts is complicated by the inherent aleatoric, or 'chance', uncertainty of these phenomena. Although specific weather extremes are unpredictable in a deterministic sense beyond a few days in the future, they can be estimated in probabilistic terms: for example, the 100-year flood which is an event with an average return period of 100 years, or an annual probability of one per cent. Natural disaster risk is commonly defined as the probability multiplied by the potential impacts affecting people, assets or the environment. Framing the analysis in terms of probability is useful since risk management strategies can be based on the entire range of extreme event scenarios, which would not be possible with the use of average values only. Also, the notion of probability entails addressing *potential* impacts before they occur, rather than coping with *actual* impacts after the fact, thus signifying a shift from reacting to impacts to anticipating and managing risks.

8.3.3 Spatial explicitness

Although driven by large, even global, scale phenomena, natural disaster events are essentially local. Local communities, but also state and national governments, therefore demand knowledge on potential or real losses with as much geographical and sectoral detail as possible. This, in turn, requires a massive amount of data, such as the details of the river catchments' properties in the case of floods, and an accurate quantification of the accumulated value of the land involved (Büchele, 2006; FLOODsite, 2006). When extending the analysis to a European scale, it is impossible to treat the problem with the same approach used for local catchments, due to the lack of consistent information and the resources needed to process data at high resolution on such a large geographical scale. The methodology of flood risk assessment in monetary and economical terms, as well as that of drought and heat stress in agriculture, relies on an unavoidable trade-off between territorial resolution, data availability and precision, and amount of data processing.

8.4 Assessing direct risks

The approach used in the ADAM project was based on an analysis of the exposure to *current* climate variability in order to address the question: 'Are we already being confronted with plausible climate-related stochastic events that challenge the limits of Europe's willingness or ability to adapt?' Once hotspots have been identified, we build on these estimates of today's weather-related losses and the efficiency of prevention and risk transfer to ask how adaptation can reduce future vulnerabilities and risks. We present methods and results on mapping flood and drought risks in Europe. The methods employed differ due to data and modelling constraints.

8.4.1 Mapping riverine flood risks

The computation of monetary flood risk relies on geographical information system (GIS) processing of hazard maps and territorial databases combining the probability and severity of flooding with spatially explicit information on exposure and vulnerability. Model outcomes are maps in grid-based resolution (50 km × 50 km) and district level detail of monetary damage computed for five return periods (50, 100, 250, 500 and 1000 years). The basic maps can be further elaborated to produce an estimate of the annual expected loss value[2]. The model provides minimum and maximum values for each return period, and the expected losses, to highlight the range of uncertainty (see Lugeri *et al.*, 2007)[3]. The data can be aggregated to various administrative levels by summing up the losses over the chosen boundaries. In Figure 8.2 we show regional level estimates as a share of GDP.

At the NUTS 2 level, regions in Eastern Europe, central France and Scandinavia seem to be particularly at risk. These findings are generally in line with the priority areas for action identified in the GP. Aggregation of risks to national level (Figure 8.3) reveals that Eastern European countries seem to be under particular stress. In nearly all new EU Member States, annualised flood risk, when measured in GDP, exceeds one per cent.

As already mentioned, only a few studies exist against which to compare our results: Defra (2001) which modelled risks for Wales and England; Słota (2000) which examined the impacts in Poland; and Feyen *et al.* (2009) for Europe. In comparison to the other studies, our modelled results seem reasonably satisfactory, indicating that the computed values of damage fall within the respective uncertainty range. Overestimations from our method compared to other studies range between 10 and 80 per cent, about 30 per cent of the difference in results being attributable to exposure estimations alone.

---

[2] Also called annual average damage (AAD) in this context, and we do not differentiate between losses and damages in this discussion.
[3] For information on the flood hazard modelling underlying the hazard maps, see de Roo *et al.*, 2007.

(a)

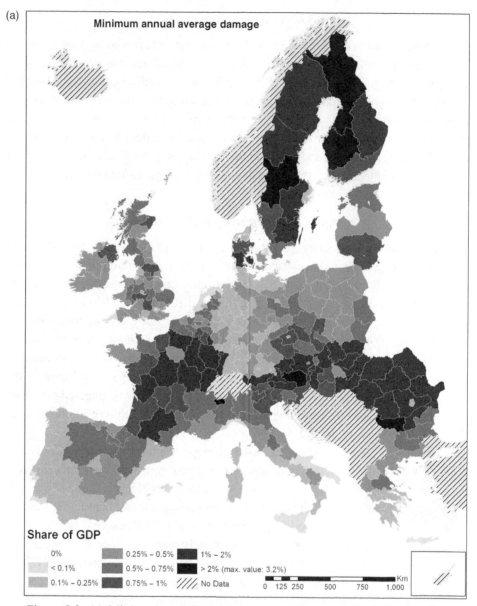

Figure 8.2. (a) Minimum and (b) maximum annual average flood damage for European provinces and regions (NUTS 2 level)[4] as a percentage of GDP for today's climate regime. (See colour plate section.)

---

[4] NUTS (Nomenclature of Territorial Units for Statistics) is the EU georeferencing standard system for hierarchical classification of administrative divisions (CEC, 2003). NUTS 3 relates to the district/province level, NUTS 2 to regional or provincial resolution and NUTS 1 to states within a country.

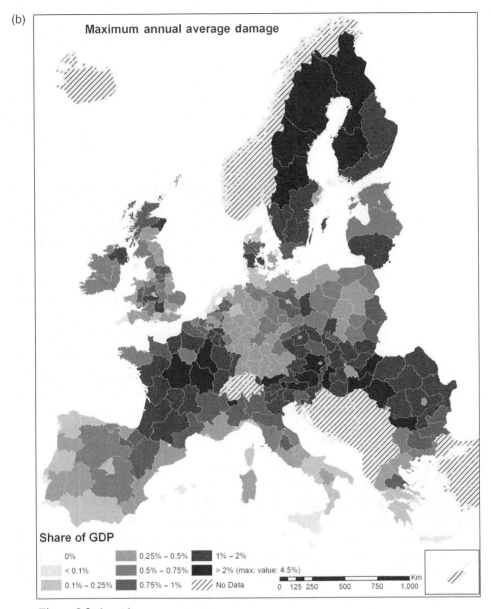

(b)

Maximum annual average damage

Share of GDP

0%	0.25% – 0.5%	1% – 2%
< 0.1%	0.5% – 0.75%	> 2% (max. value: 4.5%)
0.1% – 0.25%	0.75% – 1%	///// No Data

Km
0 125 250 500 750 1,000

Figure 8.2. (cont.)

8.4.2 *Mapping drought and heat risks to agriculture*

Agriculture is both temperature and precipitation dependent. In contrast to flooding, mean changes in these variables are decisive, and credible projections from climate modelling are at hand. In order to quantify the combined effect of drought and

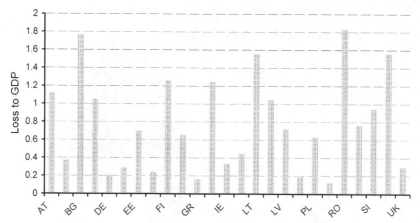

Figure 8.3. Maximum average annual flood risk across European countries (in per cent of GDP). *Note*: this is not loss *of* GDP, but loss measured *in terms of* GDP. Losses relate to assets, while GDP relates to income generated from those assets.

heatwave stress on crop yield in the European Union, we couple temperature and precipitation changes derived from global climate modelling over the near future (the period 2030–2060) with crop modelling performed using the CropSyst model (Stockle *et al.*, 2003)[5]. Climate change averaged over the period 2030–2060 represents a global warming of approximately 2 °C with respect to the pre-industrial period, which is the EU's policy target for EU wide and international negotiations. Winter and summer crop yields were simulated for the 2030–2060 time period considering both a business-as-usual scenario, with today's crop management practices continued in the future, and an adaptation scenario in which crop management practices are modified. There are a number of potential adaptation options and we considered the following: (i) shifting the sowing date to ±15 days with respect to the present period; (ii) using cultivars with a longer/shorter growth cycle of ±20 per cent for each phase duration with respect to those used in the present period.

Using spring wheat as an example[6], we compute the combined drought and heat stress risk calculations for today (Figure 8.4) and for the change in such risk for the period of 2030–2060 compared to today, with and without adaptation interventions (Figure 8.5).

As expected, Southern Europe and parts of France are today particularly exposed to drought and heat stress (Figure 8.4) and this trend is expected to worsen in the future. In a future climate scenario with a strong north–south precipitation gradient, the Mediterranean basin and Central Europe are projected to be more strongly affected by combined heat and drought stress when adaptation is not considered

[5] Specifically, mean temperature and precipitation changes derived from the global climate model HadCM3 (see Pope *et al.*, 2000) based on the SRES A2 scenario are downscaled to a regional level over the period 2030–2060 as compared to today's weather pattern, represented by the control period 1975–2005 and used as inputs for the crop model.
[6] Similar analyses were done for winter wheat, soybean and sunflower.

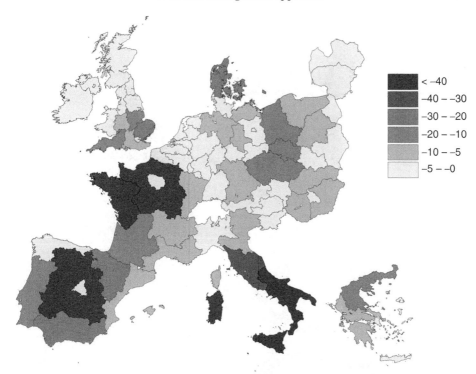

Figure 8.4. Annualised monetary risk due to combined heatwave and drought stress for spring wheat calculated for the present period (1975–2005) on a NUTS 1 level (losses in € millions). (See colour plate section.)

(Figure 8.5: panel a). When considering the two simple adaptation strategies, however (Figure 8.5: panels b and c), many regions in Europe would actually be able to benefit from a warming climate (on an annualised basis). Northern Europe, in particular, would exploit the advantage of higher precipitation by using crop varieties with a longer growing cycle: when water is not the limiting factor, a longer time for biomass accumulation results in an increased yield.

In contrast, in some regions in Southern Europe the same adaptation options would result in a negative impact on yield, since crop development shifts towards summer when longer dry spells and heatwaves may significantly affect crop growth. It should be noted that winter crops, such as barley and durum wheat, are generally not affected by water stress; in contrast to summer crops, their growth cycle is generally advanced to the autumn–winter period, which is not affected by drought or heat stress either in the present or the future.

8.4.3 Assessing indirect risks in a drought hotspot: the case of southern Spain

Flood risk predominantly causes structural damage to public infrastructure and private assets and, consequently, the costs involved relate mostly to the direct

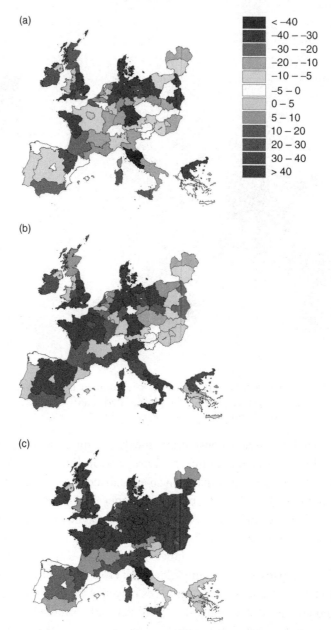

Figure 8.5. Changes in annualised drought and heatwave risks to spring wheat over a future period in 2030–2060 compared with today, without adaptation (a) and with adaptation in terms of advanced sowing (b) and longer cycle variety (c) (in € millions). (See colour plate section.)

replacement costs of the affected assets[7]. This is generally in line with our loss estimates. Drought and heat stress, in contrast, affect crops that are marketable goods. Beyond the monetary loss, of interest here are the subsequent market effects of induced climate variability and change. A study of price mechanisms and income effects, through economic modelling, sheds some light on this issue and we examine how crop loss translates into a real economic loss.

We integrate monetary crop losses from water and heat stress into a multi-market economic equilibrium model, GRACE (see also Chapter 4, Aaheim *et al.*, and Aaheim and Schjolden (2004)). The model determines how a loss in agricultural output leads to an increase in the price of agricultural products and how the loss affects other sectors of the economy[8]. We present the case of crop losses due to water and heat stress in the Guadiana River Basin (see Chapter 9, Werners *et al.*, for more details). The Cropsyst losses discussed above are used to produce an estimate of the loss of productivity in the agricultural sector, which is introduced into the GRACE model as a reduction in the productivity of land in this sector. Cropsyst losses for individual crops[9] – like the ones discussed above – are aggregated to a total yield loss based on the area of cultivation and associated market prices.

In order to assess the macroeconomic implications, we suggest that the sudden onset characteristic of extreme events implies that the scope for reallocation of resources is limited in the short–medium term. We take this constraint into account by fixing the use of natural resources in the agricultural sector and by reducing the total availability of capital and labour within the drought-affected provinces according to the losses in the agricultural sector. The model is run deterministically for a 12.5 per cent yield loss scenario, which roughly corresponds to a 20-year heat stress and/or 100-year drought event. It is run for Guadiana only (Scenario I) and across southern Spain including Guadiana (Scenario II), leading to yield losses of approximately €155 million and €340 million, respectively. Table 8.1 shows the direct effect as simply the yield loss, which is used as input to the model, and leads to losses of value added in agriculture and aggregate GDP. The model results reveal a reduction in economic output both for the agricultural sector and for the economy as a whole.

In Scenario I, in which the Guadiana Basin alone experiences a loss, the loss to farmers in Guadiana is not reduced substantially by taking market interactions into account. This is because the price of goods increases very modestly as a result of local weather events, and cannot be passed on by farmers to their customers. For the

---

[7] Flooding may also cause business interruptions, which are not considered here. Generally, they are deemed to be rather limited and short-lived (e.g. see Mechler and Weichselgartner, 2003).

[8] The model does not take into account compensation received from agricultural insurance or assistance provided by the government.

[9] The crops included are durum wheat, soft wheat, barley, maize, soya, sunflower, sugar and grapevine. These are assumed to be representative of all crops. Livestock is assumed to be unaffected.

Table 8.1. *Percentage reduction in production, sectoral and regional GDP for two agricultural loss scenarios*

Losses	Scenario (I): 12.5% yield loss in Guadiana only		Scenario (II): 12.5% yield loss in all of southern Spain	
	Guadiana Basin	Spain	Guadiana Basin	Spain
Direct effect Loss of yield	−6.0	−0.5	−6.0	−0.5
Market effect Loss of value added – agriculture	−5.5	0.0	−1.8	−1.7
Loss of value added – total economy (GDP)	−0.4	0.0	−3.1	−1.9

whole economy in Guadiana, the loss is relatively insignificant. In the rest of Spain, as the loss is relatively small, farmers will benefit from the slight price increase and the effect would be sufficient to offset the loss overall in Guadiana.

In reality, it is unlikely that extreme heat and water stress in Guadiana will occur without other parts of Spain also being affected. If other provinces of southern Spain are equally affected by the event, as assumed in Scenario II, the impacts to the Iberian economy will be more severe, with a total loss of nearly three per cent of GDP for Guadiana and two per cent for Spain. The loss to farmers is lower than in Scenario I. This is because there is a larger price increase due to a non-marginal reduction of the supply of agricultural products, which allows farmers to transfer a greater share of the loss to their customers. The loss to the local and national economy is more serious in such a larger-scale scenario when neighbouring provinces are also affected. This indicates that market effects, based on price changes strongly depend on the scale of the phenomenon affecting regions inter-linked by trade.

8.5 Assessing risk management policy responses: financing disaster risks

Not all losses can be efficiently reduced, and market effects may not efficiently help to spread losses across an entire population or an economy. For coping with the residual impacts, there are many funding modalities helping to share and finance disaster losses: private and public (tax revenue) savings, insurance and international assistance (for a discussion of risk financing options in development assistance, see Chapter 12, Linnerooth-Bayer *et al.*). Figure 8.6 shows a cross-country sample of major disasters. In addition to insurance markets, governments as 'insurers of last resort' have an important role in providing infrastructure reconstruction and relief

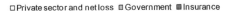

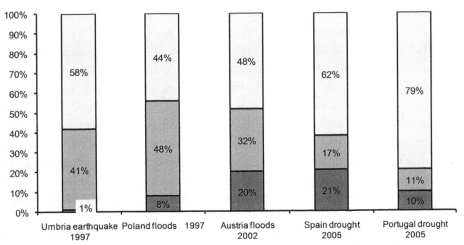

Figure 8.6. Cross-country sample of financing modalities of disaster losses by insurance, government assistance, and private sector and net loss (as a percentage of direct losses).

support for households and businesses. Spending as a share of direct losses has ranged from 11 per cent (drought in Portugal) to 48 per cent (flooding in Poland). The only outlier in Europe has historically been the UK, with a large insurance penetration and almost no relief provided to the affected[10] (Linnerooth-Bayer and Mechler, 2007)[11].

Disaster risks thus imply a contingent government liability (see also Chapter 4, Aaheim *et al.* for an extended discussion) and if governments are unable or fail to replace damaged infrastructure and to provide assistance to those in need post-disaster, longer-term consequences may ensue.

In the following, we focus on the financial implications on the public sector and options for sharing risks nationally, and EU-wide, in order to minimise the burdens. Mechanisms for absorbing risks comprise: better regulating, or even providing insurance for private sector risks to households and businesses; implementing a relief fund; or accessing the EU Solidarity Fund (EUSF), being the key EU regional disaster relief allocation mechanism. We focus on the latter two mechanisms and assess the adequacy of public sector response under a current climate regime.

---

[10] This can be attributed to an (unwritten) compact holding that the government is responsible for structural risk management while households and business need to take action in terms of insurance.
[11] Data on private sector spending are not available, and thus are lumped together with the net loss.

8.5.1 Sharing risks nationally: the case of Austria

With a relatively high flood risk primarily due to high exposure of people and assets in the Danube river basin, Austria is an important case in point. Due to relatively limited private sector insurance coverage, the public sector has played an important role in providing financial post-disaster support. Austria recently experienced a fiscal and political crisis of sorts due to the 2002 severe flooding, which caused losses of about €3 billion. The government disaster reserve fund, as the key government mechanism for financing public and private sector losses, has generally not been an effective mechanism for coping with the aftermath of disaster events. Following the severe flooding of 2002, other sources of funding, such as additional tax revenue, had to be sought, causing major frictions in the coalition government. With other issues of dispute, this finally contributed to the downfall of the government and a call for early re-elections.

Originally, the Austrian reserve fund was accumulating in nature. In 1995 it consisted of more than €190 million. Given a perceived overcapitalisation of the fund, a law was passed in 1998 decreeing that the fund capital should be limited to €29 million, at which point the accumulated capital was taken and used for other budgetary ends. Since then, at year end, the balance accumulated from tax revenue in excess of this amount has been removed and used for other purposes. This restriction led to a deficit in the fund following the severe flooding in 2002 requiring payouts lasting into 2003, which had to be refinanced by diversions from other budget items. This created important opportunity costs as other socially desirable investments were foregone.

The 2002 flood event showed that the disaster fund, the prime mechanism for public sector financial response in Austria, did not perform well for low-probability, high-consequence events. One reason for this was that the fund in fact had become an annual budget item rather than an accumulating mechanism providing more financial protection. As a counterfactual scenario, Figure 8.7 charts the funding available had the fund, in fact, been accumulating; in 2003, immediately after the large-scale flooding, a small surplus would still have been available for providing rapid post-disaster assistance.

On the other hand, there are important opportunity costs associated with 'reserving' budgetary revenue over time. In order to model this trade-off between opportunity and ability to provide assistance (in terms of availability of budgetary resources to fund relief and reconstruction), we conduct a stochastic risk assessment using a catastrophe simulation model (CATSIM) (Mechler, 2004; Hochrainer, 2006; see also Appendix 1 for further information on the model). We use CATSIM to inform contingency planning on national scales such as for national reserve funds and public–private disaster insurance pools. In order to examine the

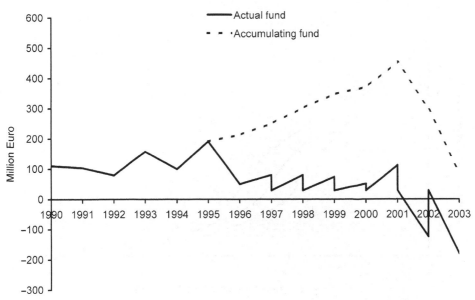

Figure 8.7. Reserves of the national disaster fund in Austria and a hypothetical simulation of an accumulating fund. (Source: Hyll, Türk and Vetters, 2004.)

fiscal repercussions of flood risk (essentially occurring in the Upper Danube basin) for Austria, we aggregate the grid-based information of the risk maps to a national level. This leads to a representation of risk by means of a loss-frequency distribution, which relates probabilities of loss to losses in terms of assets. For example, Figure 8.8 shows a distribution for today's flood risk in Austria in terms of capital stock losses, including lower and upper confidence intervals.

CATSIM modelling involves incorporating probabilistic disaster risk and potential financing gaps for funding these losses into macroeconomic projections in order to determine fiscal and economic consequences. Aggregate disaster risk is represented by the curve in Figure 8.8. Financial vulnerability can be defined as the availability of internal and external savings to refinance potential losses, as well as supporting the private sector with relief and recovery assistance. It is determined by available *ex ante* and *ex post* financing sources. Governments can raise funds *ex post* after a disaster in many ways: by diverting resources from other budget items; imposing or raising taxes; taking a credit from the Central Bank, which either prints money or depletes its foreign currency reserves; borrowing by issuing domestic bonds; and issuing bonds on the international market (Fischer and Easterly, 1990). Governments can also arrange for financing before a disaster occurs. Such financing options include reserve funds, traditional public or private insurance instruments, alternative insurance instruments, such as catastrophe bonds, or arranging a contingent credit or debt instrument.

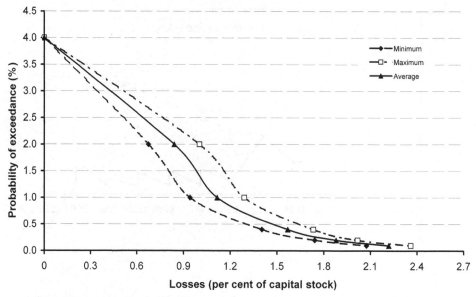

Figure 8.8. Exceedance probability curve for aggregate flood losses in 2009 for Austria.

Fiscal consequences from the perspective of the national government can be analysed in terms of Austria's financial vulnerability. A key objective of fiscal policy is to ensure sufficient *fiscal space*, which can be defined as the flexibility in a government's budget to fund new and emerging items subject to keeping the fiscal position sustainable (see Heller, 2005)[12]. One could argue that fiscal space was effectively reduced to zero in 2002, given the fiscal and political crisis ensuing. Figure 8.9 shows a selected number of scenarios of the Austrian government's fiscal space position in the presence of flood risk and with the reserve fund as a risk financing mechanism, simulated over the 10-year period from 2009–2018. In a baseline case with the disaster fund only a budget item (Figure 8.9 (a)), given assumptions on economic growth and associated increases in tax revenue, fiscal space would increase by about 27 per cent by 2018. Yet, disaster shocks may occur and necessitate disaster spending in terms of relief and reconstruction support. In the most drastic case in our analysis (lowest trajectory on the figure), fiscal space would be reduced by about 43 per cent.

In contrast, when using an accumulating fund as a true pre-disaster risk financing mechanism (Figure 8.9 (b)), potential volatility would be reduced and in the worst

---

[12] Generally, a large portion of a government's budget is tied to recurring expenditure (such as wages and salaries and debt service), and cannot be diverted for other purposes. The balance remaining is the discretionary budget, which we consider to represent fiscal space. For Austria, based on spending in the flood event of 2002 and the difficult economic environment in 2009, we estimate it broadly to amount to one per cent of the budget in 2009, i.e. about € 0.7 billion.

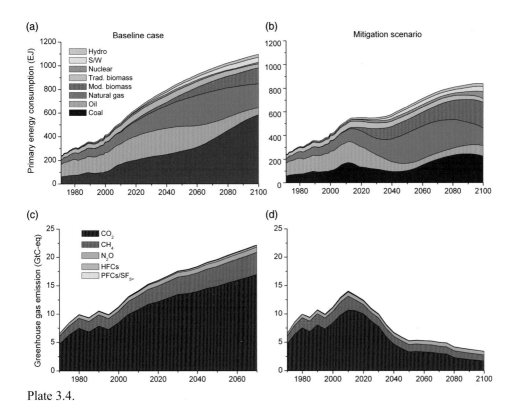

Plate 3.4.

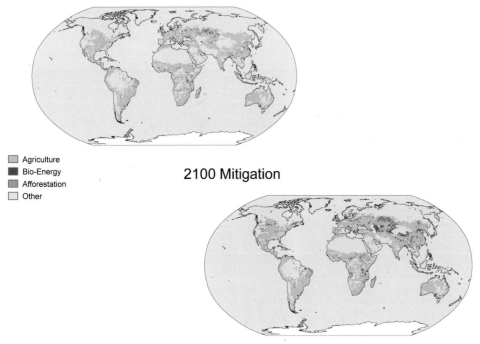

Plate 3.6.

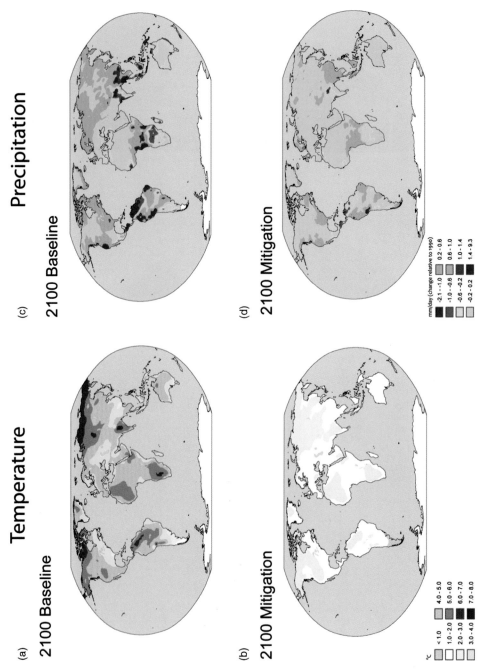

Temperature

(a)

2100 Baseline

(b)

2100 Mitigation

°C ☐ <1.0 ☐ 4.0-5.0 ☐ 1.0-2.0 ☐ 5.0-6.0 ☐ 2.0-3.0 ■ 6.0-7.0 ☐ 3.0-4.0 ■ 7.0-8.0

Precipitation

(c)

2100 Baseline

(d)

2100 Mitigation

mm/day (change relative to 1990) ■ -2.1--1.0 ☐ 0.2-0.6 ■ -1.0--0.6 ☐ 0.6-1.0 ■ -0.6--0.2 ■ 1.0-1.4 ☐ -0.2-0.2 ■ 1.4-9.3

Plate 3.8.

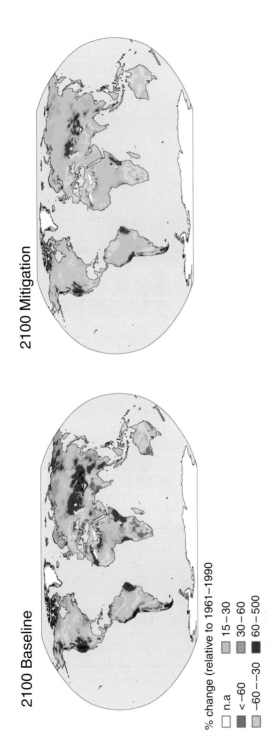

2100 Baseline

2100 Mitigation

% change (relative to 1961–1990)

n.a   15 – 30
< –60   30 – 60
–60 – –30   60 – 500
–30 – –15   >500
–15 – 15

Plate 3.11.

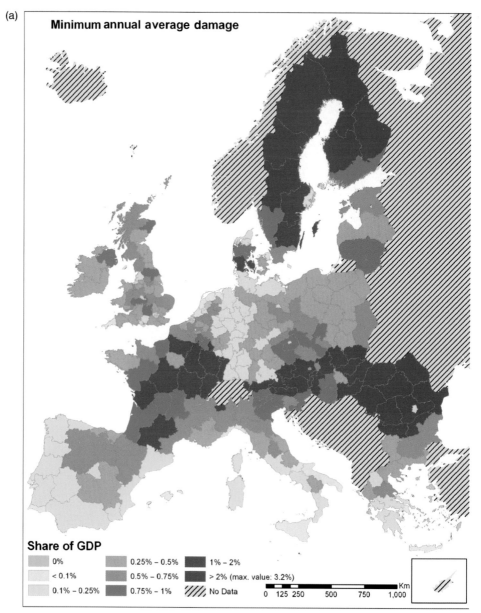

Plate 8.2a.

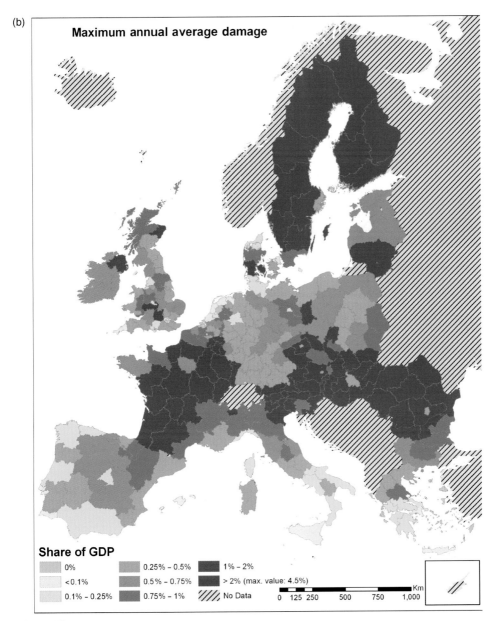

(b)

Maximum annual average damage

Share of GDP

0%	0.25% – 0.5%	1% – 2%
<0.1%	0.5% – 0.75%	> 2% (max. value: 4.5%)
0.1% – 0.25%	0.75% – 1%	No Data

0   125 250   500   750   1,000 Km

Plate 8.2b.

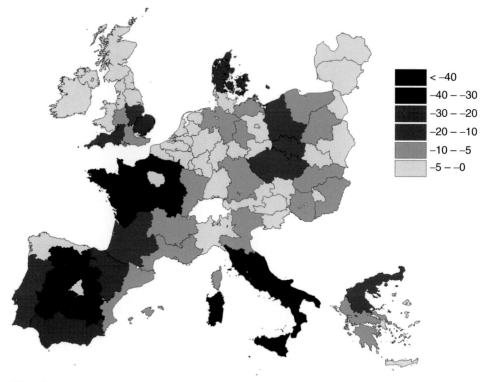

Plate 8.4.

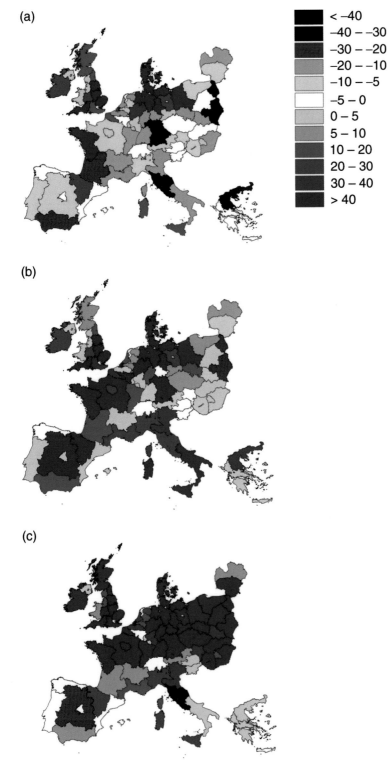

(a)

(b)

(c)

■	< −40
■	−40 – −30
▓	−30 – −20
▒	−20 – −10
░	−10 – −5
□	−5 – 0
░	0 – 5
▒	5 – 10
▓	10 – 20
▓	20 – 30
■	30 – 40
■	> 40

Plate 8.5.

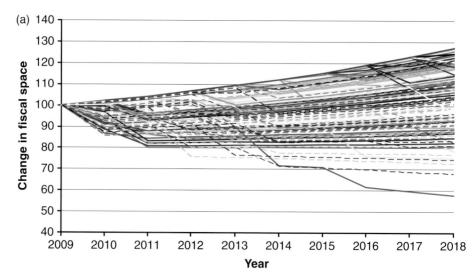

Plate 8.9a.

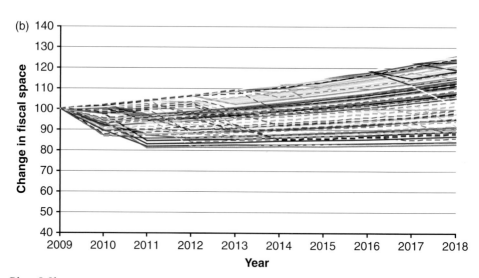

Plate 8.9b.

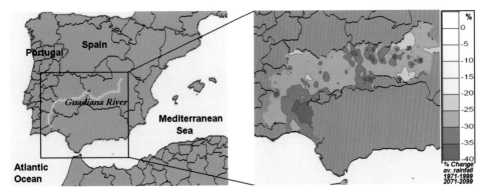

Plate 9.1.

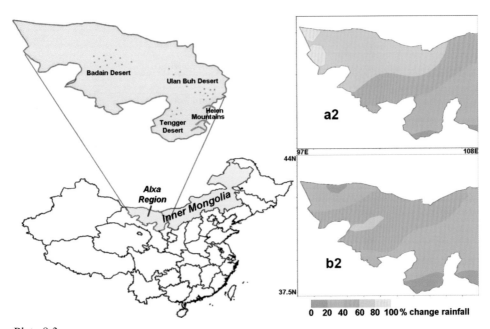

Plate 9.3.

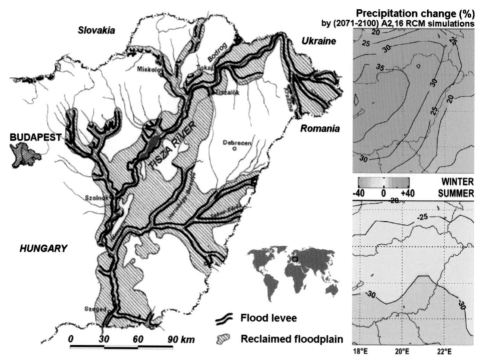

Plate 9.5.

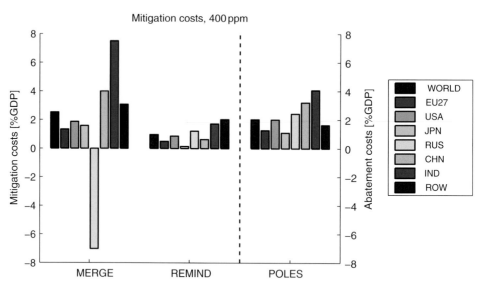

Plate 11.11.

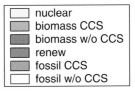

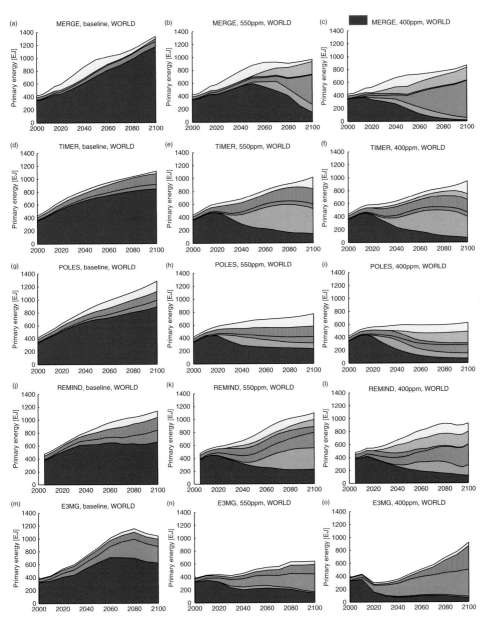

Plate 11.4.

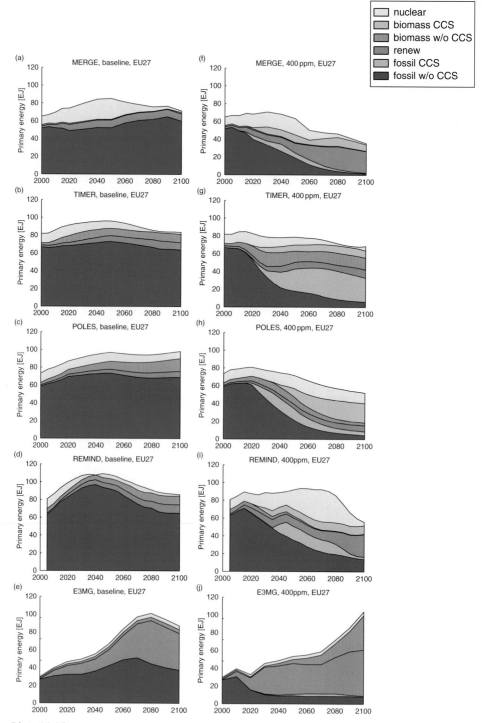

Plate 11.10.

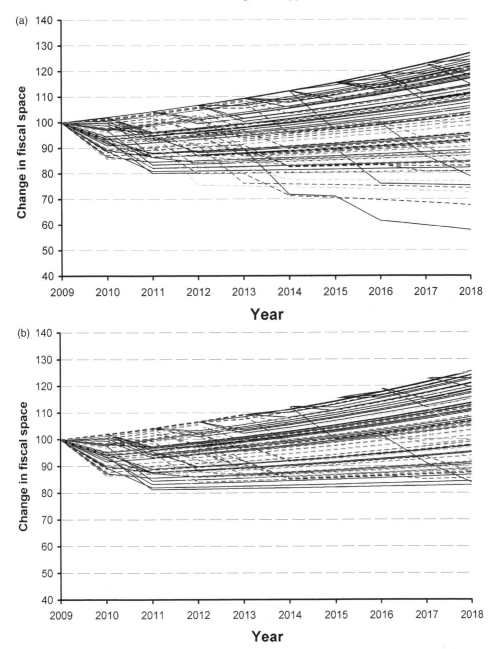

Figure 8.9. Change of the Austrian government's fiscal space for a time period of 10 years without (a) and with (b) an accumulating reserve fund. (See colour plate section.) Note: Every line represents a different future scenario dependent on the occurrence and magnitude of disaster events in the given time horizon.

case scenario the decrease would amount to 17 per cent only. Thus, on the one hand, the most debilitating down-side effects would be avoided; on the other hand, fiscal space in the scenarios without events would be slightly lower (a 25 per cent increase in 2018). This is due to opportunity costs associated with reserving budgetary funds, which otherwise could be used for (socially) profitable government investment.

This trade-off between maximising budgetary resources and reducing volatility is not a perfect one, as reduced financial vulnerability implies more sustained growth and increased revenue. Depending, however, on the exposure to contingent liabilities, such as flood risk, and financial vulnerability, it is a trade-off to be explored. Given similar and higher exposures to flood risk, there are a number of other candidate countries, also exhibiting large fiscal constraints, particularly in Eastern Europe, where the public sector seems financially vulnerable to disaster risk.

8.5.2 *Sharing risks in the EU: The European Solidarity Fund*

Funding requirements for flood and other risks may overburden national governments and necessitate international assistance, even in the EU. This was recognised by the European Union, which, after large-scale flooding in Central Europe in the summer of 2002, created the EU Solidarity Fund (EUSF). Under this mechanism, Member States and Accession Countries can request aid in the event of a major natural or technological disaster (EC, 2005). The fund provides financial aid for uninsurable losses in the event of a natural disaster causing direct damages above €3 billion at 2002 prices, or 0.6 per cent of gross national income. The payments from the Fund are limited to finance operations undertaken by the public authorities to alleviate non-insurable damages (e.g. putting infrastructures back in operation)[13]. The maximum annual budget is €1 billion per annum. Since 2002, almost all EU countries have applied for funds (mostly for flood losses) and more than half have received assistance via the Solidarity Fund. There have been instances, such as in 2002 in the aftermath of the 2002 floods, when the Fund was nearly depleted by the middle of the year. Such problems may become more pronounced if losses increase and the scope for applying for relief payments is widened, as is currently being discussed in the EU. Overall, it is not clear to what extent the EUSF is at risk of being under-funded.

When pooling our nationally aggregated risk distributions and applying the EUSF loss funding criteria detailed above, we can derive an estimate of probabilistic funding requirements (Figure 8.10). For example, today there would be about an

---

[13] The European Commission determines the amount of aid and proposes its mobilisation. The maximum annual budget is €1 billion per year. The amount annually available for extraordinary regional disasters is also limited to 7.5 per cent of the EUSF's annual budget.

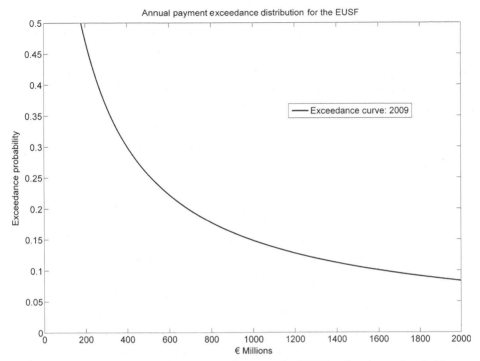

Figure 8.10. Probabilistic funding requirements for EUSF today due to flood risk.

annual 15 per cent chance that the EUSF needs to provide support in excess of €1 billion (and thus just face depletion) with average annual payments amounting to about 0.7 billion.

It is important to keep in mind that the EUSF was set up to fund other emergencies in addition to flood risks[14]. EUSF support has so far been granted only once in 2008 for drought relief, which may become more necessary and desirable in the future. Agricultural insurance, just as flood insurance, exists in many forms involving private and public players. Yet, as our research shows, uninsured losses can be quite large, necessitating the need for the government to step in with further aid. ADAM work in the Guadiana basin in the Iberian Peninsula reveals that funding for droughts under the EUSF may become desirable.

The Commission has expressed its readiness to examine all requests for EUSF aid following droughts, but will ensure that 'the request is not the indirect result of inefficient water management and that appropriate drought management plans are in place' (EC, 2007c). Moreover, the Commission has been open to assessing whether

---

[14] By statute, it may fund other emergencies such as oilspills, and has done so in the past.

changes are needed to the definition of the criteria and eligible operations, so that the EUSF may respond better to drought events (such as lowering the qualifying loss threshold per event to €1 billion). Yet, as discussed above, the rationale of funding drought losses is more difficult to establish than for flood losses, given the fact that predominant impacts are crop losses, which are traded in markets and price and compensation effects must be considered in order to determine the net impact. Overall, our analysis highlights the fact that it may be worth considering an increase in fund resources or better leveraging of resources (e.g. by re-insuring the EUSF).

8.6 Conclusions

Adaptation to climate change is emerging as an additional agenda item for climate change policy in Europe. A key area of concern is adaptation to weather extremes, for which links to the disaster risk management community are being established nationally and internationally. We presented an integrated framework for assessing extreme weather risks and adaptation options under climate change. The framework is organised around an expert-based risk-analytic approach, accounting for key uncertainties. Maps for flood and drought risk for Europe have been generated at a European scale, identifying monetary economic losses. The innovation has been that these maps are probability based, spatially explicit and indicate the probabilities of certain events and associated monetary losses occurring at different aggregation scales. They also cover all of Europe. Such maps may be useful input for policy making, because they provide a consistent framework for assessing current and future hotspots and for prioritising the allocation of funds for risk management and adaptation measures.

Future projections were judged not to be sufficiently robust for flood risk, because climate models are currently not good at projecting changes in localised climate variability. In contrast, we did feel able to project drought and heatwave risk. Our results confirm the commonly held view that regions and countries in Eastern Europe are particular hotspots in terms of flood risk.

With regard to drought and heat stress to agriculture, we find Southern Europe to be particularly vulnerable. In a future climate with a north–south precipitation change gradient, and assuming adaptation, many agricultural regions in Europe would actually benefit from a warming climate. However, some regions in Italy and Spain would not be able to do so and face continued stress and substantial associated monetary risks. Furthermore, our economic analysis showed that because crops are traded in markets, economic losses in terms of value added depend crucially on the spatial scale of the climatic extreme; price effects may importantly reduce or increase the economic effects.

We found that disasters already pose significant contingent liabilities for governments, and prudent planning is necessary to avoid debilitating consequences. This was shown by the Austrian political and fiscal crisis in the aftermath of large scale flooding in 2002. Yet, as our Austrian case study also demonstrated, there is a potential trade-off between reducing volatility and maximising budgetary resources, which has to be reconciled by decision makers.

As a consequence of budgetary pressures, many governments have been turning to the EU Solidarity Fund, which, in light of recent experience and future anticipation of severe losses from natural disasters, was set up in 2002 to provide financial aid to Member States for post-disaster emergency measures. We tested the funding status of the Fund, which holds a maximum of €1 billion per annum. We found it is sufficient to fund flood losses today; however, given other demands on the Fund, including a potentially worsening drought risk problem in Southern Europe, there might be a need to increase its funding in the future. In general, our risk assessment information may also inform a more equitable, timely, and risk-based allocation of the EU Solidarity Fund.

With limited funds for refinancing actual losses and implementing adaptation measures to prevent adverse impacts, there is increased need for adaptation support in a changing climate. It should, therefore, be an attractive option to link cohesion and adaptation policies within the EU. On the one hand, the Structural and Cohesion spending programmes concentrate on the Lisbon (innovation, growth, jobs) and Gothenburg (sustainable development) goals; but the new EU Cohesion Policy (2007–2013) also stresses the importance of climate change. On the other hand, the EU Green and White Papers on Adaptation emphasise the social dimension of an EU adaptation strategy. A linkage of both policy fields provides the chance to pursue adaptation as well as cohesion targets simultaneously. We find these to be interlinked in any case, for example, through the cohesion instrument of the EU Solidarity Fund.

Uncertainties in our assessment were notably large and, in our judgement, hindered the projection of flood risks into the future. Climate change introduced large uncertainties into the estimation of future drought risks with important implications for any assessment of adaptation practices. At this stage, the evaluation of adaptation options cannot rely on exact or fairly robust projections of future risk. It is generally difficult to assess risk-related consequences of climate policies and emission pathways with high credibility and accuracy. Clearly, more research is needed to quantify and manage the considerable uncertainties.

What is more, risk management interventions, such as flood defences, have been designed and operated on the basis of the stationarity assumption with the past being the key to the future. However, 'stationarity is dead' (Milly *et al.*, 2008) and learning therefore becomes important. Such learning should be based on continued updating of data, assessments and models, as well as by focusing on keeping adaptation decisions flexible.

These suggestions are in line with findings from the stakeholder engagement processes in Hungary and Spain conducted in the ADAM Project (see Chapter 9, Werners *et al.*), in which some of our research was used. The conclusion from these processes was that successful adaptation depends on decision makers having access to sufficient resources, including trustworthy information and guidance, and on envisaging the decision sequence as a learning process. Risk financing options, one set of adaptation measures analysed by us in detail, seem particularly flexible and can be updated over time on an almost annual basis, as soon as new information emerges. This is in contrast to many other adaptation decisions which can be characterised by long associated time horizons of up to 200 years (see Hallegate, 2009). Yet, a balance between physical adaptation measures that reduce risk, and financial adaptation options which transfer risk, needs to be struck.

Despite important limitations, it seems that using a risk-analytic approach, with the departure point being today's vulnerabilities and risks, may critically inform the identification of current and projected future risks across Europe, as well as help policy makers and stakeholders identify key adaptation decisions. Results derived in this thematic area of ADAM have been compiled in a digital compendium, which is available via the ADAM website and will hopefully provide a transparent, open-domain resource for better informing stakeholders on local, national, and EU levels on the adaptation challenges faced by Europe.

References

Aaheim, A. and Schjolden, A. (2004) An approach to utilise climate change impacts studies in national assessments. *Global Environmental Change*, **14**, 147–60.

Apel, H., Thieken, A. H., Merz, B. and Blöschl, G. (2004) Flood risk assessment and associated uncertainty. *Natural Hazards and Earth System Sciences*, **4**, 295–308.

Büchele, B., Kreibich, H., Kron, A. *et al.* (2006) Flood-risk mapping: contributions towards an enhanced assessment of extreme events and associated risks. *Natural Hazards Earth System Sciences*, **6**, 485–503.

Carter, T., Jones, R., Lu, X. *et al.* (2007) New assessment methods and the characterisation of future conditions. In *Climate Change 2007: Impacts, Adaptation and Vulnerability. Contribution of Working Group II to the Fourth Assessment Report of the Intergovernmental Panel on Climate Change*, ed. Parry, M., Canziani, O., Palutikof, J., van der Linden, P. and Hanson, C. Cambridge, UK: Cambridge University Press, pp. 133–71.

CEC (2003) Regulation (EC) No 1059/2003 of the European Parliament and of the Council of 26/05/2003 on the establishment of a common classification of territorial units for statistics (NUTS).

Christensen, J. H. and Christensen, O. B. (2003) Severe summertime flooding in Europe. *Nature*, **421**, 805.

Department for Environment, Food and Rural Affairs (Defra) (2001) *National Appraisal of Assets at Risk from Flooding and Coastal Erosion*. Final Report.

de Roo, A., Barredo, J., Lavalle, C., Bodis, K. and Bonk, R. (2007) Potential flood hazard and risk mapping at pan-European scale. In *Digital Terrain Modelling. Development and Applications in a Policy Support Environment*. ed. Peckham, R. and Jordan, G. Series: Lecture Notes in Geoinformation and Cartography. Berlin: Springer.

EC (2005) European Commission *European Union Solidarity Fund – Annual report* 2004 COM/2005/0709 final. Brussels: European Commission.

EC (2007a) European Commission *Adapting to Climate Change in Europe – Options for EU Action*, Green Paper, COM(2007) 354, Brussels: Commission of the European Communities.

EC (2007b) European Commission *Climate change: Europe must take adaptation measures to lessen impacts of current and future warming*. Press release IP/07/979, Brussels 29 June 2007.

EC (2007c) European Commission *Addressing the Challenge of Water Scarcity and Droughts in the European Union*, Communication from the Commission to the European Parliament and the Council COM(2007) 414 final, Brussels, July 2007.

EC (2009) European Commission *Adapting to Climate Change in Europe – Options for EU Action*, White Paper, COM(2009) 147/4, Brussels: Commission of the European Communities.

Feyen, L., Barredo J. I. and Dankers, R. (2009) Implications of global warming and urban land use change on flooding in Europe. In Feyen, J., Shannon, K. and Neville, M. (eds.) *Water and Urban Development Paradigms. Towards an Integration of Engineering, Design and Management Approaches*. London: Taylor & Francis Group

FLOODsite (www.floodsite.net) (2006) 6th FP Integrated Project Report Number T9–06-01 2006: Guidelines for Socio-economic Flood Damage Evaluation.

Goodess, C. M., Hanson, C., Hulme, M. and Osborn, T. J. (2003) Representing climate and extreme weather events in integrated assessment models: a review of existing methods and options for development. *Integrated Assessment*, **4**, 145–71.

Hall, J. W., Sayers, P. B. and Dawson, R. J. (2005) National-scale assessment of current and future flood risk in England and Wales. *Natural Hazards*, **36**, 147–64.

Hallegatte, S. (2009) Strategies to adapt to an uncertain climate change. *Global Environmental Change* Part doi:10.1016/j.gloenvcha.2008.12.003

Heller, P. (2005) *Understanding Fiscal Space*. IMF Policy Discussion Papers 05/4, Washington DC: International Monetary Fund.

Hochrainer, S. (2006) *Macroeconomic Risk Management Against Natural Disasters*. Wiesbaden: Deutscher Universitätsverlag.

Hulme, M., Barrow, E. M., Arnell, N. W., Harrison, P. A., Johns, T. C. and Downing, T. E. (1999) Relative impacts of human-induced climate change and natural climate variability. *Nature*, **397**, 688–691.

Hyll, W., Türk, A. and Vetters, N. (2004). *FloodRisk*. Workpackage Ökonomische Aspekte TP 05. Projektendbericht. Wien: Lebensmittelministerium.

IPCC Working Group II (2001) *Climate Change 2001: Impacts, Adaptation, and Vulnerability, Summary for Policymakers*. Report approved at the Sixth Session of Intergovernmental Panel on Climate Change WGII, Geneva, Switzerland, 13–16 February.

Jones, R. N. (2004). Managing climate change risks. In *The Benefits of Climate Change Policies: Analytical and Framework Issues*, ed. Agrawal, S. and Corfee-Morlot, J. Paris: OECD.

Kasperson, J. and Kasperson, R (2001) *Summary of International Workshop on 'Vulnerability and Global Environmental Change.'* 17–19 May 2001. Stockholm: Environment Institute (SEI).

Kelly, P. M. and Adger, W. N. (2000) Theory and practice in assessing vulnerability to climate change and facilitating adaptation. *Climatic Change*, **47**, 325–52.

Kundzewicz, Z. W., Radziejewski M. and Pińskwar, I. (2006) Precipitation extremes in the changing climate of Europe. *Climate Research*, **31**, 51–8.

Linnerooth-Bayer, J. and Mechler, R. (2007). Disaster safety nets for developing countries: Extending public–private partnerships. *Environmental Hazards*, **7**, 54–61.

Lugeri, N., Genovese, E., Lavalle, C., Barredo, J. I., Bindi, M. and Moriondo, M. (2007) *An assessment of weather-related risks in Europe*. Ispra: JRC-IES EUR 23208 EN

Mechler, R. and Weichselgartner, J. (2003) *Disaster Loss Financing in Germany – The Case of the Elbe River Floods 2002*. Interim Report IR-03-021. Laxenburg: IIASA.

Mechler, R. (2004) *Natural Disaster Risk Management and Financing Disaster Losses in Developing Countries*. Karlsruhe: Verlag für Versicherungswissenschaft.

Merz, B. and Thieken, A. (2004) Flood risk analysis: concepts and challenges. *Österreichische Wasser- und Abfallwirtschaft*, **56**, 27–34.

Mills, E. (2005) Insurance in a climate of change. *Science*, **309**, 1040–44.

Milly, P. C. D., Betancourt, J., Falkenmark, M. *et al.* (2008) Stationarity is dead: Whither water management? *Science*, **319**, 573–4.

Mizina, S. V., Smith, J. B., Gossen, E., Spiecker, K. F. and Witkowski, S. L. (1999) An evaluation of adaptation options for climate change impacts on agriculture in Kazakhstan. *Mitigation Adaptation Strategies for Global Change*, **4**, 25–41.

Nordhaus, W. D. and Boyer, J. (2000) Warming the world. In *Economic Models of Global Warming*. Cambridge: MIT Press.

Parry, M. L., Canziani, O. F., Palutikof, J. P., van der Linden, P. J. and Hanson, C. E. (eds.) (2007). *Climate change 2007: Impacts Adaptation and Vulnerability. Contribution of Working Group II to the Fourth Assessment Report of the Intergovermental Panel of Climate Change*. Cambridge, UK: Cambridge University Press.

Pope, V. D., Gallani, M. L., Rowntree, P. R. and Stratton, R. A. (2000) The impact of new physical parameterisations in the Hadley Centre climate model: HadAM3. *Climate Dynamics*, **16**, 123–46.

Rosenzweig, C. and Parry M. (1994) Potential impact of climate change on world food supply. *Nature*, **367**, 133–8.

Scawthorn, C., Blais, H., Seligson, E. *et al.* (2006a) HAZUS-MH flood loss estimation methodology. I. Overview and flood hazard characterisation. *Natural Hazards Review*, **7**, 60–71.

Scawthorn, C., Blais, H., Seligson, E. *et al.* (2006b) HAZUS-MH flood loss estimation methodology. II. Damage and loss assessment. *Natural Hazards Review*, **7**, 72–81.

Schär, C., Vidale, P. L., Lüthi, D. *et al.* (2004) The role of increasing temperature variability in European summer heatwaves. *Nature*, **427**, 332–6.

Słota, H. (ed.) (2000). *Water Management in Poland*, Cracow: University of Cracow, Institute of Meteorology and Water Management.

Smit, B., Burton, I., Klein, R. J. T. and Street, R. (1999). The science of adaptation: a framework for assessment. *Mitigation and Adaptation Strategies for Global Change*, **4**, 199–213.

Stern, N. H. (2006) *The Stern Review. The Economics of Climate Change*. Cambridge, UK: Cambridge University Press.

Stockle, C. O., Donatelli, M. and Nelson, R. (2003). CropSyst, a cropping systems simulation model. *European Journal for Agronomy*, **18**, 289–307.

Turner, B. L., Kasperson, R., Matson, P. *et al.* (2003) A framework for vulnerability analysis in sustainability science. *Proceedings of the National Academy of Sciences*, **100**, 8074–9.

UNISDR (2008) United Nations International Strategy for Disaster Reduction. *Climate Change and Disaster Risk Reduction*. Briefing Note 01. Geneva: UNISDR.

Walker, B., Carpenter, S., Anderies, J. *et al.* (2002) Resilience management in social–ecological systems: a working hypothesis for a participatory approach. *Conservation Ecology*, **6**, 14.

Wall, E. and Smit, B. (2005).Climate change adaptation in light of sustainable agriculture. *Journal of Sustainable Agriculture*, **27**, 113–23.

9

Mainstreaming adaptation in regional land use and water management

Lead authors:

SASKIA E. WERNERS[1], J. DAVID TÀBARA, HENRY NEUFELDT,
DARRYN MCEVOY, XINGANG DAI, ZSUZSANNA FLACHNER,
JENNIFER WEST, FRANCESCO COTS, GIACOMO TROMBI,
NICOLA LUGERI, PIOTR MATCZAK, GERT-JAN NABUURS

[1]Co-ordinating lead author

Summary

This chapter examines the constraints and opportunities for mainstreaming adaptation to climate change in land use and water management in three study regions of the ADAM project: the Guadiana River Basin in Spain and Portugal, the Tisza River Basin in Hungary and the Alxa region in western Inner Mongolia, China. We analyse the conditions that either facilitate or limit adaptation according to six analytical dimensions: biophysical, technical, financial, institutional, social and cognitive (the latter including informational aspects). Our research suggests that all six aspects are needed to capitalise on opportunities for successfully planning and implementing adaptation. Institutional and cognitive aspects have been identified as particularly important, but the relative weight of each aspect depends on location and will vary over time. Furthermore, we argue that, in the long term, building capacity to adapt to climate change will depend on the extent to which climate concerns are integrated into the planning and implementation of land use and water management. Based on our empirical findings, we provide recommendations that could facilitate such climate mainstreaming. We find that adaptation is enhanced by (i) adaptation pilot projects that test and debate a diverse set of new ideas in a collaboration of civil society, policy and science; (ii) open and easy access to information on climate impacts, policy and adaptation options; (iii) integration of (traditional) agro-environmental land use systems that regulate climate impacts at the local and regional scale, with new technologies, policies, organisational responsibilities and financial instruments; and (iv) flexible financial instruments that facilitate benefit and burden sharing, social learning and support a diverse set of potentially better-adapted new activities rather than compensate for climate impacts on existing activities.

Making Climate Change Work for Us: European Perspectives on Adaptation and Mitigation Strategies, ed. Mike Hulme and Henry Neufeldt. Published by Cambridge University Press © Cambridge University Press 2010.

9.1 Introduction

Since initial work for the Third Assessment Report of the Intergovernmental Panel on Climate Change (IPCC) demonstrated that adaptation to climate change is both important and complex, there has been growing attention for documenting adaptations as they happen and explaining the processes by which adaptation can occur (Adger *et al.*, 2005; Adger *et al.*, 2007). Whereas the literature on adaptation is rich in detail on impacts, vulnerability and limits to adaptation, less is known on the conditions that facilitate adaptation in practice (McEvoy *et al.*, 2008). The latter is, however, of paramount importance for developing integrated guidance on how to cope with climate change in the long term.

This chapter applies the definitions and concepts of impacts, vulnerability and adaptation as outlined by the IPCC (Adger *et al.*, 2007) and examines adaptation practices in the context of climate impacts and vulnerability. Adaptation to climate change takes place through adjustments in human and natural systems to reduce vulnerability in response to observed or expected changes in climate and associated extreme weather events. It involves changes in perceptions of climate risk, and in social and environmental processes, practices and functions to reduce potential damages or to take advantage of new opportunities. It includes anticipatory and reactive actions, as well as private and public initiatives. In practice, adaptation is an on-going process, reflecting many stresses and cross-sectoral concerns, including discrete actions to address climate change specifically. Isoard and Swart (2008) conclude that human adaptation occurs mainly at sub-national and local levels but involves many other levels of decision making from municipalities to international organisations. Adaptation is a cross-sectoral, multi-scale and transboundary issue, which requires a comprehensive and integrated response.

Societies have always had to respond to regional climate variability and extreme weather events. Many have developed ways of coping with floods, fires and droughts (for instance, the vernacular architecture of Mediterranean countries). Recent experience of weather extremes has given these efforts new impetus within countries and at the European level. Whilst climate change is a new driver for action, adaptation will in many cases be implemented by regulatory modifications of the existing policy frameworks for floods, droughts and the management of water quality. Step-wise advances in action, co-ordination and engagement of agents at the local and regional level will be needed to handle the expected level of accumulated incremental change over time, and to address the increased possibility of new extreme events (Footitt and McKenzie Hedger, 2007). Following Klein *et al.* (2007), we use the term mainstreaming for the step-wise integration of adaptation policies and measures into ongoing sectoral planning and decision making to reduce climate vulnerability and ensure long-term sustainability.

Land and water resources are directly impacted by climate change, and decisions regarding these resources affect ecosystems and human vulnerability. As such, land use planning and water management are expected to play an increasingly central role in adaptation strategy. Stern (2006) identifies land use planning as a key area for adaptation. Yet, although changing land use planning is a promising adaptation strategy to cope with climate change impacts, this strategy is not practised extensively (Footitt and McKenzie Hedger, 2007). The multitude of land uses and stakeholders means that land use and water management are complex sectors in the climate change arena that deserve special attention. So far, climate change analyses and climate policy formulations have not adequately addressed the integration of water resource issues and climate change response options, including associated synergies and trade-offs between different policy domains and scales of action (Bates *et al.*, 2008).

This chapter examines the challenges and opportunities for adaptation to climate change in three study regions of the ADAM project: the Guadiana River Basin in Spain and Portugal, the Tisza River Basin in Hungary and the Alxa region in western Inner Mongolia, China[1]. The three regions have in common that they increasingly struggle with climate impacts on land use and water resources, including desertification and the occurrence of extreme events such as floods and droughts. However, the institutional contexts and governance traditions upon which societal responses and adaptation practices have developed differ greatly in the three regions. We examine adaptation practices from a comparative, empirical and analytical perspective, before synthesising this information to sharpen our understanding of the constraints and opportunities for the successful implementation of adaptation through land use and water management policies and practices. In particular, the research in the three study regions is synthesised according to: (i) climate change and impacts on land use and water management; (ii) adaptation practices to climate variability and change; and (iii) constraints and opportunities for enhancing adaptation as a response strategy for climate change. The chapter closes with a discussion of the lessons learned from the study regions for mainstreaming and furthering adaptation through land use and water management.

9.2 Adaptation practice in three regions

In order to analyse the conditions that facilitate or limit adaptation in the three selected regions, we first briefly introduce the biophysical and institutional context in which adaptation is taking place or needs to take place. In particular, we look at the

---

[1] The ADAM project sought Chinese research co-operation to enrich the project with non-European perspectives and to acknowledge the international dimension of climate change impacts and greenhouse gas emissions. Within China, the Alxa region in Inner Mongolia was selected as a case of significant climate related land use change of great national concern.

Table 9.1. *Characteristics of the three study regions*

	Guadiana region, Spain and Portugal	Alxa region, Inner Mongolia, China	Tisza region, Hungary
Biophysical, land use	Semi-arid climate, forest, agriculture	Arid climate, desert, livestock, agriculture	Continental climate, grassland, agriculture
Area/arable land	66 800 km²/20 million ha	270 000 km² (main study area 72 000 km²)/ 30 000 ha + 9 million ha steppe	46 000 km² (Hungarian part river basin)/2.6 million ha
Technical	2000 dams. Reservoir and irrigation system	Irrigation, groundwater and water transfers	2800 km river dyke, drainage system
Economic	Participation in EU and global market. Tourism. GDP 20 000 per capita below EU average	Increasing market forces and industrialisation. GDP €2000 per capita	Transition economy. GDP: € 4500 per capita. Below country average
Institutional	EU member in 1986. EU regulation. Regional competences defined in Spain and Portugal	National (communist) interventionist state; well-defined limited regional autonomy.	EU member in 2004. Implementation national and EU regulations.
Social	4 million people. Ageing	200 000 people. Mongol minority	4.1 million people. Roma minority

potential impacts of climate change in relation to land use and water management. Second we describe adaptation practices in each region as identified by the ADAM project through workshops, interviews with stakeholders, modelling studies and the review of secondary sources. Finally, we discuss the constraints and opportunities to adaptation according to six dimensions: biophysical, technical, financial, institutional, social and cognitive[2] (cf. Smit and Pilifosova, 2001; Adger *et al.*, 2007; Bates *et al.*, 2008). Table 9.1 summarises the main characteristics of the three study regions.

9.2.1. Guadiana River Basin

Guadiana basin: climate change and impacts on land use and water management

The Guadiana is a transboundary river between Spain and Portugal. Its basin is one of the three main drainage units of the Iberian Peninsula. Eighty-three per cent of the

---

[2] The cognitive dimension includes aspects of recognition (the ability to recognise the challenge), reflexivity (the ability of people and organisations to reflect on and adapt their behaviour), perception/framing (legitimately diverse views about risks, vulnerability and adaptive capacity) as well as informational aspects.

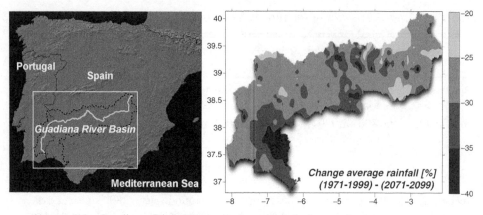

Figure 9.1. Guadiana River Basin and projected change in average summer precipitation between 1971–1999 and 2071–2099 in SRES Scenario A2 using regional climate model HIRHAM (adapted from Orioli *et al.*, 2008). (See colour plate section.)

basin is located in Spain and 17 per cent is in Portugal (Henriques Goncalves, 2005). The river basin has the typical semi-arid climate of Southern Europe with low and irregular precipitation (440 mm/year). Rainfall is greatest in the central-north, decreasing westwards towards the mouth of the river (Orioli *et al.*, 2008). Average seasonal temperature ranges from 14.5 °C (winter) to 31.9 degrees (summer). Temperature rarely drops below zero. Climate scenarios project a temperature rise in spring and summer of four to seven degrees by 2100 (Santos *et al.*, 2002), with temperature variation increasing towards the north east of the basin. At the same time, rainfall is expected to decrease in spring and in summer. Figure 9.1 illustrates this using the 'Special Report on Emissions Scenarios' (SRES) A2 Scenario results from the HIRHAM climate model (Extra High Resolution: 12.5 × 12.5 km) (Orioli *et al.*, 2008). The natural hydrological regime of the Guadiana is characterised by low flows during summer and confined high flow events in winter. Using the HadCM3 model's climate scenarios, annual runoff is estimated to fall by up to 60 per cent by 2100 (Santos *et al.*, 2002). In the last four decades, the river basin has undergone major modifications through the building of dams, tourism developments, (illegal) wells and increasing urbanization pressures. On the Portuguese side of the basin the Alqueva dam was built. This created the largest artificial lake in Europe for which water use still has to be allocated. In the Spanish Upper Guadiana basin, groundwater development substantially increased people's livelihoods, however groundwater use is not sustainable and groundwater tables are falling. The main problems associated with land use and water management are the over-exploitation of aquifers for agricultural use, agricultural contamination and river fragmentation by dams (Cosme *et al.*, 2003;

WWF, 2003; Tábara *et al.*, 2009). The combination of these developments and the semi-arid climatic conditions intensifies water scarcity along the Portuguese–Spanish border. Climate change potentially increases the pressure on available resources and the conflicts between agriculture, irrigation, tourism, nature conservation and the development of large urban areas. Furthermore, climate change impact assessments show a worsening of the current situation of desertification, particularly through the impact of forest fires (Moriondo *et al.*, 2006) and the loss of soil fertility (ECCE Project, 2005).

Land use and agricultural practice differ considerably in the region (see Figure 9.2). Going from southwest to northeast, three main zones can be distinguished. The first zone is the coastal zone primarily with tourism on the Portuguese side (Algarve) and a combination of intensive agriculture (mostly citrus and fruits) and tourism on the Spanish side. Moving inland, the next zone is hilly, forested and supports a unique ecosystem called *dehesa* (Spanish) or *montado* (Portuguese). Dehesa farmers combine cork oak (*Quercus suber*) and common oak (*Quercus robur*) with different types of tree crops (olives, figs, almond, oranges, carobs), ground vegetation (cereal,

Figure 9.2. Guadiana River Basin: (a) Guadiana River at Mertola, Portugal, (b) extensive agro-environmental production system dehesa, (c) intensive irrigated agriculture, (d) Tourism development in the estuary. (Photos: Werners, Dec 2006.)

herbs, grass) and livestock breeding (cows, sheep, red swine) (Joffre et al., 1999). Dehesa has a high landscape value and is protected as cultural heritage. The third zone is dominated by large-scale agriculture (Beja region in Portugal and Extremadura in Spain). Crops include cereals, wine, olives and citrus. Crops grown are both irrigated and non-irrigated. The share of irrigated land is expected to change in part due to the new Alqueva reservoir.

The impacts of climate change depend strongly on the cropping system. Dehesa is well adapted to the regional climatic conditions, although its economical feasibility is questioned (Joffre et al., 1999). With respect to intensive agriculture the impacts differ for summer and winter crops[3]. Climate change is not expected to significantly impact winter wheat and spring wheat. The summer crops sunflower, soybean and corn are projected to be the most adversely affected crops with consistent reduction in yield in almost the whole basin (on average 15, 30 and 35 per cent, respectively) (see also Chapter 8, Mechler et al.). By affecting crops differently, climate change will also affect the robustness of cropping patterns (Werners et al., 2007).

Guadiana Basin: assessment of adaptation practices

Stakeholders have proposed the following adaptation options in relation to land use and water management[4] (ADAM Project, 2007b; Tábara, 2007): agricultural and economic diversification, changing farm size, re-use of grey water, promotion of rural tourism and local produce, protecting as well as modernising the 'dehesa' agro-ecological production system, using (new) drought and heat-resistant crops (e.g. dates) and (changing) irrigation systems. Reforestation was welcomed as an adaptation option, though criticism was voiced with respect to the current use of fire-prone pine species. Next to biophysical interventions, stakeholders emphasised the need for building adaptive capacity at the farm and regional institutional level, through awareness raising, better co-ordination between actors and policies, and financial support for promoting alternative sustainable activities (often linked to small-scale rural tourism and businesses).

Changes in institutional arrangements relating to climate change can be observed in both countries and, in particular, at river basin scale, although these are still in an early stage (McEvoy et al., 2007). Arrangements that affect the

---

[3] The impacts of climate change on crop yields were simulated for five different crops using a crop growth simulation model (CropSyst). Crops were selected to include a winter crop (winter wheat) and four spring-summer crops (spring wheat, sunflower, soybean and corn). Present and future crop yields were simulated under the SRES climate scenario A2. Corn was the only crop that was always irrigated in the simulations. Extreme events (heat stress due to heatwaves, water stress during droughts) were included as further threats to production (for more details see Chapter 8, Mechler et al.).

[4] Interviews in Seville and Mertola, field trip and workshop held in Mertola, Portugal; 12–22 December 2006.

Guadiana River Basin include the Portuguese National Climate Change Programme (2001) and the Spanish National Adaptation Plan (2006). Implementation of these programme is the responsibility of several newly created national bodies, such as the Portuguese Climate Change Commission, the Spanish Inter-ministerial Group on Climate Change and the National Council on Climate Change. Among their duties is to promote climate change policy across government and to ensure integration of climate considerations into sectoral and regional policies in accordance with national and European regulation. These new bodies are also required to ensure the participation of public and private actors in the policy making process.

With agriculture an important economic sector in the region, the ADAM project assessed different agro-technical adaptation measures for field crops (for more details see Chapter 8, Mechler *et al.*). The results listed in Table 9.2 show that irrigation and changing crop variety for a shorter growth cycle are the most promising of the assessed measures. This concurs with opinions held by stakeholders. At the same time, Table 9.2 suggests that the effectiveness of each adaptation measure depends strongly on crop type and that some measures could even be counter-productive with respect to crop yield. Together with the potential pressure of irrigation on the already strained water resources, this highlights the potential for mal-adaptation.

Table 9.2. *Effectiveness of selected field crop adaptation measures*

Adaptation measure	Winter wheat	Spring wheat	Sunflower	Soybeans	Corn
Shift in sowing date (±15 days relative to 'no adaptation')	😐	☹	☹☹	☹	☺☺
Irrigation (of those crops that are non-irrigated under 'no adaptation')	☺	☺	☺☺	☹	*
Changed growth cycle (±20% for each growth phase duration relative to 'no adaptation')					
Shorter growth cycle (-20%)	☹	☹	☺☺	😐	☺☺
Longer growth cycle (+20%)	😐	☹☹	☹☹	☹☹	😐

☺☺: 10%–25% gain of crop yield; ☺: around 5% yield gain; 😐: no gain or loss; ☹: around 5% yield loss; ☹☹: 10%–25% yield loss relative to 'no adaptation'.
* Not included as adaptation strategy as corn is always irrigated in simulations

Guadiana Basin: enhancing adaptation – constraints and opportunities

A limiting factor in the analysis of adaptation in land use and water management is the sectoral and disciplinary perspective with which adaptation measures are often assessed. Mainstreaming requires an understanding of economic aspects, such as market trends, cost of inputs and new incentives for agriculture, as well as institutional and plant specific physiological aspects. These different dimensions of the adaptation challenge have yet to be addressed comprehensively. More complex production systems like dehesa and diverse cropping patterns offer opportunities to reduce climate risks (Werners *et al.*, 2007).

To implement adaptation practices, stakeholders underlined the importance of cooperation between administrative agencies of both countries to ensure integrated socio-ecological planning in the Guadiana River Basin (ADAM Project, 2007b). According to Cots *et al.* (2007) a promising administrative vehicle for mainstreaming climate concern into the already existing regulatory transboundary co-operation is the 'Convention on Cooperation for the Protection and Sustainable Use of the Portuguese–Spanish River Basins' (1998) and its Commission for the Convention Development and Application (CDAC). The Convention defines the regulatory framework for co-operation between the two countries to protect inland waters (surface waters and groundwater) and their dependent ecosystems and to use the shared basins waters sustainably. The experience from the Guadiana study region suggests that a transboundary organisation like the CDAC can break down regional or national divisions, while at the same time foster new co-operation and collaboration because: (i) it works on the scale of the transboundary river basin, (ii) it facilitates participation by providing a space for exchange between actors on both sides of the border, and (iii) it intervenes in a broad set of activities and regulatory areas. However, three main obstacles remain: (i) explicit climate change considerations are absent in the Convention, (ii) its implementation lacks transparency and participation, and (iii) both the Convention itself and the work of the Commission for the Implementation and Development of the Convention are almost unknown to the general public and even to the local administrative bodies in the Guadiana basin (Timmerman and Doze, 2005).

An important opportunity for adaptation in the Guadiana basin is the harmonisation and implementation of the European regulatory framework. The implementation of the Water Framework Directive (Commission of the European Communities, 2000) and the new EU regulatory processes in the field of river basin and water resources such as the EU Floods Directive, the White Paper on Adaptation and the EC communication on Water Scarcity and Drought, provide added stimulus for new transboundary governance structures. Regions have gradually developed a more important role both in the European and the national arena (Jones and Keating,

Table 9.3. *Constraints and opportunities for climate adaptation in the Guadiana River Basin*

	Constraints	Opportunities
Biophysical	Water resources (especially groundwater) already over exploited	Revalorisation of landscape diversity and dehesa ecosystem services
Technical	Highly controlled river flow	Water saving technologies and crop modification
Financial	Large sunk costs of existing water infrastructure. Subsidies indirectly increase climate risks (e.g. subsidised reforestation with pine trees increases forest fires)	European funds encouraging cross-border regional co-operation and economic diversification, e.g. rural tourism to make farmer's income less climate dependent
Institutional	Existing procedures for water allocation are contested. Adaptation policies largely unknown to general public, with limited participation in implementation	Implementation of European guidelines and directives are a driver for transnational networks of different national and sub-national agencies
Social	Pressure on available water resources increases conflicts between land and water users	Tradition of living in drought prone area. Tourism market for local products and customs
Cognitive and informational	Climate change not main concern. Suitability of traditional agro-environmental systems under climate change contested. Limited access to information on climate scenarios and adaptation options	Regional institutes start to embrace climate change and adaptation as a topic and provide contextualised climate information. European policy, directives and research encourage discourse on climate, land use and water management

1995). Today, regions interact directly with EU bodies to obtain resources from the distributive policies and further their interests. As a transboundary region, the Guadiana River Basin is eligible for the European Community Initiative INTERREG and the Structural Funds that promote transnational networks and the definition of common interests and co-operation (Tábara *et al.*, 2009). This said, the review of the Spanish National Adaptation Plan in 2008 revealed that adaptation policies were still underdeveloped, particularly at river basin scale, while regional policy makers argue that a more harmonised EU policy framework is needed to take decisive adaptive measures in the various potentially affected sectors.

Changes in policies and institutional designs are not only motivated by European policies. Evidence from the study region also suggests greater 'bottom-up' influence as indicated by a gradual increase in the involvement of local actors who were typically excluded from the policy making process. Programmes and policies that deal with adaptation to climate change are emerging at the regional and national level. The inclusion of other relevant policies, actors and governmental levels can strengthen these programmes and help with the co-ordination of objectives and policies aimed at long-term sustainability. The wide range of policies and actors that directly or indirectly influence adaptation in the Guadiana River Basin indicate the importance of building capacity among local, regional and national actors, both public and private, for mainstreaming.

Table 9.3 summarises constraints and opportunities for climate adaptation that were identified in the Guadiana River Basin study region.

9.2.2. Inner Mongolia Alxa region

Alxa region: climate change and impacts on land use and water management

The Alxa region in Inner Mongolia covers more than $270\,000\,\text{km}^2$ of Northern China and is characterised by a continental arid to semi-arid temperate desert to steppe ecosystem. The average temperature is $7\,^\circ\text{C}$ with extremely cold winters and very hot summers. Precipitation ranges from 50 mm in the north-west to 200 mm in the south-east with high spatial and interannual variability. Hence water availability is a critical environmental factor. Temperature has risen since the 1950s, and regional climate models project a further rise ranging from $3\,^\circ\text{C}$ to $5\,^\circ\text{C}$ above pre-industrial values by 2100, depending on the scenario (Xu *et al.*, 2006). As a result, surface evapotranspiration will increase, placing extra pressure on water availability. Projected precipitation changes are more uncertain and differ largely between climate models as rainfall is scarce and small absolute changes may correspond with large relative changes (Xu *et al.*, 2005) (Figure 9.3). In addition, some models suggest an increase in precipitation and a cooling trend for some areas (ADAM Project, 2008) adding to the uncertainty with which adaptation policy has to cope.

Traditionally, the Mongol lifestyle was one of nomadic pastoralism (Dickinson and Webber, 2007). The immigration of Han Chinese and Mongolian farmers brought farming practice from outside Inner Mongolia to the region. Since then, many herders have given up their nomadic lifestyles and established permanent dwellings, as recommended by government policy in view of desertification and population growth. This means that pastures that were previously grazed in rotation, are now used intensively all year round (Brogaard and Zhao, 2002).

Nowadays, most of Alxa's nearly 200 000 inhabitants live in small towns in close proximity to the hundreds of desert oases. Water availability is low ($110\,\text{m}^3/$ capita year, five per cent of the national average). The main land use is irrigated agriculture and extensive rangeland (see Figure 9.4). Currently, rangeland covers

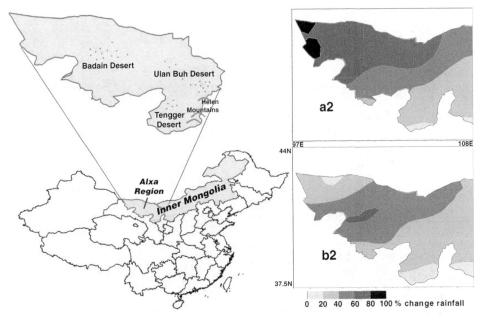

Figure 9.3. Alxa region in Inner Mongolia, China and projected change in average precipitation between current and SRES scenario A2 and B2 (2071–2100) using regional climate model PRECIS (after Xu *et al.*, 2005). (See colour plate section.)

one third of Alxa with an official livestock population of 2 200 000, managed by 30 000 herdsmen. Use of lands at the margin of deserts, conversion of rangeland, increasing animal numbers in the 1980s and 1990s, and changing herd composition[5] are officially considered to be the key human driving forces of recent desertification (Dickinson and Webber, 2007). More than 65 per cent of the rangeland in Alxa is currently classified as degraded, resulting in an estimated productivity reduction of 25 per cent in the past 20 years. The reasons interviewees gave for desertification included both climatic variation (reduced precipitation and warmer winters) and increasing livestock numbers due to general government reforms implemented after 1978 that encouraged private livestock ownership (*cf.* Brogaard and Seaquist, 2005; West, 2009). Over 80 per cent of the farmers and herdsmen interviewed in the Alxa region have noticed a decline in water quantity and quality as well as an increase of severe droughts and heatwaves. Interestingly, only a few of those interviewed associated these changes with global climate change (Dai *et al.*, 2008).

---

[5] Primarily due to the economic value of cashmere, the composition of herds has shifted from a mix of horses, camels, cattle, sheep and goats, to overwhelmingly sheep and goats. While having lower water requirements, by exposing or pulling out roots they are also more destructive to the grassland.

Figure 9.4. Alxa region in Inner Mongolia: (a) landscape of sand dunes and steppe desert along edge of desert; (b) enclosed herd at house; (c) oasis agriculture with shelterbelt woodland; (d) local style greenhouses developed by governments for migrants. (Photos: (a),(b): Werners, May 2007; (c),(d): Jia, May 2008.)

Desertification has connected three formerly distinct deserts (Badain Jaran, Tengger and Ulan Buh), resulting in more intense sandstorm events[6] and the abandonment of approximately half the human settlements in the affected region (Du and Huang, 2005). The sandstorms cause serious economic loss and ensure that impacts are not confined to the region but also adversely affect other areas (including countries as far away as Japan and South Korea), giving the problem an international dimension (ADB, 2005).

Alxa region: assessment of adaptation practices

Although the key drivers of the desertification problem are disputed, it is acknowledged that climate change can amplify existing problems (Reynolds *et al.*, 2007). Analysis of measures and policies to deal with desertification and environmental degradation can therefore provide useful insights into the constraints and opportunities for adaptation practice in China. While the intensification of

---

[6] Trends in sandstorm frequency are contested. Over the last two decades strength and frequency rose. Since 2006 the strength intensified further, but the frequency declined.

sandstorms caused concerns in the region, it also raised government awareness about desertification (Kar and Takeuchi, 2004). Policies aimed at rangeland rehabilitation have sought to address the biophysical, as well as the social and economic aspects of desertification. Examples of land use policies are grassland enclosure, grazing bans, promotion of land conversion on sloped lands (the so-called 'Grain for Green' policy (Du, 2006)), logging bans, and tree planting. Socio-economic policies include general poverty alleviation strategies (e.g. the 'Eight-Seven' Poverty Alleviation Reinforcement Plan (1994–2000)[7]) and herdsmen resettlement programs, in China referred to as ecological migration[8]. Ecological migration in Alxa has resulted in the migration of 6624 households (approximately 25 000 peasants and herders) in five years (Du and Huang, 2005). The Alxa development plan aims to reduce by 2010 the population living on extensive livestock grazing by two-thirds. It is intended that this goal will mainly be achieved by providing job opportunities in the coal mining and chemical industry sectors that have developed rapidly in north-east Alxa in recent years under the Grand Development in West China programme (Yan and Quian, 2004). Since the introduction of rangeland rehabilitation in 2001 about 20 per cent of the rangeland has been closed throughout the year and another 30 per cent has been put under rotational or seasonal grazing. In the areas under protection there is evidence of vegetation recovery (Dai *et al.*, 2008).

Recent climate policy developments are the recognition of climate change adaptation and mitigation as expressed by the central government's 'Eleventh Five-Year Plan' (NDRC, 2006) and the launch of a national climate change programme in 2007. Although the Alxa development plan makes no reference to changes in temperature, precipitation or severity of extreme weather events (Dai *et al.*, 2008), the regional government has responded to climate change by providing more specified weather forecasts[9] and by subsidising agricultural insurance in two pilot programmes.

At the farm or household level, farmers apply different water-saving techniques and many have experimented with more drought-resistant crops. Other adaptation options that are considered include the introduction of greenhouses, more wide-spread use of drip-irrigation and no-till cropping. The integration of fruit trees with annual crops in oasis settings in order to reduce denudation, raise water-use and

---

[7] In 1993, the Chinese central government launched the 'Eight-Seven' Poverty Alleviation Reinforcement Plan, which aimed to solve the poverty problem of 80 million residents in the whole country within seven years from 1993 to 2000.

[8] Ecological migration (*shengtai yimin* in pinyin Chinese) is a government policy to re-settle herders and farmers living in ecologically fragile and degraded regions to new build or existing communities. Ecological migration in China began in the 1980s in the Southern Ningxia Hui region and continues today in Inner Mongolia and other Western provinces (Du, 2006).

[9] In 2006 Alxa's meteorological bureau initiated a cell phone warning service for three Yuan/month (0.30 euro/month). Text messages are sent ahead of torrential rain, strong wind, sandstorms and heatwaves. It has been effective so far because most farmers and herdsmen in Alxa use a cell phone (Dai *et al.*, 2008).

nutrient efficiency, and support economic diversification, is also considered. The climatic and other challenges associated with farming have led some families to abandon agriculture to find another living in industry or the larger urban centres as a type of livelihood adaptation. With few exceptions, migrants see the policy of ecological migration as a successful adaptation to what they themselves described as a situation in which vegetation and grasslands had degraded to a point where livelihoods could no longer be sustained (West, 2009).

Alxa region: enhancing adaptation – constraints and opportunities

The future vulnerability and adaptation of the region will not only depend on the direction and magnitude of expected climate change, but also on the policies to deal with socio-economic and environmental change in the region. The rangeland rehabilitation policies in Inner Mongolia, in particular, have the potential to create new opportunities as well as constraints. One key challenge is to create balanced and flexible policies that can, if necessary, be altered after evaluation of longer-term effects. An example of this challenge is the widely implemented grazing ban. With the recovery of the rangelands, the accumulation of dry material adds to bush fire risk. This suggests that a certain amount of grazing might be necessary to keep dry material levels low while maintaining sufficient plant cover. Moreover, while the grazing ban has reduced human induced pressure on the rangelands, it has indirectly increased pressure on the scarce water resources as migrants have been moved to villages located in or beside the few oases remaining in the Alxa region. The continuing conversion of oases into settlements and farmland will have major long-term consequences for the regional climate and hydrology, now regulated by oasis ecosystems. Modification of the grazing ban could allow some herdsmen to return to their trade, combining use and protection of grassland.

If planned carefully, ecological migration will continue to provide alternative livelihoods for the rural poor as well as release direct pressure on marginal land. While many interviewed migrants express the view that their lives have improved in key areas such as access to healthcare, goods and services, and schooling for their children, challenges remain to ensure that the move is sustainable. These include ensuring adequate and affordable access to clean forms of electricity, providing water for irrigation, drinking and household use, promotion of new cultivation techniques and training to facilitate sustainable transitions from herding to farming. Another challenge is the equity and long-term financial sustainability of existing compensation programmes for ecological migrants. While continuation of compensation programmes for rehabilitating grasslands is desirable from the perspective of the beneficiaries – individual herdsmen and farmers – it would arguably require more central government funds than politically feasible (China Council Task Force on Forest and Grasslands, 2001). Beneficiaries will reconsider their activities when the compensation programme ends.

Table 9.4. *Constraints and opportunities for climate adaptation in the Alxa region, Inner Mongolia*

	Constraints	Opportunities
Biophysical	Large inter-decadal climate variability. Precarious balance between grassland deterioration, recovery and adding to fire risk	Sandstorms trigger national and local attention. Grassland ecosystems are recovering, stabilising the soil and reducing erosion in some areas
Technical	Technical solutions tend to be large-scale and unequal implementation is contested at a local level	Improve irrigation efficiency. Implement sustainable local electricity solutions
Financial	Current government compensation programmes may be financially unsustainable in the long term	Economic growth provides financial resources to invest in adaptation policies. International willingness to invest in climate policy and research
Institutional	National climate change concerns not reflected in local government policy documents. Inflexible policies and institutions	Stakeholder workshops and forums could help to overcome some of the gaps and engage local decision-makers in formulating and implementing climate policy
Social	Social sustainability of government policies aimed at rangeland rehabilitation	Government policies combine elements of poverty alleviation and economic development, with scope for addressing adaptation
Cognitive and informational	Little information on climate impacts, the relation to land degradation and policy development is currently available for use by local officials or citizens	People are eager to get more information. Promotion of climate and environmental consciousness at the local level through increasing number of bilateral projects

Illicit economic activities, like the extraction of medicinal plants for local and regional markets, have added to landscape degradation. However, cultivation of high-value medicinal plants also creates opportunities for alternative livelihoods and is seen as a promising measure to raise the resilience of the local population and reduce pressure on vulnerable ecosystems. Future policies therefore need to build on successful experiments, such as public–private partnerships, in which marketable products obtain additional public funding in exchange for providing environmental services and maintaining social and environmental standards (China Council Task Force on Forest and Grasslands, 2001).

Perhaps the most pervasive constraint to successful adaptation is related to local people's perceptions of the government and its policies. While government policies are important and can provide stimulus and support for successful adaptation at the local level, local people in Inner Mongolia are often ill-informed of the goal of grassland policies that have been conceived, implemented and scaled up rapidly (Runhong, 2001). In addition, there was very little awareness among those interviewed about what they themselves either as individuals or communities could do to adapt to long-term climate and environmental change (West, 2009). This highlights the importance of including the perspective of local people in policies that affect them, and of building local adaptive capacity through environmental education and training.

Table 9.4 summarises constraints and opportunities for climate adaptation that were identified in the Inner Mongolia study region.

9.2.3. *Tisza River Basin*

Tisza Basin: climate change and impacts on land use and water management

The Tisza River is the largest tributary of the Danube, receiving water from the Carpathian Mountains in Romania, Slovakia and Ukraine (see Figure 9.5). Climate change projections suggest more irregular rainfall and a warmer climate in the Carpathian basin (Láng, 2006; Bartholy *et al.*, 2007), aggravating the three main water-related problems of the Tisza region: floods, inland water stagnation and droughts (Barta *et al.*, 2000; ADAM Project, 2007a). Between 1998 and 2006 there has been at least one severe flood each year of a magnitude that was previously associated with the 100-year flood (Timár and Rácz, 2002). Paradoxically, the plains between Danube and Tisza are especially drought prone and Hungarian agriculture and forestry has suffered from extensive droughts in successive years.

Until the eighteenth century, socio-economic activities in the Tisza valley were predominantly organised around the operation of a system of creeks and channels regulating the water flow between the main river bed and the floodplain (Balogh, 2002). Inundation frequency determined land use. The Tisza floodplain provided a secure income for communities along the river, with a floodplain production system that combined plough land, forest, floodplain orchards, meadows, fish and cattle (Andrásfalvy, 1973; Bellon, 2004). Since then the Tisza River has been heavily modified to cater for large-scale mono-agriculture and river transport. Dykes were built, one-third of the river length was regulated and the floodplain was drained, decreasing the total naturally flooded area by over 80 per cent. These changes put an end to the traditional water management system and the related production systems (Bellon, 2004).

At present, the main land uses are intensive agriculture, wetlands and meadows (Hungarian *puszta*). Vulnerability of rural production systems has increased through

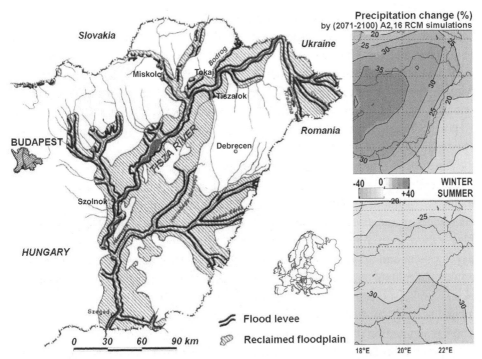

Figure 9.5. Hungarian Tisza River Basin and projected change in average precipitation by 2071–2100 in SRES Scenario A2 using 16 regional climate models (after Szlávik and Ijjas, 2003; Bartholy *et al.*, 2007). (See colour plate section.)

the loss of local flood- and drought-resistant crop varieties and processing techniques (Sendzimir *et al.*, 2008). Flood damages are expected to increase under climate change (Koncsos and Balogh, 2007, see also Chapter 8, Mechler *et al.*). Socioeconomic problems that add to vulnerability include a high unemployment rate, an ageing population, migration and minority issues (Sendzimir *et al.*, 2004; Linnerooth-Bayer *et al.*, 2006). More promisingly, the region has a large undeveloped potential for recreation and nature conservation (Vári *et al.*, 2003).

Tisza Basin: assessment of adaptation practices

Key drivers of the adaptation discussion in Hungary are climate-related extreme events, especially recurring floods and droughts, in combination with national research and institutional change. Three major vehicles for mainstreaming adaptation into land use and water management are: (i) the national research project VAHAVA, which led to the National Climate Change Strategy; (ii) the new water management plan for the Hungarian Tisza River; and (iii) the 'SZÖVET' Living Tisza Association with its micro-grant scheme and the promotion of regional markets and products.

The first of these, a research programme titled VAHAVA (VAltozas-HAtas-VAlaszadas: change–impact–response) was launched by the Ministry of Environment and Water and the Hungarian Academy of Sciences in 2003 (Láng, 2006). The VAHAVA programme aimed to support the national climate policy as well as introduce climate change issues to the public. To this end, the programme had an explicit outreach strategy. A major outcome of the programme was the development of the National Climate Change Strategy approved by the Hungarian parliament in 2008 (www.klima.kvvm.hu). The follow-up project KLIMAKKT pays more attention to the assessment of adaptation options, especially for agriculture (Erdélyi, 2008).

Secondly, recurring flooding events and the recent predictions of an increased incidence of floods and droughts (Láng, 2006) have been a driving force behind a new water management plan for the Hungarian Tisza River: the New Vásárhelyi Plan (www.kvvm.hu/vizeink). Stakeholders value the plan for its participatory development and approach to integrated land and water management. After initial criticism, regional and national actors collaboratively prepared the implementation plan, with key objectives being to establish improved socio-economic and environmental security through land use and water management (Werners *et al.*, 2009a). The key innovation of the plan is the combination of water retention and floodplain rehabilitation as measures to replace or complement prevailing engineering approaches, favouring flood levee construction (Figure 9.6b and c). Albeit behind schedule, the first polders are under construction. The late realisation of the polder scheme exposes issues relating to the organisation of water management, benefit transfer, financing and land use change continue to be problematic. In the Bodrogköz area, where the first retention reservoir of the new water plan has been built (Figure 9.6d), stakeholders debate the need for a flexible institutional set-up to manage both extreme floods and annual shallow flooding of the landscape.

Finally, the participation of regional stakeholders in the implementation of the water plan and national climate policies is supported through the UNDP-GEF supported 'Living Tisza Alliance' project[10] that aims to harmonise local initiatives in flood plain revitalisation and sustainable rural development. The alliance stimulated the debate on adaptation measures in the region. Its 'Tisza völgy' magazine has facilitated outreach and social involvement, in which local stakeholders emphasised the need to link climate change adaptation, the new water plan and agricultural subsidies in the region. It is intended that raising the income from local floodplain production systems will help to reduce vulnerability of local communities. In 2007 a

---

[10] The Living Tisza Alliance (SZÖVET) was established in 2006 as part of the 'Conservation and Restoration of the Globally Significant Biodiversity of the Tisza River Floodplain Through Integrated Floodplain Management' project (2005–2008) supported by the Global Environment Facility through the United Nations Development Programme (UNDP/GEF). The Alliance is open to private persons, NGOs, municipalities, farmers and researchers (www.elotisza.hu).

Figure 9.6. Hungarian Tisza River Basin: (a) Tisza River at Tiszadada; (b) oxbow lake: traditional water management used oxbows and creeks for water regulation. This inspired the new water management plan; (c) bicycle lane on retention reservoir dyke for rural development; and (d) construction of the first retention reservoir. (Photos: Werners, 2007.)

micro-grant scheme for climate-proofing was introduced. Granted projects include facilities for village tourism that aim to reduce climate dependency in agriculture and increase income security (Kajner *et al.*, 2008). Another initiative of the Living Tisza Alliance together with other non-governmental organisations (NGOs) is the marketing of regional products through the introduction of the 'Living Tisza' eco-label and online retail that can help producers to change production as a response to climate change.

Tisza Basin: enhancing adaptation – constraints and opportunities

While the VAHAVA project initiated a national debate on climate adaptation, many issues remain unresolved. These include the harmonisation with other policies, a coherent and financially robust adaptation programme and the clarification of the role of different actors and agencies. Regionally, public local action groups and NGO initiatives (like the Living Tisza Alliance) can carry the adaptation debate forward.

The implementation of the new water management plan holds many lessons for mainstreaming adaptation. At present, water infrastructure is in the planning stage, with the first retention area constructed. However, measures that focus on social adaptation, like rural development and support of floodplain production systems (e.g. extensive fishing, utilising biomass of water related areas, floodplain orchards), are subject to significant delays. The causes for the delays were discussed in a series of interviews and workshops (Werners *et al.*, 2009b). Direct obstacles to implementation were said to be the required land acquisition and authorisation for the measures. The fragmentation of responsibilities and the lack of strong management were considered a barrier, resulting in a comparatively low ranking of floodplain revitalisation and rural development measures in national development priorities. Institutional complexities were also cited, with EU accession and the accelerated implementation of European directives and legislation heightening the complexity of organisational responsibilities and financial flows. For example, EU accession offers access to new funds, as well as demanding the Hungarian government to reduce the budget deficit. At the same time, EU standards and practices for agriculture and water management offer an opportunity to make existing governance and planning practices more effective, participatory and sustainable (Commission of the European Communities, 2001).

Actors in the Tisza region are keen for a clarification of the roles of different parties involved in the new water management plan and expressed a strong interest in establishing a multi-stakeholder implementing agency with national government and other stakeholder representatives. In addition to formal co-ordination, actors identified options to strengthen informal relations and cooperation between non-governmental agencies (Matczak *et al.*, 2008). This includes awareness raising about the impacts of climate change and associated flood risks, amongst others. Actors also stressed the importance of local and regional markets and private sector initiatives, for example, by encouraging local investment in flood protection and floodplain rehabilitation, and by collaborating with actors in downstream urban areas to support the development of water retention.

Given the shortage of financial resources, actors suggested integrating adaptation, the new water plan and the Hungarian regional development programmes. The costs and benefits of adaptation and floodplain management have to be shared between many parties at different scales. A re-evaluation of resource allocation is suggested, including: (i) subsidies for sustainable agriculture and land use management, (ii) subsidies for renewable energy, levelling the price difference between renewable energy and average energy price, (iii) agro-environmental schemes and removing damaging current subsidies like compensation schemes for farmers in areas at risk of inundation, (iv) property rights, and (v) regulation of the use of EU funds in several sectors of the Hungarian economy (e.g. Common Agricultural Policy Support Schemes, EU cohesion funds and Natura 2000). Insurance is only developing at a slow rate as actors expect state support in emergency situations (Vári *et al.*, 2003).

Table 9.5. *Constraints and opportunities for climate adaptation in the Tisza River Basin*

	Constraints	Opportunities
Biophysical	Floods are infrequent and alternate with droughts, shifting policy attention	Using floodplains for water retention restores ecosystem services. Extreme events trigger awareness
Technical	Old infrastructure	Water retention
Financial	Severe budget constraints. High dependency on European funds. Low priority for sectoral ministries	European and international funds support regional coalition. Micro-grant scheme. Re-evaluation of resource allocation
Institutional	Harmonization with other policies poor. Fragmented responsibilities. Property rights rigid and ill-defined	Regional coalition negotiating regional solutions and policy change. Adaptation can be integrated in ongoing discourse on water policy
Social	Allocation of retention areas problematic. Dominant land use not conducive to water retention	Promotion of more climate proof local products. Learning from traditional land use and water management
Cognitive and informational	Adaptation options are contested. Limited access to socio-economic and climate data	Recognition of the limits of prevailing flood levee dominated water management. Regional coalition debating climate impacts and adaptation options

Socio-economic constraints for adaptation are the economic vulnerability of local production systems and social problems such as poverty, migration and segregation of minorities (Roma minority). The level of education is low, both in general and in vocational training, contributing to high unemployment rate and high social vulnerability (*cf.* Fekete, 2006). The initiatives of the Tisza Alliance show that the informal networks and local production systems have degraded, but can recover. Small-scale bio-energy schemes and regional markets are promising new developments that will have to prove themselves sustainable in the future. Traditional land use systems hold lessons for the future, or as one actor put it 'stay where the old village used to be'.

Table 9.5 summarises constraints and opportunities for climate adaptation that were identified in the Tisza region study region.

9.3. Discussion

In its assessment of adaptation practices, Chapter 17 of the IPCC's Fourth Assessment Report concluded that: (i) adaptation to climate change is already taking place, but on a limited basis, (ii) adaptation measures are seldom undertaken in response to climate change alone, (iii) adaptation and adaptive capacity are uneven across and within societies, and (iv) there are substantial limits and barriers to adaptation (Adger *et al.*, 2007). The results from our three study regions support these insights. In addition, we observe opportunities for mainstreaming adaptation in land use and water management. Below, we elaborate on the commonalities and contrasts in the three regions to learn about climate change impacts, adaptation practice, and constraints and opportunities for mainstreaming adaptation.

Climate change and impacts on land use and water management

In all three regions, local populations are already experiencing the impacts of a changing climate. The role of context-based science is being recognised increasingly and supported by national and regional institutes (cf. Weaver *et al.*, 2006). In particular, new climate projections are being made available, contributing to regional and political awareness of potential climate risks. Yet, improving projections, especially of precipitation and climate-related risks, remains an important challenge. These projections depend critically on the scale and the resolution of the data used, as well as on integration of social and political aspects in the evaluation of potential adaptation strategies. Furthermore, there is a need for integration of agro-ecological and economic data into regional climate risk assessments.

Traditional agro-environmental land use systems in all three regions reflect the way local populations adapted to the region's climate variability. These traditional systems have subsided under competition from the global economy and changing institutional contexts. New global market production systems tend to be less adapted to regional climate change and variability and often respond less efficiently to local demands for quality and diversity of services provided. Thus climate impacts cannot be dealt with separately from the increase of scale of economic human activities. Together, these aggravate existing challenges for sustainable land and water resources use at the local and regional level.

Assessment of adaptation practices

Adaptation to regional climate variability has always taken place. At present proactive adaptation actions are planned for future climate change, yet on a limited basis. Adaptation is mostly in the planning stage or implemented through pilot projects. There is no clear connection between regional climate impact studies and adaptation planning. Adaptation planning typically accounts for more general climate trends and scenarios, partly because detailed assessments of climate impacts have only recently become available. Although climate change has encouraged dialogue between different actors and policy communities (e.g. water and agriculture), actual adaptation planning and implementation remain largely sectoral. So far, the impact of adaptation mainstreaming on the integration of non-climate policies is limited in our study regions. With respect to adaptation planning there is a gap between ambitious policy goals and policy implementation. Whereas adaptation policy goals often refer to transitions to adaptive systems, the instruments selected in the European study regions focus on gradual change and well-established existing practices (dykes, dams, irrigation, risk management, setting targets). In the Inner Mongolia region, land-use policies have been implemented and scaled up rapidly (Runhong, 2001), leading to new adaptation challenges for policy makers and local people.

Enhancing adaptation: constraints and opportunities

While there are substantial constraints to adaptation, there are also opportunities emerging. We use six dimensions to discuss these opportunities and constraints: biophysical, technical, financial, institutional, social, and cognitive and informational aspects.

Biophysical aspects: In all three regions, ecosystems have degraded and water resources are over-exploited. Traditional landscape and resource use practices, such as the dehesa in Spain and Portugal or the traditional floodplain production systems in the Tisza, seem to have been better prepared to respond to changes in the climate and had an active role in regulating climate extremes. This regulating service has motivated local populations, scientists and policy makers to explore the traditional agro-ecological production systems. Our research in the Tisza and Guadiana river basins shows that preserving and managing diversification of land uses has a great potential for reducing climate related risks. In the Tisza region diversification of land-use in relation to the hydrological conditions is explicitly supported by the current water management plan.

Technical aspects: Existing technical solutions, like building dykes, run into limits or add to undesirable or longer-term effects. Pilot projects and demonstration activities have started to test the feasibility of new technologies for sustainable land use and the development of natural resources. There is scope for the development and exchange of more sustainable technologies and information systems, including early warning systems (e.g. the cell phone-based warning service in Inner Mongolia). Models currently available are not parameterised for assessing new technologies and more complex and innovative adaptation strategies, creating a barrier for the appraisal of mainstreaming.

Financial aspects: Financial resources are limited in each of the study regions and adaptation is often considered too costly and uncertain compared to expected benefits. Whereas there is a pressure on existing financial services (e.g. insurance) to become more expensive, new financial instruments are also emerging (like micro-grants, see also Chapter 13, Bayer *et al.*). The implementation of adaptation strategies is constrained by unequal distribution of costs and benefits. For instance, measures taken to reduce land degradation and sand storms may conflict with rural livelihoods and be financially unsustainable, and water retention increases the risks for those who store the water for the benefits of others. At a smaller scale, the difference between those who are, and who are not, included in adaptation pilots or support programmes can increase tensions. The perception of fair sharing of costs and benefits between actors is central to the successful implementation of adaptation and has to be addressed in adaptation planning. In all study regions, European and/or national government financial support is sought for the implementation. However, mainstreaming adaptation complicates existing relations with donors or subsidies. The European agro-environmental schemes, for instance, are not designed for inter-annual land use change depending on water availability. Creating markets for adaptation is a key challenge for the Tisza region (e.g. encouraging cities and industries to buy in on upstream flood water storage and floodplain management) as well as for national and international adaptation to climate change. All three regions identified opportunities for public–private partnerships in which marketable products obtain additional public support in exchange for providing social and environmental services that support adaptation.

Institutional aspects: Divided and unclear responsibilities are key constraints for adaptation actions in the Guadiana and the Tisza river basins. By contrast, in Inner Mongolia, the rigidity of the strictly defined roles of different organisations is a constraint, as is the limited communication of intended policy goals to beneficiaries. In all three regions we saw a lack of co-ordination between agencies and tensions between actors at different scales. Mainstreaming adaptation and embedding it in existing national policy and institutional frameworks allows for addressing trade-offs and synergies that are crucial for 'selling' adaptation. Yet it complicates the implementation of the original policy and diffuses the responsibility for implementing the adaptation agenda. There is a clear call on central governments to delineate and communicate the roles and responsibilities for implementation of adaptation strategies at national, regional and local levels. To achieve more adaptive governance structures capable of dealing with new risks and uncertainties, different scales of governance need to work together to make policies, plans and programmes more coherent. In this regard, there is emerging evidence of new coalitions of government and non-government actors in the Tisza and Guadiana river basins that are helping to put the adaptation agenda in a regional context and encouraging action in the region. These coalitions often have close connections to academics who act as brokers for climate risk and adaptation information.

Social aspects: Adaptation can fail or be counterproductive because social processes and structures are understood imperfectly. Some adaptation options have consequences that are socially unacceptable. In the Tisza Basin, for example, sites for water retention were rejected. In the Alxa region, the enclosure of livestock conflicts with traditional lifestyles. The Tisza study region shows that informal social networks around local production systems have degraded, but are remediable. Local populations hold a wealth of knowledge on how to cope with climate variability, which deserves to be taken into account, while developing new policies and measures.

Cognitive and informational aspects: In the Alxa and Guadiana regions, in particular, people struggle to connect regional trends to global climate change. The causes of trends in desertification and reduced water availability are heavily contested. Adaptation policy so far does not address the diverse perceptions of risks and their causes. The Tisza region shows benefits of debating climate-related risks and how best to respond; after various discussions on adaptation options, actors were quick to take advantage of a micro-grant scheme for implementing local solutions. This supports the notion of adaptation as a social learning process (see also Chapter 2, Russel *et al.* and Chapter 5, Hinkel *et al.*). Another lesson is to implement cost-effective and flexible adaptation frameworks, which can be modified as personal and scientific understanding changes. All three regions suffer from a lack of (access to) information about new adaptation options and policies. Knowledge integration can take place through 'issue-linking' in debates (e.g. in the case of linking climate change to desertification). Newly emerging forums for debating adaptation strategies may prove to be valuable in this regard. At the regional level these are often associated with internationally funded projects. An opportunity for new planning processes is to focus on the transfer of knowledge relevant for adaptation decisions including early warning systems.

Yet, a gap remains between scientific adaptation theory and adaptation practice on the ground. There is a mismatch between model assessments of impacts and adaptation on one hand and 'real' adaptation options as discussed by people in the region or in the policy plans on the other. For example, existing models in the Tisza and Guadiana basins do not include resource conflicts resulting from multi-stressors, or win–win opportunities from the integration of adaptation and longer-term sustainable land and water use planning. As such, there is an observable tension between the information demands of the implementation process and the support that scientific modelling frameworks can offer (*cf.* Vogel *et al.*, 2007).

9.4. Conclusions and recommendations

This chapter examines constraints and opportunities for mainstreaming adaptation to climate change in land use and water management. Whereas the literature on adaptation is relatively rich in detail on impacts, vulnerability and constraints to climate adaptation, less is known about the conditions that facilitate adaptation in practice. Yet, opportunities for using land use and water management planning to support adaptation and climate-proof regional development have started to emerge. In all three regions we identified institutional and cognitive aspects as crucial for successfully implementing adaptation. The degree of success critically depends on the capacity to participate in adaptation and the distribution of that capacity across different actors and governance levels. This includes, for example, access to information, governance and financial services. In all three regions, lessons can be learned from integrating traditional agro-environmental land use systems with new technologies and institutional designs, for example, to preserve diversity in landscape, ecosystem services and land uses. The study regions suggest that it is important to balance formal regulatory rules and informal social factors in planning and implementation. Informal networks are crucial for social learning and adaptive capacity and may be particularly useful in times of crisis. At the same time, formal rules are required to include adaptation in longer-term planning, investment and financial support of experimentation and adaptation.

The six dimensions of adaptation discussed above capture the constraints and opportunities for mainstreaming adaptation in land use and water management. While the relative importance of each aspect is location specific and will vary over time, our research concludes that consideration of *all* aspects is needed to capitalise on opportunities for successful adaptation planning and implementation. Our study regions therefore suggest that, next to adaptation mainstreaming, there is a need for a comprehensive and systemic adaptation agenda, reaching across sectoral non-climate policies and programmes.

Acknowledgements

We thank all interviewees and participants of the regional workshops for sharing their experience on adaptation practice. We are grateful for the valuable comments and help of research partners, especially Helene Amundsen, Marco Bindi, Maria Falaleeva, C. Fu, Zsolt Harnos, Alex Haxeltine, Levente Horváth, Mike Hulme, Francesca Incerti, Gensuo Jia, Márton Jolánkai, Marco Moriondo, István Lang, Rik Leemans, Hongtao Liu, Lorenzo Orioli, Claudia Pahl-Wostl, Anna Serra, Yanling Sun, Rob Swart, Y. Xu and L You.

References

ADAM Project (2007a) *Scoping Document P3d Tisza Case Study, Deliverable D-P3d.2.* ADAM Project (EU Project no. 018476-GOCE), Project Report www.adamproject. info, Norwich, UK.

ADAM Project (2007b) *Scoping Stakeholder Workshop and Interviews for Guadiana Region. Background report for: Scoping Document P3d Guadiana Case Study,*

Deliverable D-P3d.2. ADAM Project (EU Project no. 018476-GOCE), Project Report www.adamproject.info, Norwich, UK.

ADAM Project (2008) *Scoping Document P3d Inner Mongolia Case Study, Deliverable D-P3d.2*. ADAM Project (EU Project no. 018476-GOCE), Project Report www. adamproject.info, Norwich, UK.

ADB (2005) *Regional Master Plan For The Prevention And Control Of Dust And Sandstorms In Northeast Asia*. Asian Development Bank (ADB).

Adger, W. N., Agrawala, S., Mirza, M. M. Q. *et al.* (2007) Assessment of adaptation practices, options, constraints and capacity. In *Climate Change 2007: Impacts, Adaptation and Vulnerability. Contribution of Working Group II to the Fourth Assessment Report of the Intergovernmental Panel on Climate Change*, ed. Parry, M. L., Canziani, O. F. Palutikof, J. P. v. d. Linden, P. J. and Hanson, C. E. Cambridge, UK: Cambridge University Press.

Adger, W. N., Arnell, N. W. and Tompkins, E. L. (2005) Adapting to climate change: perspectives across scales. *Global Environmental Change Part A*, **15**(2), 75–6.

Andrásfalvy, B. (1973) Ancient floodplain and water management at Sarkoz and the surrounding area before the rived regulations (in Hungarian). *Vízügyi történeti füzetek 6*.

Balogh, P. (2002) *Basics and Method of Floodplain Management on Middle-Tisza Valley*. Budapest: VATI Kht.

Barta, K., Bódis, K. Boga T. L., *et al.* (2000) *The Hydrological Risks Influencing the Development of the Tisza Country; Resources and Opportunities (in Hungarian)*. University of Szeged, Natural Geography Department, Szeged, Hungary.

Bartholy, J., Pongrácz, R. and Gelybó, G. (2007) Regional climate change expected in Hungary for 2071–2100. *Applied Ecology and Environmental Research*, **5** (1), 1–17.

Bates, B. C., Kundzewicz, Z. W. Wu, S. and Palutikof, J. P. (eds.) (2008) *Climate Change and Water*. Geneva: IPCC Secretariat.

Bellon, T. (2004) Living together with nature. Farming on the river flats in the valley of the Tisza. *Acta Ethnographica Hungarica*, **49** (3), 243–56.

Brogaard, S. and Seaquist, J. (2005) An assessment of rural livelihood vulnerability in relation to climate – a case study in agro-pastoral northern China. In *International workshop 'Human Security and Climate Change'*, Oslo, Norway.

Brogaard, S. and Zhao, X. (2002) Rural reforms and changes in land management and attitudes: a case study from inner Mongolia, China. *Ambio*, **31**(3), 219–25.

China Council Task Force on Forest and Grasslands (2001) *Report to CCICED Annual Meeting 2001*. Beijing, China, China Council for International Cooperation on Environment and Development (CCICED).

Commission of the European Communities (2000) *Directive 2000/60/EC of the European Parliament and of the Council establishing a framework for the Community action in the field of water policy (EU Water Framework Directive)*.

Commission of the European Communities (2001) *European Governance – A White Paper*. COM(2001) 428 final.

Cosme, N., Sousa, S., Estrela, M. A., Olay, A., Álvarez, R., and Loredo, J. (2003) *Environmental Data on a Case Study From the Transboundary Catchment of Guadiana River*. Transcat project http://www.transcatproject.net/papers/Guadiana_article_final.pdf.

Cots, F., Tábara, J. D., Werners, S. E. and McEvoy, D. (2007) *Climate Change and Adaptive Water Management Through Transboundary Cooperation: The Case of the Guadiana River Basin*. In CAIWA Conference, Basel.

Dai, X., Xu, Y., Jia, G. *et al.* (2008) *Climate change impacts and adaptation strategies in Inner Mongolia. Mid-term report*. Project Report. ADAM Project (EU Project no. 018476-GOCE), www.adamproject.info, Norwich, UK.

Dickinson, D. and Webber, M. (2007) Environmental resettlement and development on the Steppes of Inner Mongolia, *PRC Journal of Development Studies*, **43**(3), 537–61.

Du, F. (2006) Grain for green and poverty alleviation. The policy and practice of ecological migration in China. *Horizons*, **9**(2), 45–8.

Du, F. and Huang, W. (2005) Approach on ecological environment problems in Alxa area (In Chinese). *Inner Mongolia Environment Protection*, **17**, 5–9.

ECCE Project (2005) *A Preliminary Assessment of the Impacts in Spain due to the Effects of Climate Change. Impacts on the agrarian sector – Final report (Evaluación preliminar de los impactos del cambio climático en España (ECCE))*. Spain, Madrid, Ministério de Medio Ambiente.

Erdélyi, É. (2008) The potential impacts of climate change on main field crops and their yields, case studies in Hungary. *'Klimá – 21' Füzetek*, **55** (English Special Edition), 53–79.

Fekete, É. G. (2006) A situation of disadvantage turned into an advantage? Convergence opportunities for backward small regions in the region of Northern Hungary. *European Integration Studies*, **5** (1), 71–89.

Footitt, A. and McKenzie Hedger, M. (2007) *Climate Change and Water Adaptation Issues*. No 2/2007. EEA Technical report. Copenhagen, Denmark: European Environment Agency.

Henriques Goncalves, A. (2005) IWRM Applied to the Guadiana River Basin. In Joint Workshop 'Bridging the Gap between Scientists and River Basin Organisations', UNESCO HELP-ISARMROSTE.

Isoard, S. and Swart, R. (2008) Adaptation to climate change. In Jol, A., Šťastný, P. Raes, F. *et al.* (eds), *Impacts of Europe's Changing Climate – 2008 Indicator-based Assessment*. Copenhagen, Denmark: European Environment Agency. **4**, pp. 161–6.

Joffre, R., Rambal, S. and Ratte, P. (1999) The dehesa system of southern Spain and Portugal as a natural ecosystem mimic. *Agroforestry Systems*, **45**, 57–79.

Jones, B. and Keating, M. (eds.) (1995) *The European Union and the Regions*. Oxford, UK: Oxford University Press.

Kajner, P., Farkas, G., and Nagy, R. (eds.) (2008) *A Chance Along the River Tisza. Biodiversity Micro Grant Find – Report and prospects*, Violet Bt., Hungary.

Kar, A. and Takeuchi, K. (2004) Yellow dust: an overview of research and felt needs. *Journal of Arid Environments*, **59**(1), 167–87.

Klein, R., Eriksen, S., Næss, L. *et al.* (2007) Portfolio screening to support the mainstreaming of adaptation to climate change into development assistance. *Climatic Change*, **84** (1), 23–44.

Koncsos, L. and Balogh, E. (2007) Flood damage calculation supported by inundation model in the Tisza Valley. In 32nd Congress of the International Association of Hydrolic Engineering and Research (IAHR), Venice, Italy.

Láng, I. (2006) *The Project 'VAHAVA', Executive Summary*. Ministry for the Environment and Water Management (KvVM) and the Hungarian Academy of Sciences (MTA), Budapest.

Linnerooth-Bayer, J., Vári, A. and Thompson, M. (2006) Floods and fairness and Hungary. In *Clumsy Solutions for a Complex World: Governance, Politics and Plural Perceptions*, ed. Verweij, M. and Thompson M. Houndmills, Basingstroke, UK: Palgrave-Macmillan, pp. 181–204.

Matczak, P., Flachner, Z. and Werners, S. E. (2008) Institutions for adapting to climate change in the Tisza river basin. *'Klimá – 21' Füzetek*, **55** (English Special Edition), 87–100.

McEvoy, D., Cots, F., Lonsdale, K., Tábara, J. D. and Werners, S. E. (2007) The role of institutional capacity in enabling climate change adaptation: the case of the Guadiana River Basin. In *Proceedings International Symposium on Transborder Environmental and Natural Resource Management*, Kyoto, Japan.

McEvoy, D., Lonsdale, K. and Matczak, P. (2008) *Adaptation and Mainstreaming of EU Climate Change Policy: An Actor-Based Perspective.* CEPS Policy Brief No.149. Brussels, Belgium: Centre for European Policy Studies (CEPS).

Moriondo, M., Good, P., Durao, R., Bindi, M. Giannakopoulus, C. and Corte-Real, J. (2006) Potential impact of climate change on fire risk in the Mediterranean area. *Climate Research,* **31**, 85–95.

NDRC (2006) *The outline of the Eleventh Five-year plan for National Economic & Social Development of The People's Republic of China.* National Development and Reform Commission (NDRC). Beijing, China: Chinese Market Press.

Orioli, L., Moriondo, M., Brandani G. and Bindi, M. (2008) *Combining Climate Change, Land Use and Some Economic Aspects for an Adaptation Strategy in Guadiana River Basin, Iberian Peninsula.* ADAM project, Tyndall, Norwich.

Reynolds, J. F., Stafford-Smith, D. M., Lambin, E. F. *et al.* (2007) Global desertification: building a science for dryland development. *Science* (**316**), 847–51.

Runhong, G. (2001) *Case Study on Conversion of Farmland to Forest and Grass in Zhuozi County, Inner Mongolia – Prepared for CCICED Western China Forest Grasslands Task Force.* Institute of Forestry, Inner Mongolia Agricultural University, Hohhot, China.

Santos, F. D., Forbes, K. and Moita, R. (Eds.) (2002) *Climate Change in Portugal. Scenarios, Impacts and Adaptation Measures,* Portugal: Gradiva, Lisbon.

Sendzimir, J., Balogh, P., Vári, A. and Lantos, T. (2004) The Tisza River Basin: slow change leads to sudden crisis. *NATO Science Series – Sub series V Science and Technology Policy,* **41**, 261–90.

Sendzimir, J., Magnuszewski, P., Flachner, Z. *et al.* (2008) Assessing the resilience of a river management regime: informal learning in a shadow network in the Tisza river basin. *Ecology and Society,* **13**(1), **11** (1), 11.

Smit, B. and Pilifosova, O. (2001) Adaptation to climate change in the context of sustainable development and equity. In *Climate Change 2001: Impacts, Adaptation and Vulnerability. IPCC Working Group II,* ed. McCarthy, J. J., Canziani, O. Leary, N. A. Dokken, D. J. and White K. S. Cambridge, UK: Cambridge University Press, pp. 877–912.

Stern, N. H. (2007) *The Economics of Climate Change: The Stern Review.* Cambridge, UK: Cambridge University Press.

Szlávik, L. and Ijjas, I. (2003) Action plan on flood prevention for the Tisza River. In *Precautionary Flood Protection in Europe, International Workshop,* 5 – 6 February 2003, Bonn.

Tábara, J. D. (2007) A new climate for Spain: a late accommodation of environmental foreign policy in a federal state. In *Europe and Global Climate Change: Politics, Foreign Policy, and Regional Cooperation,* ed. Harris, P. Cheltenham, UK, Northampton, MA, USA: Edward Elgar, pp. 161–84.

Tábara, J. D., Cots, F., Dai, X. *et al.* (2009) Social learning on climate change among regional agents – insights from China, Eastern Europe and Iberia. In *Interdisciplinary Aspects of Climate Change,* ed. Leal Filho, W. and Mannke, F. Frankfurt, Germany: Peter Lang Scientific Publishers.

Timár, G. and Rácz, T. (2002) The effects of neotectonic and hydrological processes on the flood hazard of the Tisza region (East Hungary). *EGU Stephan Mueller Special Publication Series,* **3**, 267–75.

Timmerman, J. G., and Doze, J. (2005) *Transboundary River Basin Management Regimes: the Guadiana basin case study – Background report to Deliverable 1.3.1.* Report of the NeWater project – New Approaches to Adaptive Water Management under Uncertainty, www.newater.info. Osnabrück / Lelystad: Institute for Inland Water Management and Waste Water Treatment (RIZA).

Vári, A., Linnerooth-Bayer, J. and Ferencz, Z. (2003) Stakeholder views on flood risk management in Hungary's Upper Tisza Basin. *Risk Analysis*, **23**(3), 537–627.

Vogel, C., Moser, S. C., Kasperson, R. E. and Dabelko, G. D. (2007) Linking vulnerability, adaptation, and resilience science to practice: Pathways, players, and partnerships. *Global Environmental Change*, **17**(3–4), 349–64.

Weaver, P. M., Haxeltine, A. Kerkhof, M. v. d. and Tábara, J. D. (2006) Mainstreaming action on climate change through participatory appraisal. *International Journal on Innovation and Sustainable Development*, **1** (3), 238–59.

Werners, S. E., Incerti, F., Bindi, M., Moriondo, M. and Cots, F. (2007) Diversification of agricultural crops to adapt to climate change in the Guadiana River Basin. In *Proceedings ICCC 2007*, Hong Kong.

Werners, S. E., Matczak, P. and Flachner, Z. (2009a) The introduction of floodplain rehabilitation and rural development into the water policy for the Tisza River in Hungary. In *Water policy entrepreneurs: A Research Companion to Water Transitions Around the Globe*, ed. Huitema, D. and Meijerink S. Camberley, UK – Northampton, USA (in press): Edward Elgar Publishing.

Werners, S. E., Flachner, Z., Matczak, P., Falaleeva, M. and Leemans, R. (2009b) Exploring earth system governance: a case study of floodplain management along the Tisza River in Hungary. *Global Environment Change*. doi:101016/j.gloenvcha.2009.07.003

West, J. (2009) *Report on Fieldwork in Bantanjing village and Luanjingtan town, Alxa League, Inner Mongolia, China. November 13–19, 2008.* Oslo, Norway: Center for International Climate and Environmental Research – Oslo (CICERO).

WWF (2003) *Results Overview for the Guadiana River Basin (Spain), WWF Water and Wetland Index – Critical Issues in Water Policy across Europe.* www.assets.panda.org.

Xu, Y., Huang, X. and Zhang, Y. (2005) Statistical analyses of climate change scenarios over China in the 21st century *(in Chinese)*. *Advances in Climate Change Research*, **1**, 80–3.

Xu, Y., Zhang, Y., Lin, E. *et al.* (2006) Analyses on the climate change responses over China under SRES B2 scenario using PRECIS. *Chinese Science Bulletin* **51**(18), 2260–7.

Yan, T. and Quian, W. Y. (2004) Environmental migration and sustainable development in the Upper Reaches of the Yangtze River. *Population and Environment*, **25**(6), 613–36.

Part III

Strategies beyond Europe

10

Global climate governance beyond 2012: architecture, agency and adaptation

Lead authors:

FRANK BIERMANN[1], PHILIPP PATTBERG, FARIBORZ ZELLI

[1]Co-ordinating lead author

Contributing authors:

HARRO VAN ASSELT, INGRID BOAS, CHRISTIAN FLACHSLAND,
NITU GOEL, ANDRIES HOF, ANNE JERNECK, ERIC MASSEY,
LENNART OLSSON, JOHANNES STRIPPLE

Summary

This chapter reports the core findings of the research group 'Post-2012 Climate Governance' of the ADAM project. The group has focused on three crucial aspects of future climate governance: (i) the relative performance of different *architectures* of global climate governance; (ii) the relative performance of new forms of *agency* (in particular, beyond the state), including the role of business and environmentalist organisations in governance arrangements; and (iii) policy options for the *adaptation* of regions, countries and international institutions to the consequences of climate change. Each research domain was assessed by three sets of methodologies, namely: qualitative policy analysis, formal modelling, and participatory forms of assessment. Policy recommendations concerning governance architecture include: (i) strengthening dialogues among environment, trade and development ministries; (ii) widening the scope of the EU Emissions Trading System and linking it with other schemes; (iii) initiating formal co-operation between the UN climate regime and the Asia–Pacific Partnership and other multilateral partnerships; (iv) agreeing on science-based sustainability criteria for removing trade barriers for climate-friendly goods and services; and (v) considering climate-related issue links and package deals in the World Trade Organization Doha Round. Concerning the role of agency beyond the state and of market approaches, it seemed important to: (i) create or strengthen public funds to stimulate private research and development; (ii) differentiate among Clean Development Mechanism target countries, project types and technologies; (iii) establish reliable, uniform sectoral emissions registrations on a country level; and (iv) agree on science-based sustainability standards for CDM projects. For global adaptation governance, further institutionalisation appeared crucial. This could include a legally binding agreement on the recognition,

Making Climate Change Work for Us: European Perspectives on Adaptation and Mitigation Strategies,
ed. Mike Hulme and Henry Neufeldt. Published by Cambridge University Press © Cambridge University Press 2010.

protection and resettlement of climate refugees under the climate convention; a climate refugee protection and resettlement fund; and a legally binding agreement on adaptation and food security. The complete findings of this study programme are presented in a separate volume, *Global Climate Governance Beyond 2012: Architecture, Agency and Adaptation* (Biermann *et al.*, 2010).

10.1 Introduction

Many observers have hailed the entry into force of the 'Kyoto Protocol' to the United Nations Framework Convention on Climate Change in 2005 as a landmark achievement in combating global climate change. However, this treaty is but a first step, and its core commitments will expire in 2012. Even full compliance with the Kyoto agreement will not prevent 'dangerous anthropogenic interference with the climate system' – the overall objective of the climate convention. This situation has led to wide-ranging debates among policy makers, academics and environmentalists on the future of climate governance after 2012 (for overviews see Baumert *et al.*, 2002; Bodansky *et al.*, 2004; Aldy and Stavins, 2007; Kuik *et al.*, 2008).

This quest of finding stable, effective and equitable solutions for long-term climate governance stands at the centre of this chapter. It has also been the focus of the comprehensive research programme reported here: the research group 'Post-2012 Climate Governance' of the ADAM project. The complete findings of this study programme are presented in Biermann *et al.* (2010).

This assessment of options for long-term climate governance has been unique in its systematic and comprehensive integration of different disciplinary bodies of knowledge and of different methodological tools and approaches, from international law, political science and global governance studies, to place-based development research and computer-based scenarios and modelling exercises. While core elements of this research drew on local facts and findings, for example in studies on vulnerabilities of the poorest of the poor, the focus remained at the global level and at the most important elements of an overarching governance architecture for mitigating, and adapting to, global climate change.

Our research has been academic in nature, yet policy-relevant in orientation. Most efforts were directed at scoping or developing policy options that could provide a basis for future climate governance, and at appraising these options through multidisciplinary assessment methodologies. While many of these policy options are derived from current debates, their appraisal took a much broader, long-term perspective, in search of solutions that may be relevant and viable long after the current negotiations have been brought to an end.

The study programme has been organised around three research domains, each having one central research question, and each being assessed by three assessment methods. This approach also structures this chapter: we first introduce the three core research domains of our project, that is, the *architecture* of global climate governance; *agency* in climate governance that goes beyond the central nation state; and *adaptation* to climate change at the level of global institutions and organisations. Section 10.3 then establishes the three main sets of methodologies applied in the analysis of the three research domains, namely: qualitative policy assessment, formal modelling, and participatory forms of assessment. By analysing each research domain from three different methodological viewpoints, we conducted a comprehensive appraisal that included criteria of policy feasibility, effectiveness and equity. Section 10.4 presents our major findings from all three domains. The concluding section summarises the policy options and identifies commonalities that could facilitate integrated policies towards effective global climate governance beyond 2012.

10.2 Research domains: architecture, agency and adaptation

We have focused on three crucial aspects of future climate governance: (i) the relative performance of different *architectures* of global climate governance; (ii) the relative performance of new forms of *agency* (in particular beyond the state), including the role of business and environmentalist organisations in governance arrangements; and (iii) the relative performance of different possible global governance arrangements for *adaptation* to climate change.

Research on each of these domains centres on unique, clear-cut research questions.

Architecture. Which type of global governance architecture promises a higher degree of institutional performance in terms of social and environmental effectiveness? In particular, is an almost universal, strongly integrated governance architecture likely to be more effective than a heavily fragmented, heterogeneous governance architecture? How can the increasing fragmentation of global climate governance be addressed?

Agency. What is the role and relevance of an increasing trend towards privatised and market-based governance mechanisms for climate change mitigation? How do the host of private actors, from non-governmental organisations to business actors, that surround these new mechanisms in global climate governance relate? To what extent, and under what conditions, do private or public–private transnational governance mechanisms produce policy outcomes that are comparable, or even superior, to traditional forms of intergovernmental co-operation?

Adaptation. What are the policy options for the adaptation of regions, countries and international institutions to the impacts of climate change? To what extent do effective adaptation policies require global regulatory mechanisms, as opposed to local policy making? To what extent does effective adaptation governance require the integration of adaptation policies in the overall climate governance architecture, and/or in other policy domains?

These three domains are not mutually exclusive. Questions of architecture are also relevant when developing institutions for future adaptation governance, and non-state actors are important for adaptation. Rather than providing clear-cut distinctions, the three domains serve as different lenses that, together, advance understanding of the complexity of global climate governance. Furthermore, this link in our research with broader theoretical debates in the social sciences, such as on governance architectures or on the role of the state versus non-state actors, increases knowledge of contemporary climate governance while also contributing to theory consolidation within and across disciplines. In particular, the selection of the three themes has been influenced by current debates in international relations and international law on globalisation, transnationalisation, fragmentation and legitimacy (Ruggie, 2001; Rosenau, 2003; Hafner, 2004; Börzel and Risse, 2005). Last but not least, the three research domains reflect the Science and Implementation Plan of the Earth System Governance Project, a new long-term global research effort under the auspices of the International Human Dimensions Programme on Global Environmental Change, which will last from 2009 through 2018 (Biermann *et al.* 2009). This chapter is one of the first publications that respond to the science plan of this new global research programme.

10.3 Methodologies: policy analysis, modelling and participatory assessment

These three research domains and their core research questions have been analysed from the perspective of three methodological approaches, each contributing to a comprehensive integrated examination.

10.3.1 Policy analysis

Firstly, we analysed each domain by means of policy analysis, including legal analysis. These methods advanced understanding of opportunities and barriers for policy making at different stages of the policy process, as well as of institutional interlinkages and barriers to rule-making. We covered criteria of inclusiveness and legitimacy (regarding the participation of different types of actors), social acceptability and political feasibility. These methods helped determine the viability and the legal and political effectiveness of policy strategies, that is, their chances to materialise as concrete legal provisions (for example new rules under a post-2012 climate regime) and to change the compliance incentives of actors. Theoretical approaches applied in our research include institutional theory and global governance research, bargaining and game theory, international law analysis, and economic analysis.

10.3.2 Modelling

The use of modelling tools helps to create a structured and quantitative framework for analysis. These methods focus less on political or legal implications but rather on criteria of long-term effectiveness and efficiency of policy options. They assist in determining the structural effects of selected strategies on both the global climate and social systems, for example, regarding long-term emission reductions or effects on national incomes. Methods applied in this research include the FAIR meta-model, developed by the Netherlands Environmental Assessment Agency (for further references, see Hof *et al.*, 2010a). FAIR is a stylised multi-region formal model that integrates modelling of the climate system (the relation between greenhouse gas emissions, concentrations and temperature) with the socio-economic system (costs of mitigation, emissions trading and effects of climate change on national income). A second model employed is REMIND, developed by the Potsdam Institute for Climate Impact Research (Leimbach *et al.*, 2009). REMIND is a hybrid model that is designed to integrate macroeconomic, energy system and climate modules. It is a multi-region endogenous economic growth model, which can focus on regional interactions such as trade flows, foreign investments or technological spill-overs.

10.3.3 Participatory approaches

In addition, this research employed participatory assessment approaches. Such tools give voice to stakeholders' perspectives. They allow a critical examination of policy recommendations against the interests and concerns of key stakeholders, and can assist in refining recommendations into feasible and socially robust strategies. Participatory assessments hence complement the examination of political feasibility criteria provided by policy analysis. Participatory methods applied here include: (i) a series of structured international workshops with experts and policy makers; (ii) regular consultations with an advisory group of senior experts and policy makers;

Table 10.1. *Research domains and research methods*

	Architecture	Agency beyond the state	Adaptation
Policy analysis	Institutional fragmentation *(institutional theory, bargaining theory, international law)* UN climate regime and Asia-Pacific Partnership *(international law)* UN climate regime and biodiversity convention *(international law)* UN climate regime and world trade regime *(institutional theory, bargaining theory, international law)*	Transnational climate governance *(institutional theory)* CDM reform *(institutional theory)* Research and development, and technological change *(economic analysis)*	Climate refugees *(institutional theory, international law)* Food insecurity *(institutional theory)* Adaptation funding *(qualitative economic analysis)* Interests and perspectives of developing countries *(institutional theory, international law)* Vulnerability of the poorest of the poor *(socio-economic analysis)*
Modelling	Institutional fragmentation *(FAIR meta-model)* Linking of emission trading systems *(REMIND model)*	Sectoral mitigation *(FAIR meta-model)*	Cost-benefit interlinkages between adaptation and mitigation *(FAIR meta-model)*
Participatory approaches	Institutional fragmentation *(side-events at conferences of the parties, UNEP workshop, policy workshop in Brussels, developing country conference in Delhi, interviews, survey)* UN climate regime and world trade regime *(UNEP workshop, policy workshop in Brussels, interviews)* Architecture and equity *(developing country conference in Delhi)*	Transnational climate governance *(interviews, survey)* Reform of CDM *(policy workshop in Lund, policy workshop in Brussels)* Market-based mechanisms and developing countries *(developing country conference in Delhi, survey)*	Climate refugees *(side-events at conferences of the parties, policy workshop in Brussels, interviews)* Food insecurity *(side-events at conferences of the parties, developing country conference in Delhi, policy workshop in Brussels, interviews)* Adaptation in developing countries *(developing country conference in Delhi)* Adaptation funding *(policy workshop in Brussels, interviews)*

and (iii) a major survey of Southern policy makers, academics and representatives of non-governmental organisations. The participatory appraisal exercises were held in: (i) New Delhi, India, on developing country perspectives; (ii) Geneva, Switzerland, jointly with the Economics and Trade Branch of the UN Environment Programme, on climate and trade policies; (iii) Lund, Sweden, on the reform of the Clean Development Mechanism (CDM); (iv) Brussels, Belgium, on adaptation funding; (v) Brussels, Belgium, jointly with the Centre for European Policy Studies, on the overall research results; and (vi) Bali, Indonesia, a side event at the thirteenth conference of the parties.

Table 10.1 summarises the application of these different methods to the three research domains.

As Table 10.1 illustrates, our original research did not apply all the methods to the same degree in each research domain. In some cases, we also relied on additional literature. For instance, estimates of climate change-induced migration are not based on our own models, but are drawn on a meta-analysis of other modelling and scenario exercises.

10.4 Research findings

10.4.1 Architecture: analysing the increasing fragmentation of global climate governance

A core element of the quest for long-term stable and effective climate governance is the overall institutional architecture. The term 'global governance architecture' is now widely used in the literature. It has been employed to describe the broader institutional complex in areas such as international security, finance, trade and the protection of the environment. We define the term here as the overarching system of public and private institutions: that is, principles, norms, regulations, decision-making procedures and organisations that are valid or active in a given issue area of world politics. Architecture can thus be described as the *meta-level* of governance (Biermann *et al.*, 2010c).

In policy and academic debates, there is increasing concern for widespread fragmentation of such global governance architecture. This is especially the case for climate governance, which is marked by a plethora of institutions that are not always effectively related to the overarching climate convention (see also for example Haas *et al.*, 2004; Kanie 2008). Regarding intergovernmental institutions, there are four different spheres of fragmentation in international climate politics, which can be arranged concentrically from 'purely' climate-specific institutions towards regimes and organisations with universal or crosscutting portfolios (see Figure 10.1 for an overview). If private and public–private initiatives are also considered, the global climate architecture appears even more fragmented (see Section 10.4.2 on agency beyond the state in more detail).

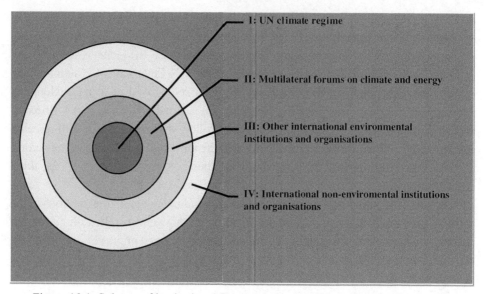

Figure 10.1. Spheres of institutional fragmentation in global climate governance.

We examined this fragmentation and its implications by the three methodological approaches of policy analysis, modelling, and participatory assessment. Based on this research, we developed novel policy options for addressing fragmentation especially between the UN climate regime and new multilateral climate partnerships such as the Asia–Pacific Partnership on Clean Development and Climate; the Convention on Biological Diversity; and the world trade regime. It may also occur between different (future) emissions trading schemes.

One benefit of institutional fragmentation is that it may facilitate *getting laggards to the negotiation table*. For instance, internal fragmentation or duplication in the UN climate regime, with various parallel tracks for discussing a future regime, allows for the direct involvement of countries that have not ratified the Kyoto Protocol to participate in discussions about a successor agreement. Notably, the United States participated in the Convention Dialogue in 2006 and 2007 and afterwards in the *Ad Hoc* Working Group on Long-Term Co-operative Action under the Convention. A fragmented governance architecture may also provide more *platforms for including non-state and sub-state actors*. For instance, major businesses are involved in multilateral technology initiatives such as the International Partnership for the Hydrogen Economy. Another advantage of fragmentation is the potential for a *meaningful division of labour* among institutions. Instead of overburdening the UN climate regime, other institutions can take over certain functions. Fragmentation might also allow for *deeper or faster agreements* by circumventing deadlocks in larger forums. For instance, the 2007 meeting of the Group of Eight was the first multilateral arena where major developed country emitters made (soft) commitments to reduce greenhouse gas emissions by at least 50 per cent by 2050. This agreement also helped *reinvigorate debates in other institutions*, by providing a major impetus on the road to the conference of the parties in 2007 in Bali.

Yet there are also many, and possibly severe, costs involved with heavy fragmentation of a governance architecture. Firstly, it gives room to many initiatives that *serve only particular interests*. The bulk of multilateral partnerships on climate and energy do not include least developed countries or small island states. They largely focus on the interests of the participating industrialised or newly industrialising countries, while side-lining preferences of poorer countries. Notably, adaptation has marginal roles in the Asia–Pacific Partnership and in the first session of the United States-initiated Major Economies meeting. Moreover, fragmentation might increase *co-ordination gaps among institutions*. At present, co-ordination on adaptation is poor between the climate convention and other institutions, for example the United Nations Food and Agriculture Organisation or the desertification convention. *Regulatory uncertainty* is another severe downside of fragmentation, especially where clear price signals and investment security are important. For example, the variety of unlinked emission trading schemes yields a patchwork of different conditions for the generation and transfer of emission credits and permits. Scholars have also pointed to the imminent danger of *'chill effects'* (Eckersley, 2004). In light of the strong dispute settlement system under the World Trade Organization (WTO), parties might have been reluctant to include further trade-restrictive measures in the UN climate regime, let alone strengthening the regime's own dispute settlement system. Finally, institutional diversity implies *the risk of 'forum shopping'* (Raustiala and Victor, 2004: 280). The Asia–Pacific Partnership for instance has provided a forum for the United States (and initially also Australia) to circumvent the UN climate regime. In the same vein, the success of such initiatives might reduce compliance incentives for parties of the Kyoto Protocol (van Asselt, 2007: 23ff.).

Following this policy assessment of institutional fragmentation in global climate governance, the question arises how the mid and long-term emission-reducing effectiveness of different scenarios of institutional fragmentation can be determined. We have quantified this for the two inner spheres of Figure 10.1 above; that is, for the UN climate regime and other institutions predominantly addressing climate change, based on the FAIR meta-model (Hof *et al.*, 2010b). This research builds on earlier projections made with the meta-model for different levels of institutional co-operation among countries, with the preferred option being a 'Grand Coalition' (Boeters *et al.*, 2007). As shown in Table 10.2, a joint 'broad-but-shallow' approach under the umbrella of a universal post-2012 regime appears far more effective than a patchwork of different approaches, notwithstanding incentives for free riding.

Table 10.2. *Emissions impact of different institutional scenarios by 2020*

Scenario:	Increase of fossil carbon dioxide emissions compared to 1990 levels	Increase of all greenhouse gas emissions compared to 1990 levels
'Grand coalition'	+28%	+22%
'Fragmented '	+45%	+43%
'Largest common denominator'	+48%	+45%
'Impasse'	+56%	+52%

Source: *Boeters et al.*, 2007: 20.

Such model-based projections for the next hundred years even suggest that a fragmented scenario with small but stable coalitions is only slightly more effective than no coalition at all (Hof *et al.*, 2010b).

In the light of our findings from both qualitative and quantitative research, a strongly integrated climate governance architecture appears to be the most effective solution. However, in current climate governance, as well as in many other areas of world politics, such integrated architectures are not always realistic. The second best solution may thus be a well co-ordinated 'web of institutions' (see also Gupta *et al.*, 2007: 791) that ensures an enhanced division of labour not only among climate-related institutions, but also with institutions from different issue areas, including the world trade regime.

Building on these findings, we have researched in detail specific institutional overlaps around the UN climate regime. We have selected one case study for each overlap of the UN climate regime with one of the four spheres identified in Figure 10.1 above.

One case study analysed internal fragmentation within core climate institutions with regard to emissions trading and prospects for a global carbon market (Flachsland *et al.*, 2010). Article 17 of the Kyoto Protocol as specified in the later Marrakesh Accords implies a top-down approach, that is, the implementation of emissions trading through multilateral negotiations. On the other hand, so-called bottom-up approaches that are considered by members to the International Carbon Action Partnership – including the EU Commission and several EU Member States, Australia, New Zealand, and some US states – emphasise the implementation and linking of domestic emissions trading schemes on the national or sub-national level. However, the latter approach implies a stepwise implementation of a global carbon market, as compared to the instantaneous implementation of a Kyoto-type trading system. Using the REMIND model, we analysed the economic costs of delaying the implementation of a comprehensive global trading system and found that when a global carbon market is implemented by 2020 instead of 2010, global mitigation costs rise from 1.3 to 2.8 per cent of global discounted consumption. While a comprehensive global top-down trading approach is the best solution to control global emissions but not realistic in the short term, the second-best option is to combine elements of different carbon market architectures. For instance, governments could agree on a system where a group of countries that are willing to adopt binding economy-wide caps continues the inter-governmental cap-and-trade system implemented by the Kyoto Protocol after 2012. By linking their domestic trading systems within this government-level framework, they can devolve the trading activity to the level of companies, which will enhance the efficiency of the international carbon market. This architecture could be designed as an open system that enables other countries to join later with some or all sectors of their economy. This approach could be environmentally and economically more effective than pure bottom-up approaches, and less prone to political stalemates and high transaction costs than the top-down approach (Flachsland *et al.*, 2010).

A second case study addressed fragmentation between the UN climate regime and the Asia–Pacific Partnership on Clean Development and Climate (van Asselt, 2007). Unlike the climate convention or the Kyoto Protocol, the Asia–Pacific Partnership does not differentiate between responsibilities of state parties and does not address adaptation or interlinkages with other regimes. Moreover, the Partnership has no systematic procedure for stakeholder participation. This example of governance fragmentation could thus undermine effective climate policies because actors receive different signals. There are several options for co-ordinating the two systems. These include (i) mutual support in treaty implementation, regarding data collection and capacity building; (ii) co-operation on flexible mechanisms, for example obtaining CDM credits for projects under the Partnership; and (iii) technology transfer, for example by using the Asia–Pacific Partnership as a testing ground for bridging diverging positions and practical barriers.

A third case study examined the relationship between the climate regime and another multilateral environmental agreement, the Convention on Biological Diversity (van Asselt *et al.*, 2008). Although both treaties are broadly compatible, some observers fear lack of respect for biodiversity protection owing to the prominent role of cost-effectiveness in the climate regime (van Asselt *et al.*, 2005: 259). Critics also argue that the CDM does not sufficiently protect biodiversity, since it allows for large-scale, monoculture plantations; lacks protection measures for old-growth forests; and fosters use of invasive alien species and genetically modified organisms (Meinshausen and Hare 2003). In response, governments agreed in 2003 on procedures for forestry projects under the CDM (UNFCCC, 2004). However, this agreement does not alleviate all concerns (UNFCCC, 2004: Annex, paragraph 12.c; see Sagemüller, 2006: 221).

A fourth case study analysed fragmentation between the UN climate regime and a non-environmental institution, namely the world trade regime. There are various overlapping policies in both regimes (Biermann and Brohm, 2005; van Asselt and Biermann, 2007; Zelli, 2007), including trade in emission allowances, unilateral policies and measures to level the playing field (for example, border tax adjustments, subsidies and technical standards), as well as the transfer of climate-friendly goods, services and technologies. We conducted a theory-guided policy analysis of overlaps, along with a major international stakeholder workshop jointly organised with the Economics and Trade Branch of the UN Environment Programme in Geneva (Zelli and van Asselt, 2010). One policy option that emerged is to better integrate scientific expertise, for example in the Committee on Trade and Environment of the World Trade Organization (WTO), the major forum where environment–trade overlaps are discussed. Another option to involve expertise is science-based sustainability criteria for the removal of trade barriers for climate-friendly goods and services. A third policy recommendation is to broaden co-ordination across institutions to overcome negotiation deadlocks in this committee. Such a dialogue could cut across ministries instead of continuing separate ministerial gathering. Moreover, at the governmental level, strategic issue-linkages could lead to package deals. One option would be to link positions on farm subsidies, trade barriers for environmental goods and services, and trade barriers for biofuels. Concessions on bio-fuels or environmental goods and services might help reinvigorate the larger debate on farm subsidies.

10.4.2 *Agency beyond the state: analysing the increasing role of privatised and market-based climate governance*

Climate governance is no longer the domain of governments and intergovernmental co-operation alone. Instead, scholars observe a growing relevance of non-state actors, such as industry and environmentalist groups, as well as public actors other than central governments, such as cities, local communities or international bureaucracies (Benecke *et al.*, 2008; Kern and Bulkeley, 2009; Kolk *et al.*, 2008; Okereke *et al.*, 2009). Increasingly, such actors also assume a role in rule-setting institutions that regulate certain sectors, or in market-based mechanisms, such as emissions trading. This emergence of 'transnational' and often 'privatised' climate governance required, firstly, a detailed conceptualisation of this new phenomenon (Pattberg and Stripple, 2008; see also Jagers and Stripple, 2003), which drew on political science and international relations studies of the public/private divide and different spheres of authority (for example Börzel and Risse, 2005).

The starting point has been the observation that 'An increasingly pertinent feature of the global public order in and beyond environmental protection and

Table 10.3. *Sites of global climate governance*

Authority mode of governance	Public	Hybrid	Private
Hierarchical	National policy; supra-national organisation		
Market	EU Emissions Trading System	Compliance market in carbon (CDM)	Carbon neutrality; company- and industry-wide emission trading
Networks	C40; Cities for Climate Protection Campaign	Partnerships for Sustainable Development (for example Renewable Energy and Energy Efficiency Partnership)	Corporate social responsibility and business-NGO self-regulation (for example Carbon Disclosure Project)

Source: *Pattberg and Stripple, 2008.*

sustainability is the dynamic mixing of the public and the private, with state-based public power being exercised by state institutions alongside and along with the exercise of private power by market and civil society institutions and other actors committed to the public interest and public weal' (Thynne, 2008: 329). Especially in climate governance, a number of actors deliberately form social institutions to address the problem of climate change without being forced to, persuaded or funded by states or other public agencies. This transnational institutionalisation of climate governance is in line with what Ruggie (2004) has called the reconstitution of a global public domain. As a domain, it does not replace states but 'embed[s] systems of governance in broader global frameworks of social capacity and agency that did not previously exist' (Ruggie, 2004: 519). The original claim about 'agency beyond the state' concerns the role and relevance of different actors. The power of individual and collective actors to change the course of events lies increasingly in sites beyond the state and its international organisations.

Based on this conceptualisation of the emergent transnational climate governance arena and agency beyond the state in climate governance, Pattberg and Stripple (2008) developed a typology that distinguishes different approaches. These range from governance through markets – including the Clean Development Mechanism (CDM) and voluntary offsets – to networked governance, which includes public non-state actors such as cities, along with transnational corporations and non-governmental organisations (see Table 10.3).

Research that is more detailed subsequently focused on particular elements of the emergent transnational climate governance arena.

Firstly, we analysed *public-private climate governance partnerships* (Pattberg, 2010). Public–private partnerships – that is, networks of different societal actors, including governments, international agencies, corporations, research institutions and civil society organisations – are cornerstones of current global environmental governance, both in discursive and material terms. Within the United Nations, partnerships have been endorsed through the establishment of the Global Compact, a voluntary partnership between corporations and the United Nations, as well as through the 'partnerships for sustainable development' (also known as 'type-2' outcomes) concluded by governments at the 2002 World Summit for Sustainable Development in Johannesburg. Both the 'partnerships for sustainable development' and the Global Compact have been criticised for privatising parts of the policy response to global change (Biermann *et al.*, 2007; Rieth *et al.*, 2007). We analysed public–private partnerships in the field of global climate governance based on three evaluation criteria: problem-solving capacity; participation and inclusiveness; and synergies or dysfunctional linkages with international climate governance.

Several obstacles prevent the realisation of the full potential of partnerships' problem-solving capacity. In particular, the geographical bias towards global partnerships instead of local or regional ones indicates that partnerships reflect pre-existing interest structures and therefore seldom deliver additional benefits that may not have been realised in more traditional multilateral or bilateral implementation arrangements. Regarding increased participation through public–private partnerships, our analysis highlights the over-representation of governments in climate partnerships as compared to the total sample of all partnerships for sustainable development registered with the United Nations. Climate partnerships are also largely dominated by states, both in terms of leadership and membership. This finding is in line with the expectation that politically contested areas such as climate politics remain overall under the control of governments. Finally, it appears that a stronger link with the UN climate regime may benefit both the 'partnerships for sustainable development' – by giving them guidance and a clear goal – and the climate regime, by assisting its implementation.

Secondly, this programme explored the CDM within the larger context of agency beyond the state. We first analysed the costs and benefits of its governance structure and then reform options through a participatory appraisal exercise in Lund, Sweden. This exercise covered a wide range of issues relating to carbon offsetting on both the regulated market of the CDM as well as in the voluntary carbon market. Major problems of the CDM include (i) the unequal geographical distribution of projects, (ii) the lack of sustainable development benefits from many projects, and (iii) complex bureaucratic processes. The market structure of the CDM and the resulting focus on cost-efficient emissions reductions are probably at the root of these problems. Reform options are many (Stripple and Falaleeva, 2008; see also van Asselt and Gupta, 2009). Firstly, the currently inequitable regional distribution of projects under the CDM requires institutional capacity building in many countries, including least developed countries and small island developing states. They could also be supported by further adapting the levies, discounting credits from richer developing countries or even through quotas for disadvantaged developing countries. To work better towards sustainable development within the CDM framework, increased differentiation between project types and technologies might be a way forward. One option is to favour projects with clear sustainable development co-benefits and to discount for projects with no or few sustainable development contributions.

A more radical option is to separate the two objectives of the CDM and to leave the achievement of sustainable development to other mechanisms. This option would focus on the CDM as an instrument for cost-effective emission reductions and create a fund for sustainable development outside of it. Such a fund could be specifically aimed at funding projects with high sustainable development benefits, but with high costs and questionable additionality, such as some renewable energy and energy efficiency projects. Regarding the third problem, sectoral approaches may promise to address some of the bureaucratic complexities of the CDM. Sectoral approaches would require the development of different methodologies for additionality and baseline emissions compared to the current project-based CDM. One could launch a pilot phase with discounted sectoral credits to examine further the potential for sectoral projects and programmes.

In addition to the appraisal of options for CDM reform, we analysed in more detail the processes that drive the current transformation of current carbon markets (Stripple and Lövbrand, 2010). Rather than asking *who* governs the carbon market domain, this study addressed the question of *how* and by which procedures carbon markets are rendered thinkable and operational in the first place. To this end, the study analysed baseline and credit markets in particular, where a complex measurement of counterfactuals (current emissions *vis-à-vis* a business-as-usual scenario) enables reductions of carbon dioxide equivalents to be assigned market value and transformed into various offset currencies.

While sectoral approaches under the CDM focus on public–private mitigation in developing countries, another case study on agency beyond the state has modelled mitigation from sectors in both industrialised and developing countries (den Elzen *et al.*, 2010). This study drew on the 'Triptych approach', a method for allocating future greenhouse gas emission reductions among countries under an international climate mitigation regime that may follow the Kyoto Protocol and be based on technological criteria at sector level. Targets are defined for the following sectors: industry (manufacturing and construction), domestic (including carbon dioxide emissions from the residential, commercial, agriculture and inland transport sectors), power production, fossil fuel production, non-carbon dioxide emissions in agriculture, and waste. Defining targets for separate sectors allows the linkage of real-world emission reduction strategies and takes into account diverse national circumstances of countries. The major advantage of this sectoral approach is that internationally competitive industries are put on the same level playing field. However, one of the major challenges is establishing reliable, uniform, sectoral emissions registrations for all countries, as this is lacking especially in developing countries.

A fourth case study analysed the role of non-state actors with regard to research and development and technological change (Alfsen *et al.*, 2010). The study argues that international agreements are best suited to boost research and development on climate friendly technologies. Such agreements and cap-and-trade systems are mutually supportive because research and development reduces future abatement costs and thus makes it feasible for politicians to agree on tighter caps.

10.4.3 *Adaptation: analysing governance arrangements for adaptation to global climate change*

It becomes increasingly clear that despite all mitigation efforts, some degree of global warming cannot be prevented, and impacts of climate change will become a reality of the twenty-first century. This poses the question of optimal adaptation governance. While a number of research programmes have addressed adaptation governance at local and national levels (see this volume Chapter 5 Hinkel *et al.*, Chapter 8 Mechler *et al.*), this research group ventured into a largely unexplored research terrain: *global* adaptation governance. How can we build global governance systems over the course of the next decades that will cope with the adaptation for global impacts of climate change required? What institutions are in need of redesign and strengthening? To what extent, and in what areas, do we need to create new institutions and governance mechanisms from scratch?

Global adaptation governance will affect most areas of world politics, including many core institutions and organisations (Biermann and Boas, 2010a). The need to adapt to climate change will influence, for example: (i) the structure of global food regimes and the work of the UN Food and Agriculture Organisation; (ii) global health governance and the agenda of the World Health Organisation; (iii) global trade in goods, the production of which will be harmed or helped by climate change; (iv) the world economic system and the ability of the International Monetary Fund to address climate-related shocks to national and regional economies; (v) the World Bank and bilateral and national agencies in raising and distributing funds to support adaptation; and (vi) many other sectors from tourism to transportation or even international security.

In line with the research domains of architecture and agency, we first conceptualised 'global adaptation governance' and then focused on specific cases and elements from the perspectives of international relations (regarding the institutional implications of global adaptation governance) and integrated assessment models (regarding global and regional damage and adaptation costs).

The quantitative research on adaptation costs (Hof *et al.*, 2010a) underscored the urgency for multi-institutional international action (see Figure 10.2). We combined here the FAIR meta-model and the AD-RICE model (de Bruin *et al.*, 2009) for analysing the mitigation costs, adaptation costs and residual damages of climate change on a global as well as regional scale. For a Contraction and Convergence emission allocation regime (with per capita emissions converging in 2050, a climate sensitivity of 3.0 degrees Celsius and the UK Green Book discounting method), the projected global adaptation costs are of the same order of

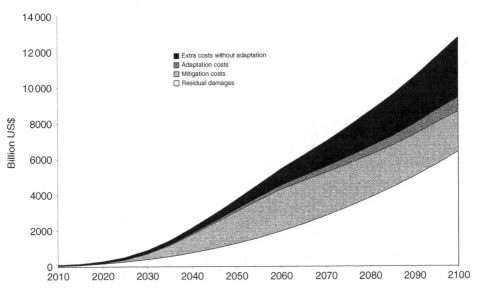

Figure 10.2. Global mitigation costs, adaptation costs, residual damages and extra costs if no adaptation is undertaken. (Source: Hof *et al.*, 2010a)

magnitude as the recent adaptation cost estimates of the World Bank (2006) and UNFCCC (2007). Yet when looking at detailed predictions, the model's division of damages into residual damages and adaptation costs reveals an intriguing finding; adaptation costs could in fact amount to only a fraction compared to damages and mitigation costs, especially in the short to medium run (see also Chapter 3 van Vuuren et al.).

Even though the share of adaptation costs in the total climate change costs is relatively small, adaptation plays a major role by reducing potential damages. The extra costs if no adaptation measures are taken (defined as the increase in residual damages minus the decrease of adaptation costs) are projected to amount to US$ 30 billion globally in 2010 and increase sharply to US$ 3.4 trillion in 2100. Investment in adaptation is therefore very effective: residual damages are on average reduced by about five dollars for every dollar invested in adaptation. Furthermore, adaptation and mitigation cannot be regarded as substitutes, but rather complement each other. Adaptation can effectively reduce climate change damages in the shorter run, but is much less effective in the end since it does not reduce climate change itself. Mitigation is very effective in reducing climate change damages in the end. Implementing both adaptation and mitigation gives the best results according to the FAIR meta-model.

Building on these insights, we analysed three challenges for future global adaptation governance: climate change-induced migration; climate change-induced food insecurity; and the need for co-ordinated adaptation funding. We further added two specific analyses, one from the perspective of developing countries as a group of nations, and from the perspective of the poorest of the poor.

It is likely that climate change will fundamentally affect the lives of millions of people who may be forced over the next decades to leave their villages and cities to seek refuge in other areas. We define these people as 'climate refugees'. They are people who have to leave their habitats, immediately or in the near future, because of sudden or gradual alterations in their natural environment related to at least one of three impacts of climate change: sea-level rise, extreme weather events, and drought and water scarcity (see Biermann and Boas, 2008, 2010b on the details and operationalisation of this definition). The exact numbers of such future climate refugees are unknowable and vary from assessment to assessment depending on underlying methods, scenarios, time frames and assumptions, and we agree with Black (2001: 2–8) and many other scholars that estimation methods and underlying assumptions are complex and controversial. Yet despite these remaining uncertainties, a meta-analysis of all available studies indicated that the climate change-induced refugee crisis is most likely to surpass all known refugee crises in terms of the number of people affected (Biermann and Boas, 2008, 2010b).

The current refugee protection regime of the United Nations is poorly prepared, and does not cover climate refugees in its mandate. At a meeting in the Maldives in 2006, delegates therefore proposed an amendment to the 1951 Geneva Convention Relating to the Status of Refugees that would extend the mandate of the UN refugee regime to cover also climate refugees. We argue, however, that such an amendment leads in the wrong direction. Firstly, the political feasibility of this proposal is highly uncertain. The UN refugee regime is already under constant pressure from industrialised countries seeking restrictive interpretations of its provisions. It is unrealistic for governments to extend the same level of protection to potentially 20 times more climate refugees. More importantly, the proposal of an extension of the UN refugee regime misses the core characteristics of the climate refugee crisis. The protection of climate refugees is essentially a development issue. It requires large-scale, long-term planned resettlement programmes for groups of affected people, mostly within their country. Often, this will be in concert with adaptation programmes for other people who are not evacuated but can be protected, for instance through strengthened coastal defences. It is therefore not the UN High Commissioner for Refugees but other international agencies such as the UN Development Programme or the World Bank that are called upon to deal with the emerging problem.

We therefore argue for a separate regime: a legally binding agreement on the recognition, protection and resettlement of climate refugees under the climate convention. This could be a separate protocol under the convention ('climate refugee protocol'), but also an integral part of a larger legal instrument, such as a protocol on adaptation, or even a single undertaking that regulates all future measures on climate governance (see Biermann and Boas, 2008, 2010b in more detail on this proposal). Importantly, the protection of climate refugees must be seen as a global problem and a global responsibility. In most cases, climate refugees will be poor, and their own responsibility for the past accumulation of greenhouse gases will be small. By a large measure, the rich industrialised countries have caused most emissions in the past and present, and it is thus these countries that have most moral, if not legal, responsibility for the victims of global warming. Industrialised countries should therefore do their share in financing, supporting, and facilitating the protection and the voluntary resettlement of climate refugees.

A second case study focused on a related challenge – food security (Massey, 2008). A changing climate will significantly affect many communities that today face hunger and malnutrition. Key impacts on agriculture are a depletion of ground water, reduced precipitation and changes, primarily a shortening, of the growing season. These may all reduce yields. For example, the Intergovernmental Panel on Climate Change (IPCC) Fourth Assessment Report suggests that a 2 to 3 °C range of warming by 2020 could decrease agricultural yields in Africa by as much as 50 per cent (Boko *et al.*, 2007: 447–448). Therefore, some form of adaptation must occur to ensure greater food security in the most vulnerable regions. Our research indicates that there needs to be a mechanism that allows for adaptation at the local level to help farmers and communities. At the same time, there needs to be a well functioning global institutional system that supports the financing and implementation of adaptive measures, including improved farming techniques and technologies.

One potential means of adaptation to meet this challenge could be improved access of farmers in developing countries to state-of-the-art research on farming technologies. Developing countries are at a competitive disadvantage as a result of the allocation of funding for agricultural research in general, including the protection offered to more adaptive crop seeds due to international intellectual property rights. Developed countries and the private sector may thus have a special role in aiding the farming sector in developing countries to adapt. This support could come in the form of an adaptation levy to fund agricultural research in developing countries as well as a renegotiation of international intellectual property rights in the domain of agriculture. The overall institutional context could be strengthened through a legally binding agreement on adaptation and food security under the climate convention (Massey, 2008). This could be a single agreement – such as a protocol to the climate convention – but could also be integrated (possibly with the agreement on climate refugees outlined above) into a larger legal instrument, such as an adaptation protocol to the climate convention. In addition, as discussed earlier under the 'architecture' domain, discussions on farm subsidies and transfer of technologies could be coupled with adaptation-related concerns, for example through sustainability criteria for trade barrier removals.

Adaptation is clearly a key priority for most developing countries, many of which have contributed only marginally to the build-up of greenhouse gases in the atmosphere but which will be especially affected by climatic change. Alam *et al.* (2010) thus examined the current discourses and negotiations on adaptation to climate change from the perspective of developing countries. Their analysis also took into account debates on a major workshop on Southern perspectives that the ADAM project organised in 2008 in New Delhi, India. Alam *et al.* (2010) concluded that, although significant progress has been made on empowering the adaptation agenda within the climate governance architecture, this resulted in a framing of adaptation that is inappropriate for addressing the many developing country concerns. First they argue that, under existing frameworks, adaptation remains an undervalued policy option relative to mitigation. Secondly, they see the type of adaptation favoured by the climate convention as not conducive to building the broader resilience that is necessary to reduce the vulnerability of developing countries. Thirdly, they view the adaptation discourse under the climate convention as largely technical and not open to alternative types of expertise that are locally generated and non-technical. In summary, Alam *et al.* (2010) suggest that it is both necessary and possible to refine the adaptation agenda under the climate convention. According to them, more deliberative policy making processes must be created for adaptation that are better able to engage with vulnerable communities and citizens to create bottom-up, locally meaningful adaptation strategies. This would require a reframing of the adaptation discourse that is more open to non-technical expertise generated from indigenous and locally based knowledge.

In addition to a comprehensive analysis of the perspectives of the developing countries, this research programme explored also the special situation of the poorest people in these countries (Jerneck and Olsson, 2010). In the context of the poorest of the poor, mitigation is not a priority because their contribution to the global emission of greenhouse gases is minuscule and their capacity to reduce emissions is low. This makes adaptation their main priority. Today, there are 923 million hungry people worldwide, who are, in general, also extremely vulnerable to climate change impacts. This large number of poor people is expected to increase further and remain large for a long time while people exposed to climate change are expected to become even more vulnerable due to increasing incidence of extreme climate events. In relation to the poorest of the poor, adaptation to climate change should thus be seen as a process of profound social change away from livelihoods threatened at their roots by climate change.

Several policy options were considered to increase the adaptive capacities of the poorest of the poor. These include mainstreaming climate change into development assistance (see also Chapter 12 Gupta *et al.*); identifying synergies with other mechanisms, such as climate change mitigation, biodiversity or desertification; as well as a number of stand-alone adaptation policies, such as special support for climate refugees (see Jerneck and Olsson, 2010). Regarding new norms and institutions, the study argued to rethink development from a sustainability perspective rather than mainstreaming climate change and adaptation into the narrower paradigm of development, even though mainstreaming may be the only option for the medium term.

The integrated assessment modelling of adaptation costs and our studies on climate refugees, food insecurity, the perspectives of developing countries and the needs of the poorest among the poor signal the need for an enhanced and targeted set of funding mechanisms for adaptation. It is thus not only important to better endow existing funds and to add new funds, but to co-ordinate the various financial mechanisms in order to reach a meaningful division of labour. We therefore also studied adaptation funding, including a participatory appraisal exercise with stakeholders and experts from developing and developed countries in Brussels (Klein and Persson, 2008).

10.5 Conclusions: mapping of policy options

This chapter has summarised a 3-year research effort, carried out by seven research institutions in Europe and India, on policy options for stable, long-term climate governance. The research focused on three areas of rapid political development which are also areas of increasing concern: the increasing fragmentation of the overall architecture of global climate governance; the increasing privatisation and marketisation of global climate governance; and the research problem of developing new mechanisms for global adaptation governance. Despite this organisation into three research domains, all domains are interlinked. For instance, most options discussed under agency and adaptation include elements of a future climate architecture, for example reform of the CDM, or protocols on climate refugees and food security. Options discussed under the 'architecture' theme involve non-state actors, for example the linking of emissions trading schemes, or may be relevant for adaptation to climate change, for example technology transfer.

This concluding section highlights connections between the various policy options. Table 10.4 restructures the options in terms of the international institutional environment where they could be pursued: the UN climate regime, in other international organisations and forums, or in cross-institutional collaboration. Moreover, the table distinguishes options according to their political and legal dimension; either they suggest new political 'hardware', that is, new norms, treaties or institutions, or they propose specific policies, measures or standards. These two dimensions take into account two crucial aspects to be considered when feeding recommendations into the negotiation process: *where?* (institutional setting) and *what?* (nature of proposal, level of ambition). These criteria are more suitable to structure policy-relevant findings, while the three domains have helped structuring and guiding research.

Through these two dimensions, Table 10.4 highlights commonalities among policy options that were analysed under the three research domains. The columns show to what extent some options can be pursued in the same institutional arena and might hence be linked in a comprehensive negotiation approach (for example protocols on climate refugees and food security). Most suggestions fall under the UN umbrella or in the middle column that at least involves the UN regime. This is in line with our general finding that, in spite of some benefits of institutional fragmentation, it is pivotal to strengthen the UN regime as the chief institution to address global climate change.

All policies, measures and standards listed in Table 10.4 relate to different institutional settings (inside and outside the UN system), with some sharing features, such as sustainability criteria based on scientific advice for both the CDM and trade barrier removals. There is an obvious potential to link issues; a scientific body such as the IPCC could for instance provide broad expertise to develop criteria across different topics. The distinction between institutional and policy-based options also points to

Table 10.4. *Overview of policy options*

	UN climate regime	Other international institutions and forums	Cross-institutional collaboration (between UN climate regime and others)
Norms and institutions	Legally binding agreement on the recognition, protection and resettlement of climate refugees under the climate convention Climate refugee protection and resettlement fund Legally binding agreement on adaptation and food security under the climate convention	Cross-ministerial dialogue among environment, trade and development ministries Opening WTO Committee on Trade and Environment for regular scientific inputs on climate-trade overlaps Public funds to stimulate private research and development Multilateral agreements on research and development of climate-friendly technologies	Open EU Emissions Trading System and link emissions trading schemes bottom-up and top-down Co-operation agreement or Memorandum of Understanding with Asia–Pacific Partnership and other multilateral partnerships for mutual support in treaty implementation, technology transfer, and so on
Policies, measures and standards	Explicit reference to biodiversity convention and biodiversity-standards for forestry projects under CDM Differentiation among CDM target countries, project types and technologies Sectoral CDM pilot phase with discounted sectoral credits Sectoral mitigation targets Science-based sustainability standards for CDM projects	Science-based sustainability criteria for removal of trade barriers for climate-friendly goods and services Issue-linking and package deals on related discussions in the WTO Doha Round (for example farm subsidies, transfer of environmental goods and services, biofuels) Deliberative adaptation policy-making processes	Dovetailing climate-related funds within and outside the UN climate regime Focused national, regional and local policies targeting the poorest of the poor – incentivised by international framework

the variant political feasibility of options. All things being equal, agreement on new policies could be expected to be easier than on new institutional instruments, for example an open emissions trading scheme or a food security protocol.

One could also combine the dimensions according to technical or material commonalities, in the attempt to advance options in parallel during negotiations. Examples are (i) issues of funding: climate refugees funds, public research and development funds, dovetailing climate-related funds; (ii) scientific advice, for sustainability criteria for the CDM and technology transfer and for the World Trade Organization Committee on Trade and Environment; (iii) trade (linkage of emissions trading schemes, issue-linking in the Doha Round); (iv) technology (research and development funding, CDM reform proposals, adaptation food security protocol, technology transfer); and (v) sectoral approaches (a sectoral CDM, sectoral mitigation targets or sector-based emissions trading schemes as part of an open trading system).

In the final analysis, and in light of the complexity of climate negotiations and the multitude of players involved, it will be important, however, not to 'over-integrate' options before communicating them in the policy process. 'Optimal' yet highly complex and demanding combinations might overburden negotiations. The potential for concrete combinations of options in the governance process will depend on political bargaining as well as on *ad hoc* opportunities of daily politics. Future climate policy does not only need well-designed strategies for long-term effective, equitable and efficient governance architectures, but also a high degree of flexibility in actual utilisation and implementation. For better or for worse, climate governance, as with most areas of policy making, will always combine long-term visioning with short-term incrementalism.

Acknowledgements

This chapter draws on the work and input of many colleagues and stakeholders. First of all, we wish to thank for their various contributions and insights all our other team members in the Post-2012 Case Study: Mozaharul Alam, Knut H. Alfsen, Jessica Ayers, Karin Bäckstrand, Lavinia Baumstark, Kelly de Bruin, Rob Dellink, Ottmar Edenhofer, Michel den Elzen, Gunnar S. Eskeland, Alex Haxeltine, Saleemul Huq, Richard Klein, Marian Leimbach, Kristin Linnerud, Eva Lövbrand, Robert Marschinski, Åsa Persson, Manish Kumar Shrivastava, Detlef van Vuuren and Harald Winkler. In addition, we are grateful to the members of the ADAM Contact Group who provided an invaluable 'reality check' for the proposals that have been developed in this research programme: Marcel Berk, Daniel Bodansky, Chandrashekhar Dasgupta, Dagmar Droogsma, Bo Kjellen, Benito Müller, Lars Müller, Willem Thomas van Ierland and Michael Wriglesworth. In addition, Simon Tay, Youba Sokona and Sebastian Oberthür provided important comments at conference presentations of this work package. Last but not least, this chapter has benefited substantially from the useful suggestions and critique from Steinar Andresen, Mike Hulme, Norichika Kanie and Henry Neufeldt.

References

Alam, M., Ayers, J. and Huq, S. (2010) Adaptation in the post-2012 architecture: Developing country perspectives. In *Global Climate Governance Beyond 2012: Architecture, Agency and Adaptation*, ed. Biermann, F., Pattberg, P. and Zelli, F. Cambridge, UK: Cambridge University Press.

Aldy, J. and Stavins, R. N. (eds.) (2007) *Architectures for Agreement: Addressing Global Climate Change in the Post-Kyoto World*. Cambridge, UK: Cambridge University Press.

Alfsen, K. H., Eskeland, G. S. and Linnerud, K. (2010) Technological change and the role of nonstate actors. In *Global Climate Governance Beyond 2012: Architecture, Agency and Adaptation*, ed. Biermann, F., Pattberg, P. and Zelli, F. Cambridge, UK: Cambridge University Press.

Baumert, K. A., Blanchard, O., Llosa, S. and Perkaus, J. F. (eds.) (2002) *Building on the Kyoto Protocol: Options for Protecting the Climate*. Washington, DC: World Resources Institute.

Benecke, G., Friberg, L., Lederer, M. and Schröder, M. (2008) *From Public–Private Partnership to Market: the Clean Development Mechanism (CDM) as a new form of governance in climate protection*. SFB Governance Working Paper Series 10, Berlin.

Biermann, F. and Brohm, R. (2005) Implementing the Kyoto Protocol without the United States: the strategic role of energy tax adjustments at the border. *Climate Policy*, **4**(3), 289–302.

Biermann, F., Pattberg, P., Chan, S. and Mert, A. (2007) *Partnerships for Sustainable Development: an appraisal framework*. Global Governance Working Paper No 31. Amsterdam: The Global Governance Project.

Biermann, F. and Boas, I. (2008) Protecting climate refugees: the case for a global protocol. *Environment*, **50**(6), 8–16.

Biermann, F., Betsill, M. M., Gupta, J. *et al.* (2009) *Earth System Governance: People, Places and the Planet – Science and Implementation Plan of the Earth System Governance Project*. Bonn: Earth System Governance Project of the International Human Dimensions Programme on Global Environmental Change.

Biermann, F. and Boas, I. (2010a) Global adaptation governance: setting the stage. In *Global Climate Governance Beyond 2012: Architecture, Agency and Adaptation*, ed. Biermann, F., Pattberg, P. and Zelli, F. Cambridge, UK: Cambridge University Press.

Biermann, F. and Boas, I. (2010b) Global adaptation governance: the case of protecting climate refugees. In *Global Climate Governance Beyond 2012: Architecture, Agency and Adaptation*, ed. Biermann, F., Pattberg, P. and Zelli, F. Cambridge, UK: Cambridge University Press.

Biermann, F., Pattberg, P., Asselt, H. van and Zelli, F. (2010c) The architecture of global climate governance: setting the stage. In *Global Climate Governance Beyond 2012: Architecture, Agency and Adaptation*, ed. Biermann, F. Pattberg, P. and Zelli, F. Cambridge, UK: Cambridge University Press.

Biermann, F., Pattberg, P. and Zelli, F. (eds.) (2010d) *Global Climate Governance Beyond 2012: Architecture, Agency and Adaptation*. Cambridge, UK: Cambridge University Press.

Black, R. (2001) *Environmental Refugees: Myth or Reality?* New issues in refugee research Working Paper 34. Geneva: United Nations High Commissioner for Refugees.

Bodansky, D., Chou, S. and Jorge-Tresolini, C. (2004) *International Climate Efforts Beyond 2012: A Survey of Approaches*. Washington, DC: Pew Center.

Boeters, S., den Elzen, M., Manders, T., Veenendaal, P. and Verweij, G. (2007) *Post-2012 Climate Policy Scenarios*. MNP Report no. 500114006/2007. Bilthoven: Netherlands Environmental Assessment Agency.

Boko, M., Niang, I., Nyong, A. *et al.* (2007) Africa. In *Climate change 2007: impacts, adaptation and vulnerability. Contribution of Working Group II to the Fourth Assessment Report of the Intergovernmental Panel on Climate Change*, ed. Parry, M. L., Canziani, O. F., Palutikof, J. P., van der Linden, P. J. and Hanson, C. E. Cambridge, UK: Cambridge University Press, pp. 433–67.

Börzel, T. A. and Risse, T. (2005) Public-private partnerships: effective and legitimate tools of international governance. In Grande, E. and Pauly, L. W. (eds.) *Reconstructing Political Authority: Complex Sovereignty and the Foundations of Global Governance*. Toronto: University of Toronto Press, p. 195–216.

de Bruin, K. C., Dellink, R. B. and Agrawala, S. (2009) *Economic Aspects of Adaptation to Climate Change: Integrated Assessment Modelling of Adaptation Costs and Benefits*. Paris: OECD.

den Elzen, M. G. J., Hof, A., van Vliet, J. and Lucas, P. (2010) A staged sectoral approach for climate mitigation. In *Global Climate Governance Beyond 2012: Architecture, Agency and Adaptation*, ed. Biermann, F., Pattberg, P. and Zelli, F. Cambridge, UK: Cambridge University Press.

Eckersley, R. (2004) The big chill: the WTO and multilateral environmental agreements. *Global Environmental Politics*, **4**(2), 24–40.

Flachsland, C., Edenhofer, O., Marschinski, R., Leimbach, M. and Baumstark, L. (2010) Developing the International carbon market beyond 2012: options and the costs of delay, In *Global Climate Governance Beyond 2012: Architecture, Agency and Adaptation*, ed. Biermann, F. Pattberg, P. and Zelli, F. Cambridge, UK: Cambridge University Press.

Gupta, S., Tirpak, D. A., Burger, N. *et al.* (2007) Policies, instruments and co-operative arrangements. In *Climate Change 2007: mitigation. Contribution of Working Group III to the Fourth Assessment Report of the Intergovernmental Panel on Climate Change*, ed. Metz, B., Davidson, O. R., Bosch, P. R., Dave, R., and Meyer, L. A. Cambridge, UK: Cambridge University Press.

Haas, P. M, Kanie, N. and Murphy, C. N. (2004) Conclusion: institutional design and institutional reform for sustainable development. In *Emerging Forces in Environmental Governance*, ed. Kanie, N. and Haas, P. M. Tokyo: UNU Press.

Hafner, G. (2004) Pros and cons ensuing from fragmentation of international law. *Michigan Journal of International Law*, **25**(4), 849–63.

Hof, A. F., de Bruin, K., Dellink, R., den Elzen, M. G. J. and van Vuuren, D. P. (2010a) Costs, benefits and interlinkages between adaptation and mitigation. In *Global Climate Governance Beyond 2012: Architecture, Agency and Adaptation*, ed. Biermann, F., Pattberg, P. and Zelli, F. Cambridge, UK: Cambridge University Press.

Hof, A. F., den Elzen, M. G. J. and van Vuuren, D. P. (2010b) Environmental effectiveness and economic consequences of fragmented vs. universal regimes: what can we learn from modelling studies? In *Global Climate Governance Beyond 2012: Architecture, Agency and Adaptation*, ed. Biermann, F., Pattberg, P. and Zelli, F. Cambridge, UK: Cambridge University Press.

Jagers, S. C. and Stripple, J. (2003) Climate governance beyond the state. In *Global Governance. A Review of Multilateralism and International Organisations*, **9**, 385–99.

Jerneck, A. and Olsson, L. (2010) Shaping future adaptation governance: perspectives from the poorest of the poor. In *Global Climate Governance Beyond 2012: Architecture, Agency and Adaptation*, ed. Biermann, F., Pattberg, P. and Zelli, F. Cambridge, UK: Cambridge University Press.

Kanie, N. (2008) Towards diffused climate change governance: a possible path to proceed after 2012. In *Global Warming and Climate Change: Ten Years After Kyoto and Still Counting*, ed. Grover, V. I. Enfield (NH): Science Publishers, pp. 977–92.

Kern, K. and Bulkeley, H. (2009) Cities, Europeanisation and multi-level governance: governing climate change through transnational municipal networks. *Journal of Common Market Studies*, **47**(2), 309–32.

Klein, R. J. T. and Persson, A. (2008) *Financing Adaptation to Climate Change: Issues and Priorities*. ECP Report No 8, Brussels: CEPS.

Kolk, A., Levy, D. and Pinske, J. (2008) Corporate responses in an emerging climate regime: the institutionalisation and commensuration of carbon disclosure. *European Accounting Review*, **17**(4), 719–45.

Kuik, O., Aerts, J., Berkhout, F. *et al.* (2008) Post-2012 climate change policy dilemmas: how do current proposals deal with them? *Climate Policy*, **8**(3), 317–36.

Lembach, M., Bauer, N., Baumstark, L., Edenhofer, O., (2009) Costs in a globalized world: climate policy analysis with REMIND-R. *Environmental Modeling and Assessment*. In press.

Massey, E. (2008) Global governance and adaptation to climate change for food security. In *Integrated Analysis of Different Possible Portfolios of Policy Options for a Post-2012 Architecture*. F. Zelli (ADAM project report No. D-P3a.2b), Norwich, UK: Tyndall Centre for Climate Change Research, pp. 143–53.

Meinshausen, M. and Hare, W. (2003) *Sinks in the CDM: after the climate, biodiversity goes down the drain: an analysis of the CDM sinks agreement at CoP-9*. [http://www. greenpeace.org/raw/content/usa/press-center/reports4/sinks-in-the-cdm-after-the-cl.pdf] (last accessed 19 September 2008).

Okereke, C., Bulkeley, H. and Schroeder, H. (2009) Conceptualising climate governance beyond the international regime. *Global Environmental Politics*, **9**(1), 58–78.

Pattberg, P. (2010) The role and relevance of networked climate governance. In *Global Climate Governance Beyond 2012: Architecture, Agency and Adaptation*, ed. Biermann, F., Pattberg, P. and F. Zelli. Cambridge, UK: Cambridge University Press.

Pattberg, P. and Stripple, J. (2008) Beyond the public and private divide: remapping transnational climate governance in the 21st century. *International Environmental Agreements: Politics, Law and Economics*, **8**(4), 367–88.

Potsdam Institute for Climate Impact Research (2007) *Portfolio of Policy and Technological Options for P3a Case Study*. ADAM Report no. D-M2.1. [http://adamproject.info/index. php?option=com_docman&task=doc_download&gid=205&Itemid=68] (last accessed 20 September 2008).

Raustiala, K. and Victor, D. G. (2004) The regime complex for plant genetic resources. *International Organisation*, **58**(2), 277–309.

Rieth, L., Zimmer, M., Hamann, R. and Hanks, J. (2007) The UN Global Compact in Sub-Sahara Africa: decentralisation and effectiveness. *Journal of Corporate Citizenship*, **7** (28), 99–112.

Rosenau, J. N. (2003) *Distant Proximities: Dynamics Beyond Globalisation*. Princeton: Princeton University Press.

Ruggie, J. G. (2001) Global governance net: the global compact as learning network. *Global Governance*, **7**(4), 371–8.

Ruggie, J. G. (2004) Reconstituting the global public domain: issues, actors, practices. *European Journal of International Relations*, **10**(4), 499–541.

Sagemüller, I. (2006) Forest sinks under the United Nations Framework Convention on Climate Change and the Kyoto Protocol: opportunity or risk for biodiversity? *Columbia Journal of Environmental Law*, **31**, 189–242.

Stripple, J. and Falaleeva, M. (eds.) (2008) *CDM Post-2012: Practices, Possibilities, Politics*. Workshop report. Lund: Lund University.

Stripple, J. and Lövbrand, E. (2010) Carbon market governance beyond the public–private divide. In *Global Climate Governance Beyond 2012: Architecture, Agency and Adaptation*, ed. Biermann, F., Pattberg, P. and Zelli, F. Cambridge, UK: Cambridge University Press.

Thynne, I. (2008) Climate change, governance and environmental services: institutional perspectives, issues and challenges. *Public Administration and Development*, **28**, 327–39.

United Nations Framework Convention on Climate Change (UNFCCC) (2004) *Decision 19/CP.9, Modalities and Procedures for Afforestation and Reforestation Project Activities under the Clean Development Mechanism in the First Commitment Period of the Kyoto Protocol* (30 March 2004). FCCC/CP/2003/6/Add.2.

United Nations Framework Convention on Climate Change (UNFCCC) (2007) *Climate Change: Impacts, Vulnerabilities and Adaptation in Developing Countries* (Bonn, Germany: Climate Change Secretariat). [http://unfccc.int/files/essential_background/background_publications_htmlpdf/application/txt/pub_07_impacts.pdf] (last accessed 20 September 2008).

van Asselt, H. (2007) From UN-ity to diversity? The UNFCCC, the Asia–Pacific Partnership and the future of international law on climate change. *Carbon and Climate Law Review*, **1**(1), 17–28.

van Asselt, H. and Biermann, F. (2007) European emissions trading and the international competitiveness of energy-intensive industries: a legal and political evaluation of possible supporting measures. *Energy Policy*, **35**(1), 497–506.

van Asselt, H. and Gupta, J. (2009) Stretching too far: developing countries and the role of flexibility mechanisms beyond Kyoto. *Stanford Environmental Law Journal*, **28**(2), 311–378.

van Asselt, H., Gupta, J. and Biermann, F. (2005) Advancing the climate agenda: exploiting material and institutional linkages to develop a menu of policy options. *Review of European Community and International Environmental Law*, **14**(3), 255–64.

van Asselt, H., Sindico, F. and Mehling, M. A. (2008) Global climate change and the fragmentation of international law. *Law and Policy*, **30**(4), 423–49.

World Bank (2006) *Clean Energy and Development: Towards an Investment Framework*. World Bank Environmentally and Socially Sustainable Development and Infrastructure Vice Presidencies. Washington DC: World Bank.

Zelli, F. (2007) The World Trade Organisation: free trade and its environmental impacts. In Thai, K. V., Rahm, D. and Coggburn, J. D. (eds.) *Handbook of Globalisation and the Environment*. London: Taylor and Francis, pp. 177–216.

Zelli, F. & van Asselt, H. (2010) The overlap between the UN climate regime and the World Trade Organization: lessons for climate governance beyond 2012. In *Global Climate Governance Beyond 2012: Architecture, Agency and Adaptation*, ed. Biermann, F. Pattberg, P. and Zelli, F. Cambridge, UK: Cambridge University Press.

11

The economics of low stabilisation: implications for technological change and policy

Lead authors:

BRIGITTE KNOPF[1], OTTMAR EDENHOFER

[1]Co-ordinating Lead Author

Contributing authors

TERRY BARKER, NICO BAUER, LAVINIA BAUMSTARK,
BERTRAND CHÂTEAU, PATRICK CRIQUI, ANNE HELD,
MORNA ISAAC, MARTIN JAKOB, EBERHARD JOCHEM,
ALBAN KITOUS, SOCRATES KYPREOS, MARIAN LEIMBACH,
BERTRAND MAGNÉ, SILVANA MIMA, WOLFGANG SCHADE,
SERBAN SCRIECIU, HAL TURTON, DETLEF P. VAN VUUREN

Summary

The European Union (EU) is committed to the goal of keeping the increase in global temperatures from pre-industrial levels to no more than 2 °C with a better than even chance. Achieving this 2 °C target would require stabilising greenhouse gas concentrations at less than 450 ppm. This chapter examines whether and how this can be done by probing the technological and economic feasibility of reaching such a low level of stabilisation with acceptable means. We explore both aspects for three carbon dioxide equivalent concentration levels, set at 550, 450 and 400 parts per million (ppm) carbon dioxide equivalents, which have different probabilities of reaching the 2 °C target. To investigate the robustness of results on mitigation costs and technological options, we compare findings from different state-of-the-art energy–environment–economy models for the time horizon 2000–2100. An in-depth sectoral analysis of how the transformation of the energy system could proceed in Europe follows this global analysis.

Our results suggest that low stabilisation is feasible in terms of technologies and moderate in costs. A broad range of technologies can be used to achieve stabilisation targets such as 550 ppm that have only a low likelihood of reaching the 2 °C goal. Much more ambitious reduction targets, such as 400 ppm, however, rely heavily on the availability of carbon capture and storage (CCS) in combination with biomass as options for removing carbon from the atmosphere and on the expansion of renewable energy. This target alone has a high likelihood of reaching the 2 °C goal.

Overall, global mitigation costs, expressed as cumulative gross domestic product (GDP) losses until 2100 relative to the baseline, are found to be below 0.8 per cent

Making Climate Change Work for Us: European Perspectives on Adaptation and Mitigation Strategies,
ed. Mike Hulme and Henry Neufeldt. Published by Cambridge University Press © Cambridge University Press 2010.

for the 550 ppm target, but nearly 2.5 per cent for the most ambitious of the three stabilisation targets, 400 ppm. These costs could be twice as high if biomass availability was smaller than first assumed or if the CCS storage potential was more limited. One model reports GDP gains for all stabilisation pathways as it incorporates existing inefficiencies.

A detailed analysis of the transformation of the energy systems in Europe leads to the conclusion that improving energy productivity and substituting renewable energy for fossil fuels are the most important means for achieving the 2 °C goal.

11.1 Introduction

Reaching the target of climate stabilisation at no more than 2 °C above pre-industrial levels by the end of this century – a goal embraced by the EU – is a historic challenge for humankind. To make it likely that this challenge will be met, greenhouse gas concentrations have to be limited to well below 450 parts per million (ppm) carbon dioxide equivalents. This presupposes a portfolio of mitigation options, in particular options to remove carbon from the atmosphere as well as early and deep emission cuts.

The 2 °C target must not only be technically feasible but also readily affordable economically if it is to be acceptable to stakeholders and decision makers around the world. For this reason we estimate the economic costs of achieving different intermediate stabilisation targets that lead with different probabilities to the 2 °C final target. In addition, we evaluate the technological feasibility of reaching these stabilisation targets and explore the importance of individual technologies.

Specifically, we try to answer two key questions:

What is the most ambitious carbon dioxide (CO_2) reduction target that is economically and technically feasible with the ability to achieve the 2 °C target? We explore three different CO_2 stabilisation scenarios that have different probabilities of reaching this target, described more fully in Section 11.3.1. The probabilities of achieving the target increase from approximately 20 per cent for stabilisation at 550 ppm, to 50 per cent at 450 ppm, and 80 per cent at 400 ppm, depending on the climate sensitivity (Hare and Meinshausen, 2006). Achieving the latter level of stabilisation is especially challenging because it would involve both early and rapid decarbonisation of the world's energy system, or negative emissions by the end of the century.

What are some of the technological barriers or economic and political obstacles that could jeopardise the intended emissions stabilisation outcome? For example, what can still be achieved if some of the technology options fail or are ruled out? Moreover, some of the technologies that may be indispensable for reaching very low emission paths, such as large-scale use of biomass, CCS, or nuclear power, may be saddled with high risks and adverse side effects.

In order to assess these key questions, five global regionalised energy–environment–economy models are compared in the Regional Modelling Comparison Project, within the ADAM project. Model comparison analysis can help to identify a range of pathways to a low carbon economy and shed light on the robustness of the associated cost estimates and technology options. Recent examples of model comparisons are Edenhofer *et al.* (2006) with focus on endogenous technological change, or Weyant and Hill (1999) and other contributions to the Stanford energy modelling forum (EMF). So far, low emission pathways have rarely been subjected to such comparisons. In the Intergovernmental Panel on Climate Change (IPCC) Fourth Assessment Report (AR4), for instance, only three models were used to produce results reported for the lowest IPCC stabilisation scenario with radiative forcing of 2.5–$3.0\,\mathrm{Wm}^{-2}$ corresponding to a 445–490 ppm CO_2 equivalent level (Fisher *et al.*, 2007).

Exploring the lower limit of stabilisation is the overarching challenge and focus of the model comparison in this Chapter. Section 11.2 presents the models and scenarios used. Section 11.3 then applies the models and compares the economic and technical results for harmonised baselines with the mitigation scenarios. It discusses alternative ways to achieve low stabilisation, assuming global co-operation and participation by all major players, and explains how individual technology options can be valued. The focus then shifts to the European Union (EU) in Section 11.4 to provide a case study detailing how the required transformation of the energy system can be achieved at regional and sectoral levels. The conclusions drawn in Section 11.5 show how a top-down modelling perspective can be supplemented with a bottom-up analysis considering different sectors, technology options and policy measures for additional insights.

The results presented here are based on the work by Edenhofer *et al.* (2009) in a Special Issue of *The Energy Journal* devoted to *The Economics of Low Stabilisation*. This work provides more information on technical details of the models and analytical specifics underlying most sections of this chapter.

11.2 Models and reference scenario

11.2.1 The models

This model comparison uses the macro-econometric simulation model E3MG (Barker *et al.*, 2006; 2008), the optimal growth models MERGE-ETL (Kypreos and Bahn, 2003; Kypreos, 2005, hereinafter called MERGE) and REMIND-R (Leimbach *et al.*, 2009, hereinafter called REMIND), and the energy system models POLES (European Commission, 1996) and TIMER (Bouwman, 2006). A more detailed description of all models is given in the Appendix and in Edenhofer *et al.* (2009).

MERGE and REMIND are hybrid models with a top-down macroeconomic model and a bottom-up energy system model. Both are optimal growth models where a social planner with perfect foresight maximises global welfare over a given period. Solved at equilibrium, these optimisation models yield least-cost energy systems under a set of constraints. In contrast, the modelling approach in E3MG, also incorporating a macroeconomic and an energy system component, is based on past observations and aims to provide projections and future scenarios consistent with historical data and trends. POLES and TIMER are bottom-up energy system models with a high resolution of different technologies. They seek to minimise the costs of transforming the energy system without assuming perfect foresight. The macroeconomic part of these models is exogenous.

All scenarios are analysed for the period 2000–2100. The models provide regional and country classifications. This exercise distinguishes seven regions which together cover the global aggregate: China (CHN), Russia (RUS), Europe (EU-27), India (IND), Japan (JPN), the United States (USA), and Rest of World (ROW).

11.2.2 The baseline scenario

As far as possible, the building blocks for the baseline without climate policy were harmonised for comparability across the different models particularly with regard to population projections and economic growth. For this, we used the *ADAM baseline scenario*, the underlying assumptions for which are detailed in Chapter 3, van Vuuren *et al.*

Due to the very different modelling assumptions, full harmonisation of all variables between all models is not possible. As Figure 11.1 shows, all models use the same exogenous projections for global and regional population (based on data from the United Nations, 2003, see Figure 11.1). The economic profile is a medium growth scenario, but with high growth rates for India and China (see Chapter 3, van Vuuren *et al.*). Models with exogenous GDP profile (POLES and TIMER) use this projection directly as an input on both the global and the regional level. All other models except E3MG stay close to the reference GDP baseline in Figure 11.1.

Regardless of their adoption of common regional and global GDP baselines, the models differ in their projections of CO_2 emissions. This can be explained by large differences in fossil-based energy prices among the models (see Figure 11.2) affecting the energy mix and the CO_2 emissions in the baseline. In MERGE for instance, the CO_2 emissions increase is greater than in other models due to low fossil fuel prices encouraging continued use of coal, gas and oil. Conversely, the low CO_2 emission pathway for REMIND arises from the assumption of an expensive price path for fossils so that a switch away from coal to renewables is already captured

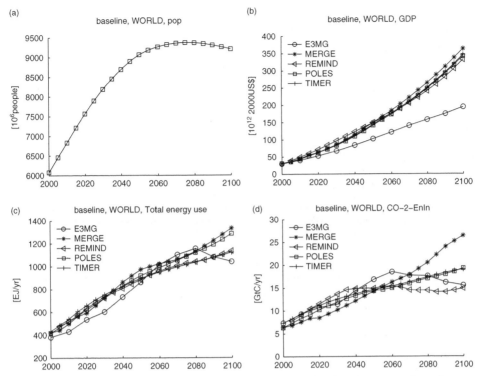

Figure 11.1. Baseline results for WORLD. Projected values aggregated to the global level are reported for (a) population, (b) GDP, (c) total primary energy use, and (d) CO_2 emissions from the energy and industry sector (CO_2-EnIn). E3MG reports lower GDP values as it does not assume long-term convergence in per capita GDP between the regions, with GDP values being reported in constant market prices and not in purchasing power parity (PPP) terms. GDP growth rates are between 2.1 and 2.4% per annum for the other models. (Source Edenhofer *et al.*, 2009.)

to some degree in the baseline. E3MG already has a large amount of renewables in the baseline and therefore shows decreasing CO_2 emissions from 2060 onwards.

11.3 Low stabilisation: opportunities and risks

11.3.1 Long-term stabilisation targets

From the range of emission stabilisation targets that have different probabilities of satisfying the objective to keep global warming to no more than 2 °C, we choose targets characterised as being 'unlikely', of 'medium likelihood', or 'likely' in IPCC terminology. These intermediate targets are associated with stabilisation at 550, 450 and 400 parts per million CO_2 equivalents, respectively, referred to as '550 ppm', '450 ppm' and '400 ppm' in this chapter. Figure 11.3 shows that the 550 ppm

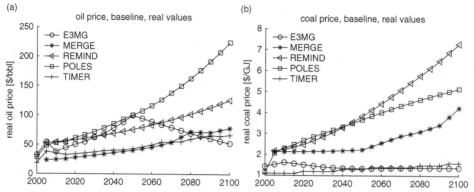

Figure 11.2. (a) Oil and (b) coal prices in the baseline scenario for the different models (in real values). Note that in MERGE, REMIND and POLES the costs are endogenous to the model; for MERGE and REMIND shadow prices are given based on resource extraction costs that are not comparable to spot market prices. In POLES, the price depends on market fundamentals, namely the differential dynamics of supply and demand and the relative amount of spare capacities. For E3MG, historical trends, not prices, are the main driver. In it the real price of oil is an input that follows the POLES price path up to 2050 and then declines by about 2% per annum. For E3MG, the coal price is the average of hard coal and other coal (for USA). (Source Edenhofer *et al.*, 2009.)

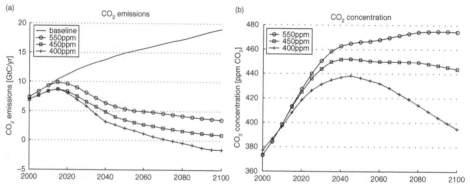

Figure 11.3. IMAGE/TIMER model: emission pathways for (a) CO_2 emissions from energy and industry for the baseline and for 550, 450 and 400 ppm, and (b) CO_2 concentration pathways for the three stabilisation scenarios (right). (Source: Edenhofer *et al.*, 2009.)

scenario (den Elzen *et al.*, 2007) yields an increasing concentration path up to (and beyond) 2100. In the 450 ppm scenario (IMAGE/TIMER 2.9, den Elzen and van Vuuren, 2007), CO_2 equivalent concentrations reach a maximum by 2045 and decline slowly thereafter. In the 400 ppm scenario (IMAGE/TIMER 2.6, van Vuuren *et al.*, 2006; 2007) concentration also peaks around 2045, but at a lower level, and then declines more rapidly than in the previous scenario.

In REMIND and POLES, the CO_2 emissions pathway from the energy and industry sector is taken as the binding cap for each scenario. Emissions from land-use and from other greenhouse gases are given exogenously in all models[1]. The data for the CO_2 emission cap from the energy and industry sector, the land-use emissions, and the emissions from other greenhouse gases are provided by the IMAGE/TIMER model (van Vuuren *et al.*, 2007) for these three stabilisation scenarios. The CO_2 equivalent emission pathway is therefore consistent with the data given by IMAGE/TIMER. MERGE is run with a climate module and binds the climate forcing for each scenario. In E3MG, the cumulated energy-related emissions by 2100 are taken as the binding limit. In POLES, the other greenhouse gases from the energy and industry sector are calculated endogenously, while land use greenhouse gases are applied from the IMAGE/TIMER model.

The first major result is that each of the models can achieve the three stabilisation targets, even in the case of the 400 ppm stringent mitigation scenario. This is a very important result because, as noted in Section 11.3, not many modelling results have been reported for such low emissions stabilisation targets. Inflexibilities in the energy systems, shortcomings in the realisation of mitigation technologies, and myopic investment behaviour are among the reasons why low concentration pathways have so far been assessed and achieved by only a small number of models. Some of the models in our analysis had to be equipped with a wider portfolio of low-carbon technologies, such as CCS and biomass in combination with CCS, to enhance their mitigation capabilities.

11.3.2 Storylines of decarbonisation

This section contrasts the mechanisms and interactions involved in the baseline with those relied upon for meeting the stabilisation targets of 550 and 400 ppm. Although some consistency is achieved through shared data on population, GDP, total energy use and CO_2 emissions (see Figure 11.1), the models reveal very different strategies for meeting future energy demands and favour different energy carriers and technologies (see Figure 11.4). Different assumptions driving the models, for example concerning the price of fossil fuels or learning rates and availability of certain technologies, lead to very different pictures of the primary energy mix in the baseline and mitigation scenarios.

The baselines for each model

The several models included in the comparison exercise span a range of possible pathways to the future. In a scenario without climate policy, fossil fuels continue to

---

[1] In TIMER this exogenous path is consistent with the optimal emission path evaluated with the FAIR model linked to TIMER.

dominate the energy system throughout the century (see Figure 11.4, left column). MERGE and TIMER rely mostly on coal; renewable energy is not important. In POLES, the extent of decarbonisation is very slight. The baseline energy mix in REMIND, however, is characterised by strong decarbonisation using biomass and the introduction of renewable energy sources[2]. In E3MG, renewables increase significantly in the baseline.

The models tell the following stories in their baselines:

In the MERGE baseline, the price of coal is low relative to natural gas and oil, which are largely exhausted in the course of this century. This leads to high levels of coal use and hence more exploitation of electricity generation from coal and of coal-to-liquids fuel production. Both technologies benefit from technological learning. Nuclear power is an important technology in the baseline scenario, particularly towards the middle of the century. This is again driven by the relatively low costs of generation. However, in a baseline in which only light water reactors and limited uranium resources are assumed available, scarcity of these resources becomes a key constraint on any longer-term role for nuclear energy. The cost of wind power technology shows moderate improvements arising from learning. Most other renewable energy sources remain uncompetitive in the baseline scenario.

In the TIMER baseline, fossil fuels remain the dominant energy carriers throughout the century, with the share of oil decreasing due to rising oil prices. The choice of energy carriers in TIMER is determined by cost and by their suitability for use in the various sectors. Costs increase as resources are used up, but decrease due to 'learning-by-doing'. The demand for 'modern' bio-fuels for both electricity and liquid-fuel production increases gradually as the costs of oil and natural gas rise. In addition, technological improvements in production also make these biofuels more competitive. Wind energy use increases steadily, although it remains a minor part of global energy use, while solar energy remains too expensive for large-scale use.

In POLES, capital and operating costs and relative prices jointly determine technological choices and the energy mix. POLES contains endogenous learning curves with a threshold that depends on technology floor costs (minimum engineering cost). The energy mix changes only slightly over time in the baseline scenario, due to inertia in capital-intensive energy production and distribution systems. The use of renewables expands, even in the baseline scenario of the POLES model, because of their cost efficiency in the long term. Wind energy is capped by its technical potential in relation to land availability and population density. For decentralised production, solar PV is constrained by the available surface space of

---

[2] Here and in what follows, renewables include solar, wind, and hydro-electric power. Biomass is reported separately.

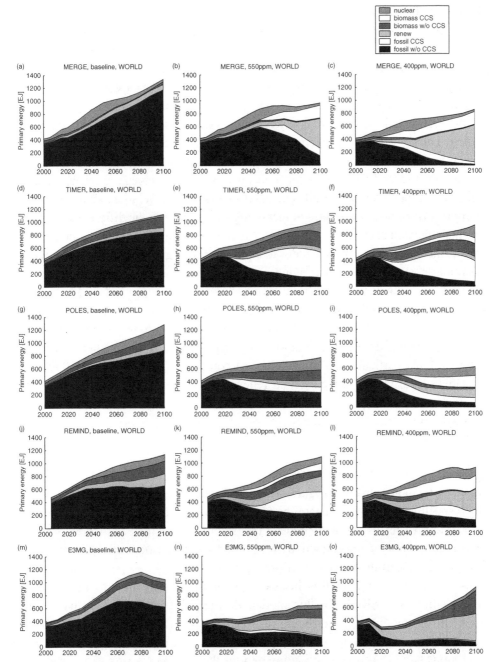

Figure 11.4. Global energy mix for the baseline, the 550 ppm and the 400 ppm scenarios (from left to right) for the five models (a–c) MERGE, (d–f) TIMER, (g–i) POLES, (j–l) REMIND and (m–o) E3MG (from top to bottom). Note that biomass is listed separately and not as part of renewables which only include energy from solar, wind, and hydro-electric power. For the balancing of renewables we apply the direct use concept. In E3MG, biomass includes combustible waste (about 80% of the total biomass use) such as primary solid biomass used for heating in the residential sector in developing countries. (Source: Edenhofer *et al.*, 2009.) (See colour plate section.)

buildings. The theoretical potential of solar thermodynamic power plants is linked to the size of sunny desert regions, but this vast potential is not usable for export because of the unavailability of transcontinental electricity grids and hydrogen transmission lines.

REMIND takes renewables, in particular biomass, into the baseline. The biomass and renewables option becomes competitive because of increasing fossil fuel prices in the second half of the century (see Figure 11.2). Biomass is a general-purpose energy carrier; it can be converted into all secondary energy carriers. Biomass-to-liquid is available at costs comparable to coal-to-liquid but helps to conserve coal for a later use and for conversion into other secondary energy carriers, such as electricity production. REMIND considers changes in the relative prices of energy carriers, driven by uneven rates of technological advance in the different sectors, as the main factor that can change the energy mix. For instance, conversion coefficients of technologies using fossil fuels tend to improve gradually over time while marginal costs of investing in wind and solar PV may be lowered dramatically through innovations resulting from learning-by-doing.

In E3MG, the baseline incorporates some decarbonisation of the global economy, projecting the historical trend of falling carbon intensity into the future. This trend combined with endogenous technological change leads to a significant replacement of fossil fuels, particularly coal, with low-carbon energy sources after 2050 following investment cycles particularly in renewables. Increasing deployment of low-carbon, rather than high-carbon, technologies further stimulates cost reduction through economies of scale in new energy-producing industries.

What the models tell about decarbonisation

In the mitigation scenarios (Figure 11.4, middle and right columns), the energy mix for a specific model is similar for the 550 ppm and the 400 ppm target, so that each model follows its own strategy under either stabilisation target. This shows that the energy mix is principally a function of each model's assumptions about the available technologies, learning rates and resource prices.

A partial exception to this insensitivity of the energy mix to the level of stabilisation is MERGE. In this model, the flexibility provided by having its own climate module and not being restricted to the prescribed CO_2 path allows the transformation of the energy system to be postponed in case of the less ambitious target. The main mitigation options that eventually start to be exercised are renewables and biomass. Hydrogen production from solar thermal, and for non-electric consumption, is an option in MERGE that becomes extremely important with stricter targets. Improvements in energy efficiency also play an important part.

In TIMER, POLES and REMIND, the use of fossil energy without CCS is very similar as their paths are constrained by the exogenous time series for CO_2 emissions. The carbon-free contributions to the energy mix, however, vary between the models. In POLES, reduction of energy use is an important strategy as POLES has demand-side energy efficiency improvements in a bottom-up approach. In general, higher energy prices can spur technological improvements that lead to energy savings in production, and they can also produce changes in behaviour, for example in residential uses and private transportation, which lead to energy savings in consumption. In TIMER, CCS is the main option, although CCS with biomass is allowed only in the most stringent stabilisation scenario. With more CCS than in any of the other models, TIMER subsumes a CCS storage potential of 470 GtC, compared to 280 GtC in MERGE. REMIND shows a steady increase of primary energy consumption because decarbonisation is available at moderate cost with CCS and renewables. Due to this low cost, energy efficiency improvements, here in a top-down representation, play only a minor role. In the policy scenarios, the primary energy consumption from biomass-CCS is connected mainly with hydrogen production for transport and not biomass-to-liquid, as in the baseline scenario.

In E3MG, the stories for 550 ppm and 400 ppm are quite different. In the former scenario the main option is increasing energy efficiency, which is an important demand-side option in E3MG. There are incentives for improving the energy efficiency of private residences and household appliances. Furthermore, regulatory policies pressing for decarbonisation of the transport sector through electrification of the vehicle fleet play a major role. In the 400 ppm scenario, the renewables and biomass options become increasingly more important. Because of learning curves, economies of scale, and Keynesian multiplier effects from the employment of resources that were unemployed in the baseline, the costs associated with increased reliance on renewables are much reduced. This induces large-scale adoption of low-carbon technologies.

Two further findings are especially noteworthy:

(i) Nuclear energy appears to be important as an interim energy source around the middle of this century in some models. The fraction of nuclear power increases in most models until 2050 and then declines, at least in some models, due partly to the depletion of uranium[3].

(ii) Concerning the CCS option, in POLES and REMIND, and to a lesser extent also in TIMER and MERGE, the total amount of CCS shows little variation with the emission stabilisation target. Rather, for the stricter target, CCS is shifted from coal combined with CCS to biomass combined with CCS. The reason is that one way to remove carbon from the atmosphere and to obtain

---

[3] In the standard policy case, fast breeders are not considered as an option in the models.

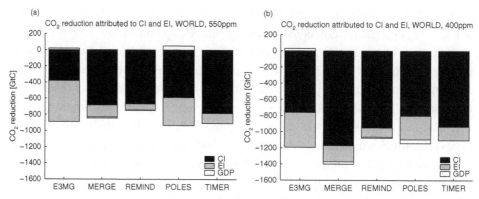

Figure 11.5. Carbon dioxide reductions attributed to lower carbon intensity (CI) and energy intensity (EI) or GDP for (a) 550 ppm and (b) 400 ppm. Reductions are always given relative to baseline. Positive values represent increases from the baseline (i.e. GDP effects in E3MG). (Source: Edenhofer *et al.*, 2009.)

negative emissions is to combine biomass with CCS. In the case of the 550 ppm scenario, negative emissions are not needed, and the use of coal and gas in combination with CCS suffices to reach the stabilisation target.

In general there are three factors that can contribute to changes in emissions according to Kaya's identity (Kaya, 1990). Any CO_2 changes from baseline that are required to achieve the mitigation target can take the form of reductions in (i) carbon intensity (CI), defined as CO_2 emissions per unit of primary energy, (ii) energy intensity (EI), defined as primary energy per GDP, or (iii) growth of GDP. A decomposition analysis enables quantification of the contributions of these different factors (see Figure 11.5). In nearly all cases reducing carbon intensity is the most important strategy, the more so the stricter the stabilisation target.

Except for E3MG, the reduction of energy intensity plays only a minor role as a mitigation option. The MERGE and REMIND do not model end-use energy efficiency technologies in the same detail as supply-side technologies. Moreover, energy intensity is already reduced in the baseline by around one per cent per annum. This is in line with the historical record (e.g. Nakicenovic *et al.*, 2000, Fig. 3–13; Fischer *et al.*, 2007, Fig. 3.6). Lowering energy intensity in response to a more stringent target is more of an option in the energy system models POLES and TIMER as they provide little flexibility for substitution on the supply-side. Moreover, POLES accounts for an explicit bottom-up representation of demand-side technologies (see Section 11.5.2).

11.3.3 Mitigation costs

This section investigates whether there are robust findings, for instance concerning the cost of mitigation or the importance of certain technologies, despite the

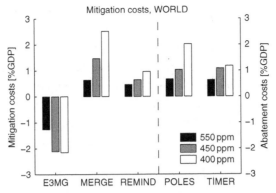

Figure 11.6. Mitigation costs for the 550, 450 and the 400 ppm scenarios. For E3MG, MERGE and REMIND, the mitigation costs (gains for E3MG) are given as cumulative GDP losses up to 2100 relative to baseline in per cent of baseline GDP. POLES and TIMER report the increase of abatement costs relative to baseline in percent GDP. The discount rate is 3%. (Source: Edenhofer *et al.*, 2009.)

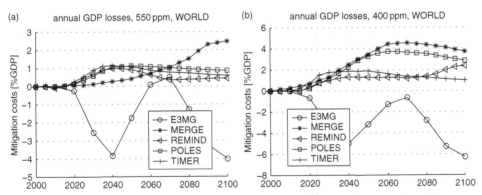

Figure 11.7. Annual mitigation costs in GDP percentage. POLES and TIMER report abatement costs. (Source: Edenhofer *et al.*, 2009.)

different model assumptions. Mitigation cost[4] percentages are net present value sums to 2100 of shortfalls in global GDP or consumption relative to the like sum of baseline values[5]. They are given in Figure 11.6 for the three reference mitigation scenarios. Figure 11.7 shows the corresponding time paths of annual mitigation costs expressed in per cent of GDP. The energy-system models POLES and TIMER report abatement costs, i.e. the sectoral costs for the transformation of the energy sector.

---

[4] In the following, we will use the phrase 'mitigation costs' simultaneously for both losses and gains due to mitigation.
[5] Unless otherwise stated, we use a discount rate of three per cent here.

Most of the models in Figure 11.6 show GDP losses that are increasing with the stringency of the stabilisation target. The only exception is E3MG which is discussed later as a special case. For the other four models, the costs for all stabilisation targets are moderate, with aggregate losses for this century below 2.5 per cent of GDP for the most stringent scenario. The annual losses displayed in Figure 11.7 are moderate until about 2040 but increase in all four models during the transition phase of the energy system and stabilise or even decline thereafter. Overall, the cost estimates are comparable to those appearing in the IPCC AR4 (Fisher *et al.*, 2007, Fig. 3.25, p. 205).

All models include endogenous technological change[6] which can be stimulated by policy measures to yield induced technological change (ITC) and make an important contribution to achieving CO_2 stabilisation targets (Edenhofer *et al.*, 2006). As additional investigations show, without the inclusion of ITC, the costs increase.

MERGE reports the highest costs for the 400 ppm scenario but the lowest, at least until the middle of this century, for the 550 ppm scenario. This is partly due to an increasing use of coal in the MERGE baseline causing much higher CO_2 emissions (see Figure 11.1) requiring increased emission reduction. Compared to the other three models, REMIND yields the lowest overall average annual mitigation costs partly because the price path for fossil energy is assumed to be high (see Figure 11.2) and renewable energy sources are already utilised to some extent in the baseline. In addition, REMIND provides a high degree of flexibility in the choice of low carbon technologies.

The fact that POLES includes only the costs for the transformation of the energy system but no macroeconomic costs might suggest that costs in POLES would be lower than in REMIND and MERGE. However, POLES reports relatively high costs of abatement. This is because the increase of the carbon price induces greater energy efficiency mostly through demand-side technological innovation but also partly through different consumption behaviour. In POLES, stepped-up energy efficiency improvements are essential to reach the set CO_2 mitigation objectives because decarbonising the supply-side alone will not be sufficient. Limiting factors on the supply side are the amount of land available for alternative energy generation and biomass production or the shortage of uranium. Moreover, the most suitable renewable energy production sites are not located where most of the energy is consumed.

Unlike the other models, E3MG reports overall gains from low-level emissions stabilisation (see Figure 11.6). In addition to the application of global carbon prices, a major driver of the mitigation strategy in E3MG is the recycling of revenues raised from auctioning carbon permits to the energy sector and applying carbon taxes for non-energy activities. Key assumptions are that 40 per cent of the revenues collected

---

[6] The models have different representations of endogenous technological change. All models include for example learning in different technologies, and some models include research and development spending. Details of endogenous technological change in the models are given in ADAM Deliverable D-M2.4.

are recycled and used for research and development investments in renewables as well as for investments in energy savings and conversion of energy-intensive sectors towards low-carbon production methods. In contrast to the other models, E3MG is not a supply-driven but a demand-driven model, where resources in a business-as-usual case are not fully employed or optimally utilised. Given the existence of worldwide idle capacities, unemployment and underemployment of resources, mitigation policies may lead to overall gains if employment of these resources is then improved. The question remains whether these gains can be attributed to climate policy or are just an effect of existing inefficiencies in the baseline.

The mitigation benefits and costs reported by E3MG vary greatly over the coming decades (see Figure 11.7) because of investment cycles in new low-carbon technologies. The wave of early investments in plug-in vehicles, greater energy efficiency in buildings, and low greenhouse gas energy supplies lead to an acceleration of GDP growth to 2040, producing negative costs. GDP then falls below baseline for two decades, yielding positive costs as the first-generation investments are replaced, before reverting to net gains in the final decades of the century.

In all models, including to some extent in E3MG, the carbon price drives investments in carbon-free technologies. The price for CO_2 is rising over time in most models, and, at any time, costs in the 400 ppm scenario are more than five times as high as in the 550 ppm scenario (Figure 11.8). However, the high price by the end of this century affects only a small amount of CO_2 emissions (see Figure 11.3(a)) and prevents fossils from re-entering the energy system.

It is important to note that despite the very different assumptions and structures employed in the models, mitigation costs fall into a limited range if results from the Keynesian aggregate-demand model E3MG are disregarded. A robust finding is that mitigation costs are moderate, independent of the energy mix and available technologies.

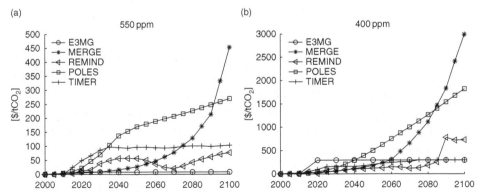

Figure 11.8. Carbon price for (a) 550 ppm and (b) 400 ppm scenario. Note the different scales. (Source: Edenhofer *et al.*, 2009.)

11.3.4 Technology options

The previous section has shown that mitigation costs will be moderate if all technology options, including nuclear energy, the use of CCS and a substantial increase of renewables and biomass are available. The values of particular technology options are explored in this section. To evaluate the 'option value' of including a particular technology in a mitigation programme, we run the models including the full range of mitigation options and then evaluate the extra costs that would arise from excluding a particular option from the portfolio. Thus, the benefits of having, for example, CCS available for use in the mitigation programme would be measured by the added costs of implementing such a programme without it (scenario name 'noccs'). We proceed in this way, one by one, fixing the deployment of renewables at baseline values ('norenew') or holding the use of nuclear power generation is at baseline levels ('nonuke'). To further explore the role of biomass and CCS, we run some additional sensitivity analyses where the biomass potential is fixed alternatively at 100 EJ yr^{-1} ('biomin') and 400 EJ yr^{-1} ('biomax') compared with the standard biomass potential of 200 EJ yr^{-1} in all models. For CCS, a constraint is set that limits the CCS storage potential to 120 GtC ('ccsmin'), compared with 400 GtC in MERGE, or no advance constraints at all in the other models in the standard mitigation scenarios. All technology options analysed are listed in Table 11.1. They are evaluated for the 550 ppm as well as for the 400 ppm scenario by the models MERGE, REMIND and POLES. The estimates of mitigation costs for the different technology options are given in Figure 11.9.

In the 550 ppm scenario, each separate technology (CCS, renewable energy, nuclear energy) can be fixed to its use in the baseline without affecting the feasibility of that scenario, but this is lost in the case of low stabilisation. When CCS is not available or when renewables are held at their baseline values, the 400 ppm target is not achievable by any of the three models. On the other hand, the 400 ppm target can still be achieved when nuclear power is kept at baseline levels.

Biomass

The results in Figure 11.9 show that the amount of biomass included in each of the three models is crucial to the level of mitigation costs. In MERGE and REMIND, costs are more than doubled when biomass potential is cut from 200 EJ yr^{-1} to 100 EJ yr^{-1} and in POLES the target cannot even be met. Conversely, a higher biomass potential of 400 EJ yr^{-1} decreases costs by almost half compared with the reference 400 ppm scenario (for MERGE and REMIND). In general, the biomass potential is not only an important determinant of mitigation costs but it also affects the energy mix because biomass is competing with other renewable energy sources (for MERGE and REMIND). With higher biomass use, reliance on other renewable

Table 11.1. *Technology options for the 550 ppm and 400 ppm target (white) and sensitivity scenarios (grey).*

Scenario name	Description	MERGE	REMIND	POLES
500 ppm / 400 ppm	All options, unlimited CCS potential,[7] biomass potential limited to 200 EJ yr^{-1}	+/+	+/+	+/+
- norenew	Amount of renewable energy is fixed to baseline values	+/−	+/−	+/−
- noccs	Amount of CCS is fixed to baseline values (to zero)	+/−	+/	+/−
- nonuke	Amount of nuclear energy is fixed to baseline values	+/+	+/+	+/+
- biomin	Biomass potential is limited to 100 EJ yr^{-1}	+/+	+/+	+/−
- biomax	Biomass potential is limited to 400 EJ yr^{-1}	+/+	+/+	+/+
- ccsmin	CCS storage potential is limited to 120 GtC	o/+	o/+	o/o

For MERGE, REMIND and POLES it is shown whether the target is achieved for 550/ 400 ppm. A plus (+) means that the stabilisation target has been met, a minus (−) means that the stabilisation target has not been met, a circle (o) means that this scenario is not run.

sources declines. For further analysis of the biomass potential scenarios we refer to Edenhofer *et al.* (2009).

It is important to add that, so far, only the technical potential has been varied in the model. Thus, for biomass production, conflicts with other types of land use, in particular, food production and biodiversity protection, as well as the question of whether a given biomass harvest can be sustained, have not been investigated. Cost effects of higher land prices due to increased demand have so far not been accounted for in the models. Furthermore, zero emissions are attributed to bio-energy use, thus neglecting emissions from direct and indirect land use changes and the biomass production process itself. Certain types of land use changes, such as converting wetlands or clearing tropical forests, lead to increased greenhouse gas emissions rather than emission reductions. Neglecting these emissions not only hides possible additional climate damage, but also yields an overly optimistic assessment of the economic potential of biomass in scenarios including carbon pricing. All these points are crucial for the assessment of low stabilisation scenarios. Indeed, it

---

[7] In MERGE, the CCS potential is limited to 400 GtC.

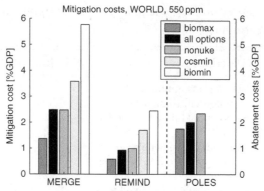

Figure 11.9. Mitigation costs as cumulative GDP losses (MERGE, REMIND) up to 2100 relative to baseline; POLES reports the increase of abatement costs relative to baseline in percentage GDP. The option values for different technologies for the 400 ppm scenario are shown (see Table 11.1). The reference case is the scenario where all mitigation options are available (in black). None of the models achieves the target in the norenew and noccs scenario, POLES does not stay below the cap for the biomin scenario. The mitigation costs of the sensitivity scenarios are always given relative to the respective baseline, which for the biomin run for instance would be a baseline with a biomass limit of 100 EJ yr^{-1}. (Source: Edenhofer *et al.*, 2009.)

could turn out that the costs of low stabilisation, incorporating all these factors, would be at the upper end of the numbers shown in Figure 11.9.

CCS

Without use of CCS, the low stabilisation target cannot be attained by any of the models. Limiting the CCS potential to 120 GtC (ccsmin) allows the target to be achieved but at high cost. In MERGE and REMIND, mitigation costs increase accordingly by about 1 percentage point which equates to a doubling of the costs in REMIND. Since additional costs of CCS, such as investments in regulatory frameworks regarding health, safety and environmental risks, are not included in the models, the full costs certainly could be even higher than shown here.

Nuclear power

When limiting the use of nuclear power to the baseline values, costs increase only moderately for REMIND and do not increase at all for MERGE and POLES. Hence, the nuclear option is less important than renewables or CCS. This is due partly to the fact that nuclear energy is already attractive in the baseline scenario, but cannot be further extended due to limited uranium resources (in MERGE and REMIND). An additional nuclear phase-out scenario, with no investment in nuclear power generation after 2000, raises costs 0.5 percentage points in MERGE, 0.1 percentage points in REMIND, and 0.7 percentage points in POLES. Therefore, at least in MERGE and POLES, the use of nuclear power is important in the baseline.

Two models also explore the option of implementing a fast breeder. With it, costs can be reduced by 0.3 percentage points for MERGE and 0.1 percentage points for REMIND compared with the standard scenario without this option. These benefits should be balanced against the increased risk of proliferation, safety and more nuclear waste.

Overall, MERGE, REMIND and POLES provide similar assessments of the value of the individual technology options. Three key findings emerge that are shared by all three models: (i) renewables and CCS are the most important options because the 400 ppm target is not feasible, and the 550 ppm target is very expensive, without them; (ii) the biomass potential dominates costs under low stabilisation; and iii) nuclear energy is dispensable as a mitigation option as the 400 ppm target remains feasible and mitigation costs increase only slightly when nuclear power is kept at its baseline level.

However, nuclear power is important up to the extent of its use in the baseline. A more detailed analysis of the option values and a comparison of the 550 ppm with the 400 ppm scenario are given in Edenhofer *et al.* (2009).

11.4 Spotlight on the EU: necessity and feasibility

Previous sections conclude that low stabilisation is feasible technologically and also economically from a global perspective. In this section we will put the spotlight on the EU-27 region to show how mitigation policy might be implemented in Europe. We first show top-down results from the model comparison introduced in the last sections for EU-27 and then we present a disaggregated sectoral analysis of the 400 ppm low stabilisation scenario.

11.4.1 Top-down analysis: energy mix and mitigation costs

We now explore the regional results from the different models. The energy mix for the EU-27 is shown in Figure 11.10. Although the energy mix differed greatly by model for the WORLD region (see Figure 11.4), for Europe the baseline is very similar for all models except E3MG. The strategies of decarbonisation for Europe, however, differ among the models. They revolve around reduction of energy use and implementation of CCS and appreciable expansion of renewables in only two of the five models, MERGE and REMIND. TIMER and REMIND concentrate on the use of CCS, whereas the mitigation solution in POLES involves greater use of biomass relative to baseline. In E3MG, the large increase in energy use is due to the assumed GDP growth rates and fossil fuel price increases raising the demand for energy so that renewables increase substantially in the baseline.

To evaluate the mitigation costs for EU-27 in relation to the other world regions, we applied an emission trading scheme to three of the models, where emission

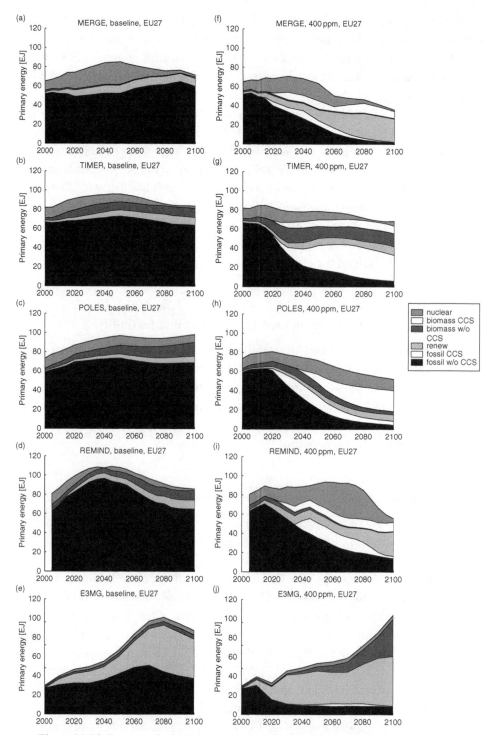

Figure 11.10. Energy mix (a)-(e) for Europe (EU-27) for the baseline and the (f)-(j) 400 ppm CO_2e scenario. Note that MERGE reports results only for EU-15. Note the different scale for E3MG. Renewables include solar, wind, and hydro-electric power. For the balancing of renewables we apply the direct use concept. (See colour plate section.)

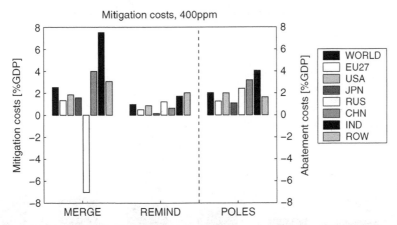

Figure 11.11. Regional distribution of mitigation costs for the 400 ppm scenario. Results are shown for a contraction and convergence allocation scheme with full emission trading. Note that MERGE reports results for EU-15. (See colour plate section.)

permits for each region are allocated according to a contraction and convergence scheme (see Figure 11.11). The associated costs for this transition in all three models are lower for EU-27 than for WORLD.

China reports much higher costs than the world average in two of the three models. This could be an important sticking point in international negotiations, as China may demand compensation before consenting to incur high mitigation costs.

India faces the highest mitigation costs in MERGE and POLES, and costs for India are higher than the WORLD average in REMIND. In MERGE, Russia benefits substantially from its large biomass potential and can therefore sell emission permits. By contrast, mitigation costs for Russia are higher than the WORLD average in REMIND and POLES. Hence, results are difficult to generalise for developing countries, including ROW.

Costs for the three developed country categories, EU-27, USA, and Japan, however, cluster closely together. The United States consistently has the highest costs of the three, but pairwise differences between their costs are distinctly less than one per cent within models and not much larger across models. By contrast, differences between the developing country groups or countries tend to show much larger variations between models and depend substantially on the target (not shown here).

11.4.2 Bottom-up analysis for residential and service, industry and transport sectors

We used POLES to provide a detailed bottom-up sectoral analysis of the 400 ppm stabilisation scenario until 2050 (Jochem *et al.*, 2009). Final demand in all energy sectors in Europe differs from the baseline to the mitigation scenario through

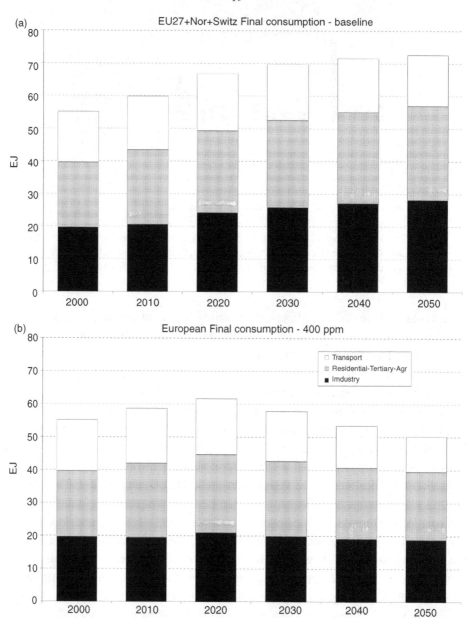

Figure 11.12. European final energy demand by sector, 2000 to 2050, for (a) baseline and for (b) the 450 ppm CO_2e scenario. (Source: Jochem *et al.*, 2009.)

additional energy conservation and more efficient energy use in mitigation (see Figure 11.12(a)). While projected European final energy demand increases by one-third to about 70 EJ in 2050 in the baseline, in the 400 ppm scenario it peaks in 2020 and then decreases slightly to 50 EJ in 2050, which is less than the level in 2000.

In the baseline scenario, final energy demand rises slowly in two of the three sectors identified in Figure 11.12. However, it declines slightly in the transportation sector through substantial improvements in energy efficiency and stagnating population. This decline arises despite continued growth in GDP

Several factors may contribute to energy demands decreasing in the second half of this century, as projected in nearly all models' baselines shown in Figure 11.10. They include saturation with energy-using equipment, a levelling-off in standards of personal comfort, limited time-budgets for personal transport, and significant oil price increases. Stricter technological efficiency standards for buildings, electrical appliances and road vehicles, and structural change toward less energy-intensive industry branches and service sectors also play a role in lowering energy demand. The absolute decline of fuel demand in the transport sector after 2020 may suggest that Europe will soon be entering a second phase of increased energy efficiency, following that for stationary energy services. Electricity is the only energy carrier which continues to grow by some 1.5 per cent annually, and still at 1.1 per cent per annum between 2030 and 2050.

In the 400 ppm CO_2e scenario, climate policies have a marked impact on European final energy demand in all sectors (see Figure 11.12(b)). Industrial energy demand continues to increase, but at a slower pace. Energy demand of the transport and residential and service sectors decreases after 2020 to 38 per cent and 74 per cent, respectively by 2050, compared with the corresponding energy demand in the baseline (see Figure 11.12). In the transport sector, new propulsion technologies and lighter vehicles advance the development of cleaner cars. The stock effects, due to economic longevity of capital assets, are less important than for buildings. Due to the impact of higher oil prices, conventional cars steadily lose market share against hybrid, electric, hydrogen, and hydrogen fuel cell technologies after 2020, even in the baseline scenario. In the 400 ppm stabilisation scenario, these changes in the market for transport fuels occur more rapidly. In the industry sector, advanced technologies such as waste heat recovery, small cogeneration, new processes based on physico-chemical techniques and biotechnology, and advances in recovering brake energy by power electronics contribute to efficiency improvements in the stabilisation scenario.

In Europe, the largest reductions in per capita energy demand occur in thermal uses in residential and non-residential buildings (see Figure 11.13). In the baseline, this particular demand trajectory is flat because the effects of increased energy efficiency driven by higher fuel prices roughly offsets the use of more floor area per capita. Price incentives are too weak to trigger much development of low energy and passive buildings, so their market share remains small. By contrast, results from the 400 ppm CO_2 equivalent scenario incorporate strong growth in the construction of low energy or passive buildings, lifting their share to 40 per cent of the building stock by 2050. Current experience in many countries, particularly

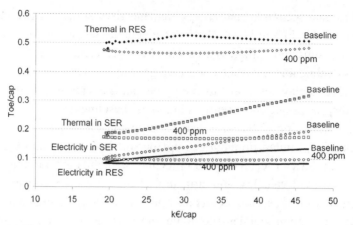

Figure 11.13. European energy demand per capita for residential and non-residential buildings and for electricity (in tonnes of oil equivalent) in relation to GDP per capita (in thousands of Euros) for the baseline and 400 ppm scenarios for 2000–2050. (Source: Jochem *et al.*, 2009.)

in the north of Europe, indicates that such buildings use only one-tenth to one-half as much energy per unit as the existing European building stock. Investments in zero-energy or even positive-energy buildings, for example, with integrated solar PV panels, are also factored in.

On the other hand, per capita demand for electricity in the European residential and service sectors increases almost proportionally with per capita GDP in the baseline and even in the 450 ppm stabilisation scenario (see Figure 11.13). In this scenario, the impact of higher energy efficiency is offset in part by additional electricity demands for ventilation systems and heat pumps for low energy and passive buildings. The net result is only a slight decrease in per capita electricity demand or electricity intensity compared to the baseline scenario (see Figure 11.13).

The transformation in the energy system directly affects the energy-related CO_2 emissions. Without any climate policy, greenhouse gas emissions would increase 1.3-fold in Europe (see Figure 11.14(a)). Regarding energy-related CO_2 emissions in the stabilisation scenario, ambitious climate policies could cause European CO_2 emissions to peak at about 4.2 Gt CO_2 around 2020 and thereafter impose a decrease of about three per cent annually to some 0.9 Gt CO_2 in 2050 (see Figure 11.14(b)). This is about one-fifth of the current level of CO_2. All sectors contribute to this decline in emissions, but the major contribution comes from electricity generation in the energy producing 'transformation' sector in Figure 11.14. This is due to the rapidly rising use of renewables and CCS technologies in power generation and coke ovens, and the use of nuclear energy in some European countries.

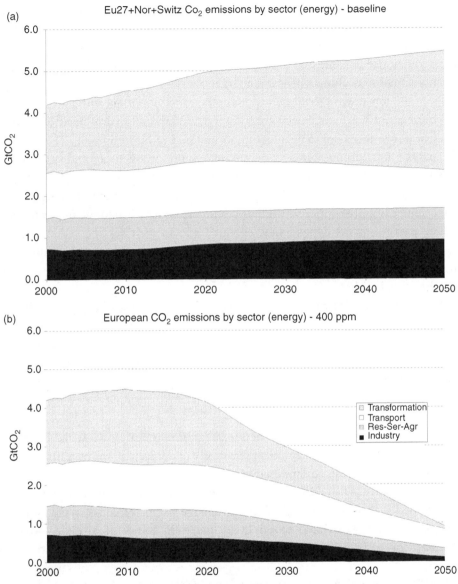

Figure 11.14. European energy-related CO_2 emissions by energy sector in $GtCO_2$, for (a) the baseline and (b) the 400 ppm scenario for 2000 to 2050. (Source: Jochem *et al.*, 2009.)

11.5 Conclusions

This chapter shows that it is likely that the 2 °C target embraced by the EU will be achievable, at moderate costs, if the full suite of technologies is available and effective policy instruments are applied. The model comparison identifies a number of different pathways by which a low stabilisation target of 400 ppm for atmospheric greenhouse gas emissions can be achieved by 2100. However, stricter mitigation targets bring greater dependence on selected technologies, such as CCS and bio-mass, giving rise to some loss of flexibility in the choice of technologies to achieve the more ambitious climate protection targets.

From the top-down global perspective, the most important mitigation options are biomass, CCS, and renewables (wind, solar and hydro-electric power). Nuclear power turns out to be a less important option for mitigation. The bottom-up analysis for Europe points to improving energy productivity and substituting renewables for fossil fuels as the most promising approaches to achieving a sustainable energy system in Europe until 2050. Improvements in energy efficiency and using fossil fuel substitutes would create new economic opportunities, reduce energy import dependency, enhance human welfare, and reduce energy costs and greenhouse gas emissions. Europe has an significant opportunity to improve energy productivity, with substantial benefits for both its own economy and the global climate.

Although some policy measures and markets function best with global participation on shared terms, there can be many pathways to a carbon-free economy. Countries or entire regions can choose those approaches to decarbonisation best suited to them. Further modelling work could usefully explore extreme scenarios that place disproportionate reliance, say, on renewables or nuclear power. This would allow deeper analysis of the national or regional context-specificity of alternative low-carbon technologies and mitigation options.

One approach would be to seek better integration of top-down with bottom-up models when assessing the technical feasibility and implications of mitigation. In addition, each portfolio of mitigation options has social and economic consequences that are not fully captured in quantitative models, and these would need to be assessed with other methods. Thus extensive use of biomass, wind, solar or nuclear power would have to be investigated in terms of attitudes to risk, other aspects of social and political acceptance, and delayed dynamic effects in society.

Finally, model results do not normally consider barriers to the timely implementation of optimal policies with the best technical instruments. Moreover, significant energy efficiency improvements would require the implementation of new technologies and need policy instruments to stimulate this development. Calculating option values for policy instruments, i.e. analysing the impact of single policy measures, would be an important next step in informing policy makers and the public about the cost of ruling out any particular measure in the efficient set. Such evaluations could enable the re-examination of pre-existing obstacles and political taboos that may conceivably be overcome.

References

Barker, T., Köhler, J., Anderson, D. and Pan, H. (2006) Combining energy technology dynamics and macroeconometrics: the E3MG model for climate stabilisation scenarios. *The Energy Journal (Special Issue: Endogenous Technological Change and the Economics of Atmospheric Stabilisation)*, 113–33.

Barker, T., Scrieciu, S. S. and Foxon, T. (2008) Achieving the G8 50 percent target: modeling induced and accelerated technological change using the macro-econometric model E3MG. *Climate Policy*, **8**, S30–S45.

Deliverable D-M2.4 (2007) *Report on Model Comparison*. Potsdam: PIK

Edenhofer, O., Lessmann, K., Kemfert, C., Grubb, M. and Koehler, J. (2006) Induced technological change: exploring its implications for the economics of atmospheric stabilisation: synthesis report from the Innovation Modeling Comparison Project. *The Energy Journal (Special Issue: Endogenous Technological Change and the Economics of Atmospheric Stabilisation)*, 57–107.

Edenhofer, O., Knopf, B., Barker, T. *et al.* (in press) The economics of low stabilization: Model comparison of mitigation strategies and costs. *The Energy Journal (Special Issue: The economics of low stabilization)*.

den Elzen, M. G. J., Meinshausen, M. and van Vuuren, D. P. (2007) Multi-gas emission envelopes to meet greenhouse gas concentration targets: costs versus certainty of limiting temperature increase. *Global Environmental Change*, **17**, 260–80.

den Elzen, M. G. J. and Van Vuuren, D. P. (2007) Peaking profiles for achieving longterm temperature targets with more likelihood at lower costs. *Proceedings of the National Academy of Sciences, USA*, **104**, 17931–17936.

European Commission (1996) *POLES 2.2*. European Commission DG XII, EUR 17358 EN.

Fisher, B. S., Nakicenovic, N., Alfsen, K. *et al.* (2007) Issues related to mitigation in the long term context. In *Climate Change 2007: Mitigation. Contribution of Working Group III to the Fourth Assessment Report of the Intergovernmental Panel on Climate Change*, ed. Metz, B., Davidson, O. R., Bosch, P. R., Dave, R. and Meyer, L. A. Cambridge, UK and New York, NY, USA: Cambridge University Press, pp. 169–250.

Hare, B. and Meinshausen, M. (2006) How Much Warming are We Committed to and How Much can be Avoided? *Climatic Change*, **75**, 1–2, 111–149.

Jochem E., Schade, W., Barker, T. *et al.* (2009) Adaptation and mitigation strategies for Europe. *Final Report of Work Package ADAM-M1*, Karlsruhe: Fraunhofer ISI, May 2009.

Kaya, Y. (1990) Impact of carbon dioxide emission control on GNP growth: interpretation of proposed scenarios. *Paper presented to the IPCC Energy and Industry Subgroup, Response Strategies Working Group*, Paris.

Kypreos, S. and Bahn, O. (2003) A MERGE model with endogenous technological progress. *Environmental Modeling and Assessment*, **8**, 249–259.

Kypreos, S. (2005) Modeling experience curves in MERGE (model for evaluating regional and global effects). *Energy*, **30**(14), 2721–2737.

Leimbach, M., Bauer, N., Baumstark, L. and Edenhofer, O. (2009) Mitigation costs in a globalised world: climate policy analysis with REMIND-R. *Environmental Modeling and Assessment*. In press.

Nakicenovic, N., Alcamo J., Davis G. *et al.* (2000) *Special Report on Emissions Scenarios: A Special Report of Working Group III of the Intergovernmental Panel on Climate Change*. Cambridge, UK: Cambridge University Press.

United Nations (2003) *Proceedings of the United Nations Technical Working Group on Long-Range Population Projections United Nations Headquarters New York*. Medium fertility version (up to 2050) and the UN Long-Range Medium (up to 2100). www.un. org/esa/population/publications/longrange/long-range_working-paper_final.PDF van Vuuren, D. P., van Ruijven, B., Hoogwijk, M., Isaac, M. and de Vries, B. (2006) TIMER 2: Model description and application. Integrated modelling of global environmental change. In *An Overview of IMAGE 2.4*, ed. Bouwman, A. F., Kram, T. and Klein Goldewijk, K. The Netherlands: Netherlands Environmental Assessment Agency (MNP), Bilthoven, pp. 39–59.

van Vuuren, D. P., den Elzen, M. G. J., Lucas P. L. *et al.* (2007) Stabilising greenhouse gas concentrations at low levels: an assessment of reduction strategies and costs *Climatic Change*, **81**(2), 119–159.

Weyant, J. and Hill, J. (eds.) (1999) The costs of the Kyoto protocol: a multi model evaluation. *The Energy Journal*, Special Issue.

12

Mainstreaming climate change in development co-operation policy: conditions for success

Lead authors:

JOYEETA GUPTA[1], ÅSA PERSSON, LENNART OLSSON

[1]Co-ordinating lead author

Contributing authors:

JOANNE LINNEROOTH-BAYER, NICOLIEN VAN DER GRIJP,
ANNE JERNECK, RICHARD J. T. KLEIN, MICHAEL THOMPSON,
ANTHONY PATT

Summary

An important means of assisting developing countries in dealing with climate change has been to mainstream climate change into development co-operation. Three supporting arguments for a mainstreaming response are: (i) ensuring that technology transfers and choice of development paths reduce the rate of growth of greenhouse gas emissions in developing countries; (ii) making development projects and their outcomes resilient to the potential impacts of climate change; and (iii) avoiding the transaction costs of setting up a new institutional framework to deal with support from developed countries for the climate change challenge in the developing world. However, there are also three sets of concerns with respect to such mainstreaming. Politically, mainstreaming undermines the promise under the climate change regime to provide new and additional resources to developing countries. It also raises questions of control, conditionality and accountability. From a development economics perspective, the concerns are that development aid has traditionally not served the poorest of the poor and, furthermore, that mainstreaming may distort existing markets, and thereby not serve the interests of the partner countries. Sustainability concerns are that mainstreaming – simply by association with the overall development paradigm it is working within – will propagate the existing flawed development paradigm to developing countries and may lead to aid dependency. The chapter concludes that ten generic conditions of success emerge from the analysis on concerns including: additionality, developing country ownership, partnership, joint developed country–partner Official Development Assistance (ODA) policy development and accountability, prioritisation of the poorest, avoidance of market distortions, limiting ODA dependency, and ensuring tripartite (private, government and civil society) decision making.

Making Climate Change Work for Us: European Perspectives on Adaptation and Mitigation Strategies,
ed. Mike Hulme and Henry Neufeldt. Published by Cambridge University Press © Cambridge University
Press 2010.

12.1 Introduction

It has become conventional wisdom that the impacts of climate change will be felt most severely by developing countries. This is for a number of reasons (IPCC, 2007; Parry *et al.*, 2007; Stern, 2007; UNDP, 2007). Firstly, these countries, compared to the industrialised North, have a greater dependence on climate-sensitive sectors (Antle, 1995; IRI, 2005). Secondly, while countries in higher latitudes may experience a 'greening' from climate change (Lucht *et al.*, 2002), the effect on ecosystems and human health in developing countries, though highly uncertain, is more likely to be negative (Boko *et al.*, 2007; Martens *et al.*, 1997; McBean, 2004). Thirdly, persistent poverty exposes people to suffer disproportionately from the impacts of climate change (Ribot *et al.*, 1996), while preventing their active engagement in successful adaptation (Thornton *et al.*, 2006; Washington *et al.*, 2006).

This disproportionate burden, recognized already in the early 1990s, led to the adoption of the notion of common but differentiated responsibilities and respective capabilities of Parties in the United Nations Framework Convention on Climate Change (UNFCCC, 1992; Rio Declaration, 1992). Developed country Parties committed to 'provide new and additional financial resources' to help developing countries comply with the Convention (UNFCCC, 1992: Art. 4(3)) and to assist the most vulnerable countries to meet the costs of adaptation (UNFCCC, 1992: Art. 4(4)). Furthermore, development co-operation agencies started examining whether their development projects influence, or are vulnerable to, climate change (CEC, 2003; Klein *et al.*, 2007), and how best to help vulnerable countries to adapt to climate change and variability (e.g. IRI, 2006). Climate change and international development are two different programmatic and institutional areas of policy concern; both areas are grappling with the question of how best to integrate the desire to reduce climate vulnerability, while promoting lasting development (Gupta, 2009).

This effort towards integrating climate change mitigation and adaptation into existing development co-operation policy is called mainstreaming. In this context, mainstreaming means using the existing bricks and mortar of ODA agencies to handle the additional challenge of climate change. This involves both designing new activities with mitigation and/or adaptation as primary purposes, and also ensuring that other ODA activities are not contradictory or counterproductive, but rather synergistic with climate-related goals.

Mainstreaming within ODA simply seems to be common sense. If ODA is meant to support development, then that development ought both to ensure reduced contributions to greenhouse gas emissions and to be resilient to the potential impacts of climate change in the future (Section 12.2). But there are a number of concerns that surround mainstreaming (Section 12.3). After elaborating these concerns, we explain some of the conditions under which mainstreaming of climate change in development co-operation could be successful (Section 12.4). The final section offers some conclusions and policy advice for the EU.

12.2 Mainstreaming: its logic, political support and justification

This section overviews the current policy and practice of mainstreaming climate change mitigation and adaptation into development policies and provides arguments in favour of doing so. Mitigation includes the adoption of measures to reduce greenhouse gas emissions and to ensure carbon sink capacity. Adaptation refers to measures taken to cope with the impacts of climate change.

12.2.1 The concept and rationale of mainstreaming

The idea of mainstreaming, often used in policy discourses, implies that the issue at hand (e.g. gender, environment) is explicitly taken into account, or integrated, in the policy formulation and implementation process (Lenschow, 2002; Persson, 2007). For example, the Treaty of the European Union stipulates that the environment should be integrated (i.e., through tools such as guidelines and checklists) in all EU sector policy processes, since this is assumed to lead to more environmentally beneficial policy outputs and outcomes. However, Picciotto (2002: 323) argues that mainstreaming goes further than technical forms of policy integration, in that it: 'suggests a deliberate perturbation in the natural order of things' and 'creates winners and losers, challenges vested interests'. Thus mainstreaming includes both technical and political aspects.

In the context of ODA and climate change, mainstreaming refers to the integration of mitigation and adaptation responses in development co-operation policy, programmes and projects. Mainstreaming *mitigation* calls for reducing (the rate of growth of) greenhouse gas emissions and enhancing carbon sink capacity through the use of ODA. Thus, where traditional ODA projects promoted thermal power plants in developing countries, mainstreaming would promote renewable energy instead. However, this may involve trade-offs with certain development priorities, as mitigation measures may have other negative environmental, economic or social costs as well as opportunity costs. Thus, promoting renewable energy may imply fewer resources for priorities such as access to drinking water and sanitation or health and education. Climate change negotiations reflect such perceptions of trade-offs between mitigation and economic development paths, or the assumption that mitigation has low opportunity costs in terms of development progress (Gupta, 1997; 2006).

Mainstreaming *adaptation* into ODA implies ensuring that development cooperation takes potential climate impacts into account. For example, if the ODA

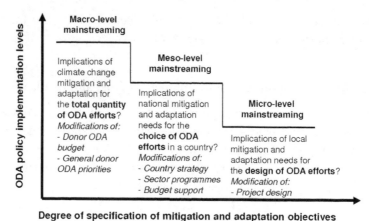

Figure 12.1. Levels of mainstreaming in ODA decision making (after Persson and Klein, 2009).

project is to set up a hydro-electric plant, the project should take into account whether water supply will be affected by climate change. Mainstreaming adaptation also means considering how adaptive capacity to climate change can be actively enhanced through development projects and policies. Often building adaptive capacity may not entail the introduction of 'new' technologies or management practices, such as renewable energy technologies when mainstreaming mitigation, but rather upscaling or modifying existing development initiatives (e.g. educational efforts to address current climate variability). Much of the adaptation literature sees a relative affinity between adaptation and 'good' development, in that the two may be mutually supportive (McGray *et al.*, 2007; Eriksen *et al.*, 2007). However, implementation of adaptation can be challenging. Downscaling climate models to the level of decision making needed for a hydro-electric plant or a reservoir may not be easy. While an adaptation strategy may call for less construction near coastlines, the government may see such construction as critical for revenue in the tourism industry or simply because no other land is available.

The technical versus political aspects of mainstreaming referred to above can be related to three levels (Figure 12.1). At project level (micro, sub-national level) it is about 'climate proofing' existing, more or less pre-defined, sectoral ODA projects to ensure that mitigation and adaptation aspects have been considered: the former to ensure that the aid project is not counterproductive to mitigation objectives and the latter to ensure that the aid project is effective even under climate impacts. This also implies that, in the future, new projects will be designed explicitly taking climate change into account, including making it the primary purpose of the project. Mainstreaming can also be seen at a strategic national policy level (meso-level) where priorities for donor country budgets are agreed between donor and partner

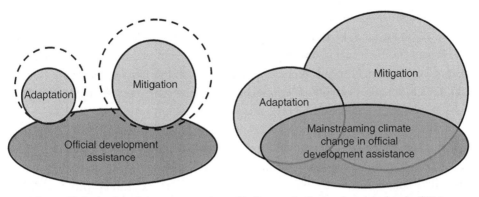

Figure 12.2. A pictorial representation of before and after mainstreaming in ODA.

countries. Mainstreaming at the macro-level entails decisions on the total size of national ODA budget and how it is to be spent, i.e. whether mitigation and adaptation should be allocated a share of the budget (and how much) or if they are funded outside the ODA budget.

The implications of mainstreaming are represented in Figure 12.2. The 'before mainstreaming' diagram shows that some ODA is used or climate proofed for adaptation and mitigation activities. The rest is not. The 'after mainstreaming' diagram shows that, through mainstreaming, all ODA becomes subject to some kind of climate proofing, potentially also catalysing more rapid non-ODA investment in both areas.

12.2.2 The evolution of mainstreaming

Initially, in the international arena, mitigation was given more importance than adaptation (*cf.* Bodansky, 1993; Gupta, 1997). As part of mitigation strategies, emphasis was placed on the need for technology transfer to enable developing countries to adopt modern low greenhouse gas technologies rapidly (SWCC, 1990). This recommendation was geared towards both the private and the public sector.

An international instrument to promote, *inter alia*, technology transfer is the clean development mechanism (CDM), which aims to engage the private sector in mitigation activities. Development aid has been used to fund pilot projects and increasingly to support capacity building in relation to the CDM in the developing countries (Yamin, 2005a, b). This shows that gradually climate mitigation objectives were being incorporated into the ODA agenda, although there have been objections to this (see Section 12.3.2).

Pressure to mainstream climate change mitigation in foreign policy can be seen, for example, in relation to export credits, which subsidise the exports of private companies to developing countries and make such products internationally competitive.

Annual Organisation of Economic Cooperation and Development (OECD) export credits of about US$ 17 billion were used to leverage USD 200 billion of annual investments in fossil fuel in developing countries (Maurer and Bhandari, 2000). Such export credits have been the subject of court cases in the US and Germany (Gupta, 2007). This puts pressure on government and aid agencies to revisit their technology transfer practices, and the OECD has adopted new export credit rules in 2007.

Unlike mitigation, adaptation has only more recently become important as a global policy objective. Lack of new and additional resources and the need to ensure effective (i.e. climate-proofed) aid activities led ODA agencies to promote mainstreaming adaptation rather than focus on stand-alone, additional projects. ODA agencies started to screen their portfolios for climate risk in the early 2000s. Subsequently, a range of 'climate proofing' tools and guidance for mainstreaming adaptation into Poverty Reduction Strategy Papers (PRSPs), country strategies, sector programmes, and projects have been developed.

12.2.3 The current practice of mainstreaming

A majority of the 23 OECD Development Assistance Committee (DAC) member countries have adopted specific climate change objectives and policies, whereas some countries include climate change in a broader environmental strategy and discuss climate change in their regular, high-level policy dialogues with partner countries. Denmark, Germany, the Netherlands, and UK are proactive in this field (OECD survey cited in Gigli and Agrawala, 2007).

Mainstreaming of *adaptation* (and less of *mitigation*) started as one-off initiatives by development banks and donor agencies (e.g. the agencies of Norway and the UK) to screen their investment portfolios (Klein *et al.*, 2007). This enables them to understand the relationship between development, poverty, climate and adaptation (Eriksen *et al.*, 2007) and to set priorities in terms of sectors and countries in need of adaptation. More recently, political milestones included the 2005 G8 Gleneagles Plan of Action, the 2006 multi-agency Clean Energy and Development Investment Framework led by the World Bank, and the 2006 Declaration on Integrating Climate Change Adaptation into Development Co-operation (OECD, 2006).

The 2006 OECD Declaration stated that 'adaptation to climate change is not a "stand-alone" agenda but needs to be integrated into development policy making and planning'. It thereby contributes to achieving development objectives (including the Millennium Development Goals) and should take 'into account the legitimate priority needs of developing countries'. The Declaration set out a systematic approach to mainstreaming for ODA agencies. Several bilateral ODA agencies have now implemented portfolio screenings as a first step to 'climate-proof' their ODA activities, although the quality is varied (Klein *et al.*, 2007). Several ODA agencies are now developing procedural guidance and tools, and general and sector-specific checklists. As ODA is moving towards a more programmatic approach, with project funding being gradually substituted by general budget support and sector-wide approaches (SWAps) (OECD, 2008), mainstreaming of adaptation calls for intervention at higher strategic decision making levels, such as the partner country-led Poverty Reduction Strategy Papers (PRSPs) or the donor-led ODA country assistance strategies.

Turning to *mitigation*, mainstreaming has been undertaken less systematically. While there is increasing attention being paid to export credits and the role of aid in the CDM, a systematic evaluation of the aid portfolio to study its climate consequences has not so far occurred. OECD DAC has focused only on ODA support for energy efficiency, renewables and forestry practices (OECD, 1992) and not on a systematic evaluation of its portfolio. The European Union (EU) adopted a strategy and action plan for 2004–2008 (CEC, 2003) prioritising the issues of raising the profile of climate change, support for adaptation and mitigation and capacity building in development co-operation. The EU has also established a Global Climate Change Alliance with a budget of 50 billion Euros (2008–2010) to systematically integrate climate change into development co-operation with least developed countries and small island developing states (CEC, 2007).

The United Nations Development Group (UNDG) has set up a United Nations Development Assistance Framework (UNDAF, 2009), which aims to provide the country teams with a framework within which development action should be taken. The framework emphasises that aid should, *inter alia*, focus on national ownership of policies, tailor-made country analysis, a human rights approach, gender equality, environmental sustainability, capacity building and disaster management. It focuses on environmental sustainability (implying both mitigation and adaptation) and refers to the climate change convention. It has adopted an Action Plan in 2005 to implement, *inter alia*, the OECD 2005 Paris Declaration on Aid Effectiveness.

Considering the prospects for progress on mainstreaming, bottlenecks hampering mainstreaming of adaptation include: a lack of awareness of climate change among development practitioners and partner country planning and finance ministries; limited resources; limited relevance of available climate information (Agrawala and Van Aalst, 2005; Tearfund, 2006); lack of targets and metrics for assessing adaptation outcomes on the ground (Klein, 2001); and sectoral compartmentalisation in aid agencies and related turf battles (Persson, 2007). Challenges hampering the effective mainstreaming of mitigation options include the lack of adequate self-reflection in the developed countries themselves as to how best to develop while minimising greenhouse gas emissions. For example, the World Bank's energy portfolio still includes several major fossil fuel projects (Mainhardt-Gibbs, 2009). Developing countries also currently see cleaner technologies as unaffordable.

Finally, domestic economic priorities of developed countries have often pushed for the exports of older technologies subsidised by export credits.

12.3 Mainstreaming: political, economic, sustainability concerns

Although there is a clear rationale in favour of mainstreaming, there are also a number of fundamental concerns, which are related both to the mainstreaming imperative itself and to the broader practice of development co-operation within which mainstreaming currently takes place. These concerns are clustered here in three groups.

12.3.1 Administrative and political concerns

Will mainstreaming climate change serve as a back door exit from the commitment to provide new and additional resources?

The problem of climate change is seen by developing countries as having been primarily caused by the developed countries during their industrialising process. Hence, developing countries call on developed countries to provide new and additional resources to help them cope with climate impacts and avoid increasing emissions to the extent possible by adopting modern technologies (UNFCCC, 1992: Art 4(3)). Underlying this call is the assumption that developed countries should compensate developing countries for the negative side-effects of their historical legacy (see Oxfam, 2007).

These new and additional resources are expected to be over and above the existing political commitment of the developed countries to provide 0.7 per cent of their gross national income to developing countries as resources for development co-operation (UNGA, 1970: para 43). This target has often been repeated in official declarations (ICFD, 2002: para. 42; WSSD, 2002: 52; G8, 2005: para. 7) and is seen as critical to meeting the Millennium Development Goals.[1]

Looking at actual performance, however, only five small developed countries have met their 0.7 per cent target. In 2006, the OECD DAC countries provided only 0.31 per cent of their combined gross national income as ODA and it has not increased since the early 1990s (OECD, 2008). Thus, the climate change funding provided so far has not been *over and above* the 0.7 per cent commitment for development assistance, except in relation to a few countries like the Netherlands and Sweden. The concerns of developing countries are thus that (i) small and decreasing ODA budgets for poverty alleviation and other relevant development issues should not be eaten up by new climate priorities, and that (ii) the nature and logic of the new and

---

[1] Some economists believe that 0.7 per cent of gross national income/product is not sufficient to address the inequity between rich and poor countries (Tinbergen, 1996).

additional funds is different from that of ODA (Government of Argentina, 2008). The Government of India (2008) calls on developed countries to provide an additional 0.5 per cent of their GDP to meet the new and additional criterion.

Furthermore, since some ODA agencies use ODA resources for capacity building for CDM, developing countries see this as subsidising private sector activity (Yamin, 2005a; Gupta, 2009), leading to policy decisions condemning this practice at national (Ming *et al.*, 2008), OECD (OECD DAC, 2004) and UN level (COP 7 – 17/CP7 2001). A politically sensitive response is to ensure that developed countries meet their financial commitments.

Control, conditionality and accountability: Will mainstreaming climate change become a new source of tied aid and conditionality?

Common concerns, for both developed and developing countries, are: Who sets the rules for how to use climate change related money? Who is controlling the process and does it lead to a de facto new conditionality? Who should be accountable for the resources? Furthermore, given control and accountability concerns, should mitigation and adaptation be pursued as stand-alone projects, outside ODA, or as mainstreamed activities within ODA?

From the developed country perspective, resources provided by aid agencies to partner countries need to be used for specific goals under specific conditions. In the past, aid has often been linked to the economic interests of donor countries, which is commonly referred to as 'tied aid' (Jepma, 1991). According to recent calculations, only 41.7 per cent of aid from OECD countries is *not* tied (OECD, 2006: 31). Another element of control can be seen in aid provided under certain conditions (e.g. structural adjustment programmes). The literature reveals that both tied and conditional aid do not lead to successful implementation and distort policy processes (Jepma, 1991; Joint European NGOs report, 2006), calling for 'developing country ownership' of projects and programmes, as exemplified by the 2005 OECD Paris Declaration on Aid Effectiveness.

From a developing country perspective, there are three points relating to control over financial resources: (i) if the financial resources are meant to compensate developing countries for the harm done to them, then it should be up to developing countries themselves to decide how the resources are ultimately spent; (ii) if the resources are part of a funding mechanism then there should be joint management of the resources; and (iii) successful design and implementation of policies, programmes and projects calls for joint design.

The second issue of joint management has been highly sensitive in the last 15 years, especially with respect to the choice of the Global Environment Facility (GEF) as the operating entity of the financial mechanism. The developed countries trust the GEF to meet their accountability goals, while most developing countries see the GEF as representing interests outside of the climate community (Mace, 2005). The tension on this subject has continued through the last two Conferences of the Parties in Bali and Poznan. The fear of the developing countries is that mainstreaming climate change will lead to new conditions, while the developed countries want to meet 'common' global goals and ensure accountability (see Levina, 2007).

12.3.2 Development economics concerns

Aid effectiveness: If ODA does not reach the poorest people, what are the prospects for mainstreaming climate change in ODA?

A key challenge is channelling adaptation assistance in such a way that it reaches the most vulnerable. These are often the poorest people with the least ability to adapt and with few resources to fall back on (see Section 12.1). This is in many ways similar to the ODA challenges, which also focus on eradicating poverty. The track record of development co-operation on this is rather poor (Isbister, 2003). The concern is: how can mainstreaming of climate change into development co-operation be justified, when ODA, in general, has a weak record of reaching the poorest?

Although many poor regions now experience large improvements in life expectancy, health status and living conditions, there are at least one billion people living in poverty and hunger who depend on degraded areas and dangerous environments for their daily survival and longer-term livelihood (UNDP, 2007). These extremely poor people live mostly in Asia, and Sub-Saharan Africa has a larger share in relation to its total population. Currently, the difference in life expectancy between the richest and the poorest regions of the world is 40 years (UNDP, 2007). Such data can be taken as strong evidence that the trickle down effect is not universally effective as a dynamic force, working its way from the richest to the poorest (Martinussen, 1997).

Poverty, social exclusion and destitution remain serious problems and as such they constitute the supreme unsolved problem in the development project. The rich countries have not fulfilled their obligations towards the poor and this is a fact that nurtures the notion of 'promises not kept'. With climate change and adaptation, there is a new obligation that has to be fulfilled through new and additional resources also to be dedicated to the poorest (UNFCCC, 1992). This gives rise to the question of whether adaptation to climate change can be mainstreamed into ODA, since ODA is designed on the basis of development theory and thinking (see Section 12.3.3).

Market distortion: To the extent that ODA-funded projects distort markets, will not mainstreaming climate change in these projects also serve to do the same?

Poorly crafted policy interventions can reduce the ability of markets to respond effectively. This critique has been levelled at some types of ODA project funding, and especially at humanitarian post-disaster assistance, where food aid has occasionally led to the collapse of local markets (Hess, 2006). To cope with climate change, the World Food Programme (WFP) and others are trying to mainstream climate change into their assistance policies. A novel idea (see Chapter 13, Linnersooth-Bayer *et al.*) is to provide security nets for at-risk communities through developed country-supported micro-insurance programmes. The WFP has pioneered a system in Malawi, where smallholder farmers can purchase affordable index-based drought insurance (Hess and Syroka, 2005). Unlike traditional claims-based insurance, indemnity is based on an index of rainfall measured at a local weather station.

While the benefits of such types of insurance are uncontested, subsidising insurance is controversial. Similar price support has led to planting crops in areas unsuited for their cultivation (e.g. rice in Texas; Skees, 2001), or crowding out food imports from the developing world, exacerbating poverty (Alston *et al.*, 2000). Finally, such subsidies may keep recipients in an aid-dependent business when they should actually shift to other livelihoods. Mainstreamed policies that distort prices can lead to maladaptation.

If subsidies are inappropriate for rendering insurance systems affordable to the poor, how can Northern countries that have contributed to creating the risks help the most vulnerable? Linnersooth-Bayer *et al.* (see Chapter 13) show that, because of market failure, catastrophe insurance premiums can be far greater than the social cost of insurance, in which case subsidies can actually set the price right. However, this is not always the case with insurance or other forms of ODA, and direct cash transfers may be a better alternative (Skees, 2007). The idea of cash transfers is not new to ODA, and many development organisations are diverting some of their project funding for this purpose (de Janvry and Sadoulet, 2004, on conditional cash supports to four million mothers in Mexico). Mainstreaming, then, might take the form of increased cash to those most vulnerable to climate change.

12.3.3 Sustainability and governance concerns

Dependency: Will mainstreaming climate change exacerbate the aid dependency problem?

One concern in the aid world is whether ODA itself is a problem because of the way in which it creates dependency (Collier, 1999). Sub-Saharan Africa has received aid that is about 10 per cent of its national income. Possibly Africa has developed slowly precisely because, relatively speaking, it received high aid volumes in relation to its income, discouraging it from developing appropriate national policies (Bauer, 1982; Heller and Gupta, 2002). Alternatively, such aid cannot be absorbed effectively by the governments' management capacity (Kanbur *et al.*, 1999). Case studies on

Ghana reveal that budget support to the country led to a diversion of resources to meet the goals of the developed countries rather than the national priorities. Sometimes countries use aid for enhancing government consumption expenditure (Khan and Hoshimo, 1992; Remmer, 2004) and become dependent on it. Moreover, countries that receive high aid often tend to be less inclined to invest in democratic institutions and accountability.

Aid that creates dependency is problematic, and in such cases mainstreaming climate change into ODA will exacerbate this problem. However, there is an important difference between ODA and climate change adaptation. Although the provision of ODA is a strong political commitment, the choice of partner countries for ODA donors is voluntary and OECD DAC countries may reform or discontinue aid to countries where a dependency problem may be created. However, the more compensatory nature of climate funding implies that OECD DAC countries may have less room for manoeuvre. This would suggest limiting aid dependency where possible by limiting aid to a percentage of partner country gross national income. However, climate assistance may be independent of such a criterion to the extent that it is either compensating for harm or else helping avoid emissions.

Development paradigm: Will mainstreaming climate change have a greater chance of leading to sustainable development than development aid alone?

It is easy to assume that development aid assists development, and that mainstreaming climate change would simply modify that aid so that it also does something positive towards managing climate change.

The paradigms of development aid theory have, however, changed practically every decade (Gupta, 2009). Its focus on 'elegant prescriptions' has mostly failed to explicitly address: (i) the empirical and theoretical links between nature and society; (ii) the gap between rich and poor countries; (iii) the distributional dilemmas between living and unborn generations, or to (iv) meet the 'requisite plurality' of decision making, leading to severe institutional distortions in the South (Thompson, 2008a, b, c). Furthermore, counter-productive trade and investment policy at global level and poor contextual relevance at local level have reduced their effectiveness.

Transdisciplinary research in Nepal, completed in the ADAM project, shows that development aid often does not stimulate development and therefore there is not much point in adjusting development projects or programmes to incorporate climate change strategy. Interviews with local actors show that the contemporary emphasis on 'the user pays', lean government, private sector participation in services and the Washington consensus have led to a 'double burden' on Nepal's poor: little in the way of resources compounded by social exclusion. The traditional development paradigm has consistently pushed for elegance, driving out the plurality, contention and noisy argumentation that are the pre-requisites for development.

Instead of a one-way policy definition, with design and checklists in developed countries, the development-and-environment agenda needs to ensure the constructive interaction of the state, market and civil society (e.g. Dixit, 2002; Sharma *et al.*, 2004; Thompson and Gyawali, 2007). Very often, one or more of the 'voices' emanating from these three institutional apices was not heard, and this lack of 'requisite variety' can be considered responsible for project failures (e.g. Finland's Bara Forest Management Plan). Where development efforts have been successful, it is thanks to other voices managing to force their way into the decision making process. These are 'clumsy solutions' (Verweij and Thompson, 2006), for example, community forestry, where each institutional apex gets more of what it wants (and less of what it does not want) than if it had managed to establish hegemony (Thompson, 2008a and b). A new development paradigm would move from elegance to clumsiness, where all three voices are both heard and responded to by the others.

12.4 Conditions for success

12.4.1 Assessing the arguments

Mainstreaming climate change into development co-operation is therefore justified by three *prima facie* arguments. Firstly, integrating climate change concerns into development projects that promote, for example, energy production, transport systems and agriculture, would imply taking greenhouse gas emissions into account in the development process. This would ensure that modern technologies are part of the technology transfer process to developing countries (e.g. SWCC, 1990) and that older environmentally harmful technologies are not exported (e.g. Werksman, 1993; Hicks *et al.*, 2008).

Secondly, integrating adaptation into existing ODA in economic sectors that are exposed to climate risk or that contribute to the vulnerability of communities to climate change impacts makes such projects more effective and resilient. Many ODA sponsored projects may be affected by climate change (Agrawala and van Aalst, 2008) and adaptation and 'good' development are often compatible objectives (McGray *et al.*, 2007). Mainstreaming and climate-proofing at the operational level, in particular, where a budget has already been allocated and the cost of mainstreaming is modest, contributes to the sustainability of ODA. It can hence be a 'no-regrets' strategy (Persson and Klein, 2009) unless ODA itself is maladaptive and hinders development progress (see above). However, moving up to more strategic levels, such as sector-wide approaches, budget support programmes, and country assistance strategies, mainstreaming may lead to lower priority and lower budgetary allocations to other more urgent development objectives, which the 2006 OECD Declaration itself recognises as 'legitimate priority needs'. Furthermore, there are also political constraints to mainstreaming in OECD DAC countries since there may be winners and losers in these countries.

Thirdly, from the developed country perspective, it has been argued that using the ODA disbursement infrastructure is more efficient and effective than setting up new and separate mitigation and adaptation funding schemes (OECD, 2005).

There are, however, also three main concerns about mainstreaming climate change. The political concern is that mainstreaming allows an escape route for the developed countries to backtrack from their financial commitments. The economic concern is that mainstreaming of adaptation in aid policies might be unsuccessful as such aid is unlikely to trickle down to the poor and most vulnerable, and may in the process also distort markets and policy processes within countries. The sustainability concern is that mainstreaming, by association with the traditional development paradigm, will not lead to sustainable development, and moreover will strengthen aid dependency.

The question now arises as to whether there are any suitable alternatives to mainstreaming climate change. The most obvious alternative options are: (i) business-as-usual, (ii) aid mechanisms serving primarily climate related objectives, or (iii) non-aid mechanisms dealing with climate change-related problems in developing countries. Business as usual, however, may imply financing counter-productive developments, and other non-aid international and market mechanisms may be too restricted in scope. The challenge of dealing with climate change is too big to allow a choice between options. Instead, pursuing a mix of options is a prerequisite for dealing with climate change in an effective manner. This could involve working simultaneously for marginal change (e.g. climate-proofing ODA projects) as well as systemic change (e.g. questioning the dominant development paradigms governing contemporary ODA). This means that mainstreaming climate change is desirable as part of a broader package, and that the main challenge is to ensure that conditions for successful mainstreaming are met.

12.4.2 *Conditions for successful mainstreaming of climate change*

Based on the material presented in this chapter, Table 12.1 presents the ten conditions of success if the three sets of concerns are to be dealt with. It should be noted, however, that the concerns are rooted in different perspectives and there can be no ideal recipe to address all these concerns.

12.5 Conclusions

In response to developing countries' call for assistance, developed countries have voluntarily signed on to two sets of commitments: (i) to provide 0.7 per cent of their GNI for key development issues and the Millennium Development Goals; and (ii) to provide new and additional resources to help tackle climate change. Although these two

Table 12.1. *Conditions for success*

Concern	Nature	Conditions of success
Political	Will mainstreaming avoid the new and additional argument?	1. *Additionality*: Increase ODA to 0.7 per cent and raise new and additional resources above this amount and mainstream all assistance subject to conditions below.
	Who controls the aid?	2. *Partnership* between OECD DAC countries and partners in all relevant forums – UNDG, OECD/DAC, EU, in Climate and Development (C&D) decision making processes.
		3. *Ownership* by developing countries of C&D agenda at programme and project level.
		4. *Joint accountability of partners*: The system should promote mutual accountability in terms of both appropriate provisions and use of climate change funds.
Development economics	If ODA does not reach the poor, will mainstreaming do so?	5. *Prioritise the poorest*: In recent years, development co-operation has tried to focus on helping the poorest.
	Will mainstreaming in ODA distort markets and create perverse incentives?	6. *Avoid market distortion* where appropriate, consider cash transfers to compensate for climate impacts as a substitute for mainstreamed ODA projects.
		7. Design mainstreamed projects with as little market distortion as possible.
Sustainability	Will mainstreaming make aid dependency worse?	8. *Limit ODA dependency*: Ensure that ODA to countries is below a certain percentage of partner gross national income (GNI).
		9. Climate aid should be seen *independent* of this.
	Will mainstreaming within existing development paradigm create structural sustainability problems?	10. Ensure *tripartite decision making* between stakeholders, private parties and governments of both ODA countries and partners to design context relevant, locally owned policies. Avoid focus on formulae, efficiency, rationality and conditionality and accept clumsy solutions.

commitments result from different sets of concerns and discourses, they have become increasingly intertwined as a result of the 'mainstreaming discourse' that has now been adopted in development co-operation policy documents (such as those of OECD, the EU and many bilateral agencies as well as the International Financial Institutions).

There are three *prima facie* reasons that justify mainstreaming of climate change mitigation and adaptation into development co-operation activities: to promote leap frog development, ensure resilience of programmes, and reduce transaction costs. These reasons are pushing developed country governments to focus on mainstreaming. However, there are also three sets of political, development economics and sustainability concerns regarding whether such mainstreaming will be successful in terms of meeting the goals of both climate policy and development co-operation policy. These main concerns are related to (i) the commitment and integrity of developed countries, (ii) the fairness of distributional patterns and whether aid actually serves the interests of the poor, and (iii) the sustainability level of current development paths and whether they indeed stimulate the resilience of systems and people. When these concerns are taken into account, then climate change and development goals may both be met. However, neither goal may be served if the resources remain too low and approaches remain self-serving.

This chapter has identified ten conditions for success that should be applied to mainstreaming approaches and ODA in general. Furthermore, we believe that, in the short-term climate proofing existing portfolios of projects can, if undertaken in collaboration with stakeholders from the partner countries, be a beneficial first step and less likely to be controversial. A more comprehensive approach to mainstreaming must avoid becoming a new conditionality and must be debated and designed in collaboration with partner countries. The tendency of the OECD DAC (and the EU) to make policies for developed countries without adequate representation and participation of developing countries will not lead to successful policies on the ground.

Mainstreaming approaches need to be carefully designed. They must build on success stories in aid strategies, whilst calling for structural changes to the development and development co-operation paradigms, and employing appropriate mainstreaming tools and techniques. Our particular advice to the EU is: (i) to work rapidly towards the 0.7 per cent target for ODA; (ii) that climate aid should be over and above this amount; (iii) that climate aid builds on the lessons from development aid; and that (iv) aid policy is not developed without the involvement of partner countries from the South. The global South is not the object of assistance, but the partner in development co-operation.

Alternatively, if we ask the question 'should climate adaptation and mitigation be addressed only within development assistance?' the answer must clearly be 'no.' In some cases climate change issues may be best addressed within other policies, for example, on trade, or even on a stand-alone basis.

References

Agrawala, S. and van Aalst, M. (2005) Bridging the gap between climate change and development. In *Bridge over Troubled Waters: Linking Climate Change and Development*, ed. Agrawala, S. Paris: Organisation for Economic Co-operation and Development, pp. 133–46.

Agrawala, S. and van Aalst, M. (2008) Adapting development cooperation to adapt to climate change. *Climate Policy*, **8**, 183–93.

Alston, L. T., Lacher, T. E., Slack, R. D. *et al.* (2000) *Ecological, Economic, and Policy Alternatives for Texas Rice Agriculture*. Report TR-181, Institute for Science, Technology and Public Policy in the George Bush School of Government of Government and Public Service to the Texas Water Resources Institute/Agricultural Program, Texas A&M University System.

Antle, J. M. (1995) Climate change and agriculture in developing countries. *American Journal of Agricultural Economics*, **77**, 741–6.

Bauer, P. T. (1982) *Equality: The Third World and Economic Delusion*. Cambridge, MA: Harvard University Press.

Bodansky, D. (1993) The United Nations Framework Convention on Climate Change: a commentary. *Yale Journal of International Law*, **18**, 451–588.

Boko, M., Niang, I., Nyong, A. *et al.* (2007) Africa. In *Climate Change 2007: Impacts, Adaptation and Vulnerability. Contribution of Working Group II to the Fourth Assessment Report of the Intergovernmental Panel on Climate Change*, ed. Parry M. L., Canziani O. F., Palutikof J., van der Linden, P. and Hanson C. Cambridge: Cambridge University Press, pp. 433–67.

Collier, P. (1999) Aid 'dependency': A critique. *Journal of African Economies*, **8**(4), 528–545.

COP 7; 17/CP7 (2001) Modalities and procedures for a clean development mechanism as defined in Article 12 of the Kyoto Protocol, Decision of the Conference of the Parties to the Climate Change Convention, available at http://unfccc.int/resource/docs/cop7/13a02.pdf#page=20

CEC (2003) *Communication from the Commission to the Council and the European Parliament: Climate Change in the Context of Development Cooperation*, COM(2003) 85 final, 11 March 2003. Brussels: Commission of the European Communities.

CEC (2007) *Communication from the Commission to the Council and the European Parliament: Building a Global Climate Change Alliance between the European Union and Poor Developing Countries Most Vulnerable to Climate Change*, COM(2007)540 final, 18 September 2007. Brussels: Commission of the European Communities.

Dixit, A. (2002) *Basic Water Science*. Kathmandu: Nepal Water Conservation Foundation.

Eriksen, S. E. H., Klein, R. J. T., Ulsrud, K., Næss, L. O. and O'Brien, K. L. (2007) *Climate Change Adaptation and Poverty Reduction: Key Interactions and Critical Measures*, GECHS Report 2007:1. Oslo: University of Oslo.

G8 (2005) *Chair's Summary*. Gleneagles: Gleneagles Summit, 8 July 2005.

Gigli, S. and Agrawala, S. (2007) *Stocktaking of Progress on Integrating Adaptation to Climate Change into Development Co-operation Activities*. Paris: Organisation for Economic Co-operation and Development.

Government of Argentina (2008) *Argentina: Views on Enabling the Full, Effective, and Sustained Implementation of the Convention through Long-Term Cooperative Action Now, Up to, and Beyond 2012*. Available online at: http://unfccc.int/files/kyoto_protocol/application/pdf/argentinabap300908.pdf.

Government of India (2008) *Submission on Financing Architecture for Meeting Financial Commitments under the UNFCCC*. Available online at: http://unfccc.int/files/kyoto_protocol/application/pdf/indiafinancialarchitecture171008.pdf.

Gupta, J. (1997) *The Climate Change Convention and Developing Countries: From Conflict to Consensus?* Dordrecht: Kluwer Academic Publishers.

Gupta, J. (2006) The global environment facility in its north–south context. In *Contemporary Environmental Politics: From Margins to Mainstream*, ed. Stephens P. H. G., Barry J. and Dobson A. London: Routledge, pp. 231–53.

Gupta, J. (2007) Legal steps outside the climate convention: Litigation as a tool to address climate change. *Review of European Community and International Environmental Law*, **16**(1), 76–86.

Gupta, J. (2009) Climate change and development (cooperation). In *Climate Change and Sustainable Development: New Challenges for Poverty Reduction*, ed. Salih, M. Cheltenham: Edward Elgar, 94–108.

Heller, P. and Gupta, S. (2002) *Challenges in Expanding Development Assistance*, IMF Policy Discussion Paper PDP/02/5, Washington, DC: International Monetary Fund.

Hess, U. (2006) *Weather Insurance Derivatives to Protect Rural Livelihoods. Presentation at International Workshop of Agrometeoroligical Risk Management*, New Delhi, 26 October 2006.

Hess, U. and Syroka, J. (2005) *Risk, Vulnerability and Development. Presentation at BASIX Quarterly Review & Insurance Meeting*, Hyderabad, 21 October 2005.

Hicks, R. L., Parks, B. C., Roberts, J. T. and Tierney, M. J. (2008) *Greening Aid? Understanding the Environmental Impact of Development Assistance*. Oxford, UK: Oxford University Press.

ICFD (2002) *International Conference on Financing for Development in Monterrey, Declaration*, Mexico.

IPCC – Intergovernmental Panel on Climate Change (2007) *Impacts, Adaptation and Vulnerability*. Cambridge, UK: Cambridge University Press.

IRI (2005) *Sustainable Development in Africa: Is the Climate Right?* IRI Technical Report 05/01. New York, NY: International Research Institute for Climate Prediction.

IRI (2006) *A Gap Analysis for the Implementation of the Global Climate Observing System in Africa*. IRI Technical Report 06/01. New York, NY: International Research Institute for Climate and Society.

Isbister, J. (2003) *Promises Not Kept*. Sterling, VA: Kumarian Press.

de Janvry, A. and Sadoulet, E. (2004) *Conditional Cash Transfer Programs: Are They Really Magic Bullets?* Department of Agricultural and Resource Economics, University of California at Berkeley. Available online at: http://are.berkeley. edu/~sadoulet/papers/ARE-CCTPrograms.pdf

Jepma, C. K. (1991) *The Tying of Aid*. Paris: Organisation for Economic Co-operation and Development.

Joint European NGO report (2006) *EU Aid: Genuine Leadership or Misleading Figures? An Independent Analysis of European Governments' Aid Levels*. Brussels: Concord.

Kanbur, R., Sandler, T. and Morrison, K. M. (1999) *The Future of Development Assistance: Common Pools and International Public Goods*, ODC Policy Essay No. 25. Washington, DC: Overseas Development Council.

Khan, H. A. and Hoshimo, E. (1992) Impact of foreign aid on the fiscal behaviour of LDC governments. *World Development*, **20**, 1481–8.

Klein, R. J T. (2001) *Adaptation to Climate Change in German Official Development Assistance: An Inventory of Activities and Opportunities, with a Special Focus on Africa*. Eschborn: Deutsche Gesellschaft für Technische Zusammenarbeit.

Klein, R. J. T., Eriksen, S. E. H., Næss, L. O. *et al.* (2007) Portfolio screening to support the mainstreaming of adaptation to climate change into development assistance. *Climatic Change*, **84**(1), 23–44.

Lenschow, A. (2002) Greening the European Union: An introduction. In *Environmental Policy Integration: Greening Sectoral Policies in Europe*, ed. Lenschow, A. London: Earthscan, pp. 3–21.

Levina, E. (2007) *Adaptation to Climate Change: International Agreements for Local Needs*. Paris: Organisation for Ecnonomic Co-operation and Development.

Lucht, W. Prentice, I. C., Myneni, R. B. *et al.* (2002) Climate control of the high-latitude vegetation greening trend and pinatubo effect. *Science*, **296**, 1687–9.

Mace, M. J. (2005) Funding for adaption to climate change: UNFCCC and GEF developments since COP-7. *Review of European Community and International Environmental Law*, **14**(3), 225–246.

Mainhardt-Gibbs (2009) *World Bank Energy Sector Lending: Encouraging the World's Addiction to Fossil Fuels*. Washington DC: Bicusa. Available online at: http://www.bicusa.org.

Martens, W. J., Jetten, T. H. and Focks, D. A. (1997) Sensitivity of malaria, schistosomiasis and dengue to global warming. *Climatic Change*, **35**, 145–56.

Martinussen, J. (1997) *Society, State and Market: A Guide to Competing Theories of Development*. London: Zed Books.

Maurer, C. and Bhandari, R. (2000) *The Climate of Export Credit Agencies*. Washington, DC: World Resources Institute.

McBean, G. (2004) Climate change and extreme weather: a basis for action. *Natural Hazards*, **31**, 177–90.

McGray, H., Hammill, A. and Bradley, R. (2007) *Weathering the Storm: Options for Framing Adaptation and Development*. Washington, DC: World Resources Institute.

Ming, L., Gupta, J. and Kuik, O. (2008) Will CDM in China make a difference? *Presentation at the Berlin Conference on the Human Dimensions of Global Environmental Change: International Conference of the Social-Ecological Research Programme on Long-Term Policies: Governing Social-Ecological Change*, Berlin, 22–23 February 2008.

OECD (1992) *OECD DAC Guidelines on Aid and Environment No. 4: Guidelines for Aid Agencies on Global Environmental Problems*. Paris: Organisation for Economic Co-operation and Development.

OECD (2005) *Paris Declaration on Aid Effectiveness*. Paris: Organisation for Economic Co-operation and Development.

OECD (2006) *Declaration on Integrating Climate Change Adaptation into Development Co-operation*. Adopted by Development and Environment Ministers of OECD Member Countries on 4 April 2006.

OECD (2008) *Development Co-operation Report 2007*. Volume 9, No. 1. Paris: Organisation for Economic Co-operation and Development.

OECD DAC (2004) Statement adopted by members of the OECD's Development Assistance Committee (DAC), High level meeting , 15–16 April 2004. Available online at: http://www.oecd.org/dataoecd/42/26/31 505731.pdf.

Oxfam (2007) *Adapting to Climate Change: What's Needed in Poor Countries, and Who Should Pay*. Oxfam Briefing Paper 104. Oxford, UK: Oxfam International. Available online at: http://www.oxfam.org/en/files/bp104_climate_ change_0705.pdf/download.

Parry, M. L., Canziani, O. F., Palutikof, J., van der Linden, P. and Hanson, C. (ed.) (2007) *Climate Change 2007: Impacts, Adaptation and Vulnerability: Contribution of Working Group II to the Fourth Assessment Report of the Intergovernmental Panel on Climate Change*. Cambridge, UK: Cambridge University Press.

Persson, Å. (2007) Different perspectives on EPI. In *Environmental Policy Integration in Practice: Shaping Institutions for Learning*, ed. Nilsson, M. and Eckerberg, K. London: Earthscan, pp. 25–48.

Persson, Å. and Klein, R. J. T. (2009) Mainstreaming adaptation to climate change into official development assistance: challenges to foreign policy integration. In *Climate Change and Foreign Policy: case studies from East to West*, ed. Harris, P. London, UK: Routledge, pp. 162–77.

Picciotto, R. (2002) The logic of mainstreaming: A development evaluation perspective. *Evaluation*, **8**(3), 322–339.

Remmer, K. (2004) Does foreign aid promote the expansion of government? *American Journal of Political Science*, **48**(1), 77–92.

Ribot, J. C., Magalhaes, A. and Panagides, S. (ed.) (1996). *Climate Variability, Climate Change, and Social Vulnerability in the Semi-Arid Tropics*. Cambridge, UK: Cambridge University Press.

Rio Declaration (1992) *Rio Declaration and Agenda 21: Report on the United Nations Conference on Environment and Development, Rio de Janeiro, 3–14 June 1992*, UN doc. A/CONF.151/26/Rev.1 (Vols.1-III).

Sharma, S., Koponen, J., Gyawali, D. and Dixit, A. (2004) *Aid Under Stress: Water, Forests and Finnish Support in Nepal*. Lalitpur: Himal Books.

Skees, J. R. (2001) The bad harvest: more crop insurance reform – a good idea gone awry. *Regulation: The CATO Review of Business and Government*, **24**, 16–21.

Skees, J. R. (2007) Presentation discussion. Expert workshop on Insurance Instruments for Adaptation to Climate Risks, International Institute for Applied Systems Analysis (IIASA), Munich Re, the German Agency for Technical Cooperation (GTZ) and World Bank, and as part of the activities of the EU Integrated Project on Adaptation and Mitigation (ADAM), Laxenburg, Austria, 24–25 September 2007.

Stern, N. (2007) *The Economics of Climate Change*. Cambridge, UK: Cambridge University Press.

SWCC (1990) *Scientific Declaration of the Second World Climate Conference*. Geneva.

Tearfund (2006) *Overcoming the Barriers: Mainstreaming Climate Change Adaptation in Developing Countries*. Climate Change Briefing Paper 1. Teddington: Tearfund.

Thompson, M. (2008a) *Organising and Disorganising*. Axminster: Triarchy Press.

Thompson, M. (2008b) Beyond boom and bust. *Journal of the Royal Society of Arts*, Winter, 24–9.

Thompson, M. and Gyawali, D. (2007) Introduction: Uncertainty revisited or the triumph of hype over experience. In *Uncertainty on a Himalayan Scale*, ed. Thompson, M. Ewarburton M. and Hatley T. Lalitput: Himal Books.

Thornton, P. K. Jones, P. G., Owiyo, T. *et al.* (2006) *Mapping Climate Vulnerability and Poverty in Africa*. Nairobi/London: United Kingdom Department for International Development.

Tinbergen, J. (1996) Is 0.7% development assistance enough? *Social Indicators Research* **39**(3), 307–9.

United Nations Development Assistance Framework (UNDAF) (2009). Available online at http://www.undg.org/docs/9879/WGPI_2007-CCA-and-UNDAF-Guidelines-FINAL-February-2009-LOCKED.doc

United Nations Development Programme (UNDP) (2007) *Human Development Report 2007–2008: Fighting Climate Change – Human Solidarity in a Divided World*. New York, NY.

UNFCCC (1992) *United Nations Framework Convention on Climate Change*, signed New York, NY (US), 9 May 1992 (entered into force 21 Mar. 1994), (1992) 31 *International Legal Materials*, 4, 849. Available online at: http://www.unfccc.int.

United Nations General Assembly (UNGA) (1970) *Second United Nations Development Decade*, UN Doc. A/8124, 16 October 1970.

Verweij, M. and Thomson, M. (2006) *Clumsy Solutions for a Complex World*. Basingstoke, UK: Palgrave MacMillan.

Washington, R., Harrison, M., Conway, D. and Black, E. (2006) African climate change: Taking the shorter route. *Bulletin of the American Meteorological Society*, **87**(10), 1355–1366.

Werksman, J. D. (1993) Greening Bretton Woods. In P. Sands (ed.). *Greening International Law*. London: Earthscan, pp. 65–84.

WSSD (2002) *Johannesburg Plan of Action of the World Summit on Sustainable Development*. Johannesburg, 2002.

Yamin, F. (2005a) *Climate Change and Carbon Markets: A Handbook of Emission Reduction Mechanisms*. London: Earthscan.

Yamin, F. (2005b) The European Union and future climate policy: Is mainstreaming adaptation a distraction or part of the solution? *Climate Policy*, **5**, 349–361.

13

Insurance as part of a climate adaptation strategy

Lead authors:
JOANNE LINNEROOTH-BAYER[1], CHRISTOPH BALS,
REINHARD MECHLER

[1]Co-ordinating lead author

Summary

The Bali Action Plan calls for 'consideration of risk sharing and transfer mechanisms, such as insurance' as a means to address loss in developing countries particularly vulnerable to climate change. This paper examines the case for insurance mechanisms by addressing two fundamental questions: *whether* support for insuring against droughts, floods, tropical cyclones and other weather extremes should be part of an international adaptation strategy, and if so, *how*? Examining recent experience, we show that catastrophe insurance can provide significant benefits in the developing world by providing security against the wholesale loss of assets, livelihoods and even lives in the post-disaster period. At the same time it engages the private sector in vast markets, ensures reliable and dignified post-disaster relief, sets incentives for prevention, and not least, spurs economic development. We also examine the many challenges of catastrophe insurance: assuring sustainability and affordability in light of covariant risks; defining an appropriate role for donors given the inefficiencies of subsidies; and ensuring that systems avoid moral hazard and contribute to adaptation. We conclude that insurance mechanisms have a promising and legitimate role in an adaptation regime, and that practical options exist for including insurance in the adaptation strategy expected to emerge from the Conference of the Parties (COP) 15 in Copenhagen, 2009. We draw this conclusion in full recognition that insurance is not appropriate in all contexts, and that it must be viewed as only part of a comprehensive risk management programme. As a practical way forward, this chapter concludes by discussing two recent proposals submitted by the Munich Climate Insurance Initiative (MCII) and the Alliance of Small Island States (AOSIS). By suggesting internationally supported facilities for enabling micro- and sovereign insurance in the developing world, these proposals are expected to play an influential role in the Copenhagen Agreed Outcome.

Making Climate Change Work for Us: European Perspectives on Adaptation and Mitigation Strategies, ed. Mike Hulme and Henry Neufeldt. Published by Cambridge University Press © Cambridge University Press 2010.

13.1 Introduction

Insurance and related financial instruments that provide economic security against droughts, floods, tropical cyclones and other weather extremes have emerged as an opportunity for developing countries to reduce their vulnerability to weather variability and adapt to climate change. Emerging financial risk management opportunities for the developing world, although not a panacea for adapting to increasing climate risks, are demonstrating their potential for reducing the effects of weather extremes on national economies and providing security for investments as an important precondition to escape poverty. Many donor governments and bodies, including the European Commission[1], are thus moving away from post-disaster assistance towards supporting pre-disaster financial instruments.

Taking stock of this opportunity, and forging an appropriate role for risk-pooling and risk-transfer mechanisms within an adaptation strategy, is timely and urgent. The recent Bali Action Plan specifically calls for 'consideration of risk sharing and transfer mechanisms, such as insurance, as a means to address loss and damage in those developing countries that are particularly vulnerable to climate change (UNFCCC, 2008a). The Plan strengthens the mandate to consider insurance instruments, as set out by Article 4.8 of the UN Framework Convention on Climate Change (UNFCCC) and Article 3.14 of the Kyoto Protocol.

If risk-sharing and transfer instruments (Box 13.1) are to be included in the post-2012 adaptation strategy, the role of insurance must be urgently established (see Harmeling and Bals, 2008). Proposals for insurance instruments have been put forward and even tabled in the negotiation process, e.g. the proposal by the Alliance of Small Island States for an International Insurance Pool (AOSIS, 1999); yet, their precise role in an adaptation strategy is still largely undetermined. The purpose of this paper is to help define this role by addressing two fundamental questions: *whether* insurance programmes that protect against weather variability and extremes should be part of an international adaptation strategy, and if so, *how*? To provide insights on these questions, it is necessary to ask further:

(i) How are developing countries affected by weather variability and extremes, and to what extent are the risks attributable to climate change?
(ii) When is it advisable for low-income households, SMEs and governments to insure against climate-related risks?
(iii) What is the experience with insurance instruments and programmes in developing countries?

---

[1] The European Commission has recently allocated €25 million for a trust fund to support the Global Index Reinsurance Facility (GIRF), which is a new risk-taking entity that will (re)insure governments, banks and primary insurers in developing countries.

Box 13.1 Risk sharing and transfer

Societies have many ways to *share* or *pool* risks. An insurance company with heavy exposure in a hurricane-prone area might form a risk sharing or pooling arrangement with an insurance company in a tornado-prone area. Should a severe hurricane season occur, the latter company would share some of the loss with the former, and vice versa. On a more informal level, a household in one village may form a risk sharing arrangement with relatives in a far-away village that is not exposed to the same hazards, or a government may form a solidarity fund to provide assistance to low-income victims after a disaster occurs. There are no premiums paid, yet arrangements are put into place before the disaster. This contrasts with (usually ad hoc*)* post-disaster arrangements, usually in the form of humanitarian assistance.

Alternatively, risks can be *transferred* through market payments. An insurance company may transfer its hurricane risk to a reinsurer, or a farmer may transfer his risk by purchasing commercial insurance. Besides insurance, there are alternative risk transfer (ART) arrangements. One such arrangement is a *catastrophe bond*, which is an instrument whereby the investor receives an above-market return when a specific catastrophe does not occur (e.g. an earthquake of magnitude 7.0 or greater in a pre-defined area), but shares the insurer's or government's losses by sacrificing interest or principal following the event. Another instrument is *contingent credit*, which requires a pre-event fee to secure a pre-defined post-event annuity. Risk pooling and transfer can be combined. For example, the Caribbean states have recently formed a catastrophe pool covering their flood and hurricane risks, and transferred part of their collective risk by purchasing reinsurance and issuing a catastrophe bond.

Although *insurance* is defined as the transfer of the risk of a loss from one entity to another in exchange for a premium, it should be recognised that risk pooling and sharing commonly form the basis of insurance contracts.

(iv) Can climate insurance be designed in a way that contributes to adaptation instead of maladaptation?
 (v) Which principles should guide outside support for insurance programmes?
(vi) What role might insurance instruments play in a climate-adaptation regime?

By addressing these questions in the sections to follow, we conclude that there is a promising and legitimate role for insurance instruments in an adaptation strategy. This conclusion is based on full recognition that insurance is not appropriate in all contexts, and that it must be viewed as only a *part of a comprehensive risk-management programme*. We also recognise that national and international support, especially premium subsidisation, has a propensity to distort price signals, promote maladaptation and crowd out private insurance initiatives; yet we argue that it may be essential to enable insurance to play its role in developing countries, especially for the most vulnerable people. As a practical way forward, this chapter concludes by discussing two recent proposals submitted by the Munich Climate Insurance Initiative (MCII, 2008) and the Alliance of Small Island States (AOSIS, 2008), both of which suggest a role for insurance as part of a multi-lateral adaptation strategy.

13.2 How are developing countries affected by weather variability and extremes, and to what extent are the risks attributable to climate change?

In the past quarter-century over 95 per cent of deaths from natural disasters occurred in developing countries, and direct economic losses (averaging US$100 billion per annum in the last decade) in relation to national income were more than twice as high in low-income as opposed to high-income countries (Munich Reinsurance Company, 2007). Over a recent 10-year period, as shown in Figure 13.1, a dis-proportionate share of the economic and human burdens from weather disasters has fallen on low-income and lower middle-income countries.

While Figure 13.1 depicts immediate direct losses, it does not show the long-term consequences of disasters on economic development, which can greatly amplify both the economic and human losses. Due to limited tax bases, high indebtedness and low uptake of insurance, many highly exposed developing countries cannot fully recover from slow- and sudden-onset disasters by simply relying on limited external donor aid (Mechler, 2004). In turn, external investors are wary of the risk of catastrophic infrastructure losses, and small firms and farmers cannot access the credit necessary for investing in higher-yield/higher-risk activities. This can lead to slowed economic recovery.

More than three-quarters of recent worldwide economic losses from all natural disasters can be attributed to windstorms, floods, droughts and other weather-related

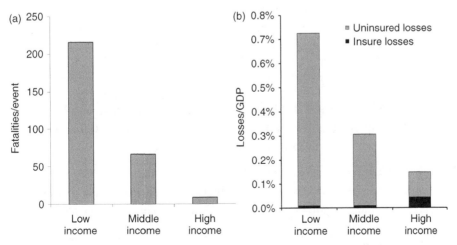

Figure 13.1. (a), (b) Differential burden of natural disasters according to country income groups (Note: country income groups according to World Bank classification) (Data source: Munich Reinsurance Company, 2005.)

hazards (UNISDR, 2007). Experts generally agree that under all linear scenarios of future climate change, increasing disaster losses in the coming decades will be dominated, not by climate change, but by changes in populations and wealth (Munich Reinsurance Company, 2006). Climate change is expected to play a role, however. The Intergovernmental Panel on Climate Change (IPCC) has predicted that climate change will magnify losses because of increasing weather variability, and that overall extreme event impacts are 'very likely' to change (Solomon *et al.*, 2007). Although no one storm or flood can be attributed to climate change, it is expected that the odds of these events occurring will increase (see Box 13.2). There is even mounting evidence of a current 'climate signal' with the IPCC reporting observations of widespread changes in temperature, wind patterns and aspects of extreme weather, including droughts, heavy precipitation, heatwaves and the intensity of tropical cyclones (Carter *et al.*, 2007).

Box 13.2 Climate change and attribution: loading the odds

Scientists cannot attribute any particular storm or other event to a changing climate, but they can present evidence on the contribution of climate change to the risk of climate variability and extremes. By way of analogy, if we throw a loaded dice and *six* appears, we cannot ascertain whether that *six* is due to loading or not. We can only say that the probability of throwing a six is increased. In the case of weather extremes, it is not possible at this time to determine precisely how much the odds are loaded, or to define the scope of damages attributable to climate change, in a scientifically sound manner. Thus, it is not possible to earmark an international fund that provides or supports insurance, or for that matter one that provides or supports any adaptation measure addressing multi-causal impacts, such that it only protects against damages that are proven to be caused by a changing climate. Although early adaptation efforts did aim at funding activities strictly related to climate *change*, there are increasing calls that adaptation should be driven by vulnerability and poverty, and that it should be mainstreamed into the development process (see Chapter 12 *et al.*; Klein *et al.*, in press).

13.3 When is it advisable for low-income households, small and medium-sized enterprises (SMEs) and governments to insure against climate-related risks?

While valuable in reducing the long-term effects of climate disasters on poverty and development, insurance instruments, particularly if left entirely to the market, cannot be a panacea for adapting to climate change. There are many reasons, including:

- insurance is generally inappropriate for very slow-onset climate impacts, such as sea level rise, ocean acidification or desertification, which are considered uninsurable; other instruments are needed in this case;

- without government or donor support, private insurance is not easily affordable by governments, households and SMEs in highly exposed and vulnerable countries, where the opportunity costs of private risk-financing instruments can be prohibitively high in terms of meeting other human needs; and
- many developing countries lack an insurance tradition and market, which will take time to develop (Mechler *et al.*, 2006).

Keeping in mind these limitations, insurance can play an important role in many contexts. By spreading stochastic losses temporally and geographically, and assuring timely liquidity for the recovery and reconstruction process (which can itself save lives and livelihoods), insurance is beneficial to those in the risk pool. Moreover, it provides the pre-disaster security essential for productive risk taking. These benefits, however, come at a cost. Insurers that operate in developing countries have high start-up and transaction expenses, which can greatly limit affordability and constrain insurance penetration. Moreover, because disasters can affect whole communities or regions (co-variant risks), insurers must be prepared for meeting large claims all at once. Their cost of requisite backup capital, diversification or re-insurance to cover co-variant claims can add greatly to the business expenses and raise the premium far above the client's expected losses (see Box 13.3).

If insurance premiums cost clients more than their anticipated losses, and some-times significantly so, what is the rationale for governments, household and farms to insure? The textbook rationale for insurance is based on the concept of *risk aversion*. Risk-averse persons prefer lower consumption, if it is steady, to higher consumption, if it is highly irregular or even subject to catastrophic shortfalls. This is particularly the case for disasters affecting the poor since even the economic outcomes can threaten their lives.

13.3.1 Prevention and non-insurance financing mechanisms

Even risk-averse parties, however, should not purchase insurance if they have lower cost alternatives for reducing their burdens through preventing disaster losses and financing the recovery process by means of family, loans and other conventional practices. It is beyond the scope of this chapter to discuss in any detail the potential of disaster prevention. We note only that (scant) studies documenting the returns on disaster mitigation show a great deal of potential across countries and disaster types. Moench *et al.* (2007), for instance, examine 27 disaster mitigation projects in developing and developed countries, ranging from polders and mango forests, to relocation of buildings, and they record high benefit/cost ratios across the board.

Moreover, the poor may have less costly ways of paying for disaster losses. Table 13.1 gives examples of non-insurance (compared with insurance) mechanisms for micro-, meso- and macro-agents. One of the most prominent strategies for coping with weather disasters, which we refer to as *collective loss sharing*, operates at all scales in the form of government assistance, humanitarian aid and other common forms of domestic and international solidarity. Solidarity can also take the form of mutual arrangements among family or community members. This can be referred to as *informal risk sharing*. Remittances, or transfers of money sent by foreign workers to their home countries, are a good example of this type of informal financial mechanism. Alternatively, or in addition, households can save cash or

Table 13.1. *Examples of insurance and non-insurance mechanisms for managing risks*

	Micro-scale risk financing	Meso-scale risk financing	Macro-scale risk financing
	Households, SMEs, farms	Financial institutions, donor organisations, etc.	Governments
Non-insurance mechanisms			
Collective loss sharing (solidarity)	Post-disaster government assistance; humanitarian aid	Government guarantees/ bail outs	Bilateral and multilateral assistance
Informal risk sharing	Kinship and other mutual arrangements; remittances		Diversions from other budgeted programmes
Inter-temporal risk spreading	Micro-savings; micro-credit; fungible assets; food storage	Emergency liquidity funds	Reserve funds, regional pools, post-disaster credit; contingent credit
Insurance mechanisms			
Risk pooling and transfer/ insurance	Micro-insurance; crop and livestock insurance; weather hedges	Re-insurance	Sovereign risk financing; regional catastrophe insurance pools
Alternative risk-transfer		Catastrophe bonds	Catastrophe bonds; risk swaps, options, and loss warranties

Source: Linnerooth-Bayer, 2008.

accumulate fungible assets to provide them with capital for investments in adaptation and for post-disaster relief; or they can make pre- or post-disaster credit arrangements for these same purposes. Not only households, but also governments save by creating catastrophe reserve funds. Savings, credit arrangements and reserve funds are all forms of *inter-temporal risk spreading* since they provide resources or require payments at some time in the future.

While non-insurance mechanisms appear to work reasonably well for low-loss events, they are often unreliable and inadequate for catastrophic events (Cohen and Sebstad, 2003). Faced with large losses, households may be forced to sell productive assets at very low prices; post-disaster inflation may greatly reduce the value of savings; money lenders may exploit their clients; entire families, even if geographically diverse, may be affected; and donor assistance rarely covers more than a small percentage of losses (Mechler, 2004). The resulting liquidity deficit can greatly aggravate poverty and reduce capacity for adapting to climate change.

While this chapter focuses mainly on insurance instruments, it should be kept in mind that many non-insurance mechanisms may be candidates for support as part of an adaptation strategy. As only one of many possibilities, a climate adaptation fund could provide emergency liquidity funds to micro-finance institutions providing loans after a disaster.

13.3.2 Insurance mechanisms

We focus on insurance mechanisms mainly because non-insurance instruments may be less appropriate for dealing with high loss events and because they have other advantages in specific contexts. As shown in Table 13.1, insurance mechanisms include a range of familiar risk pooling and transfer activities, such as micro-insurance and regional catastrophe pools. They also include alternative risk transfer instruments, such as catastrophe bonds, risks swaps and others. These alternative instruments essentially allow parties to transfer their risks by finding investors who will take them on for a price (see Box 13.2). Insurance, in contrast to many non-insurance mechanisms, can provide valuable incentives for risk-reducing adaptation measures, as well as providing security for investments enabling escape from poverty and an increased ability to cope with climate change.

Beyond post-disaster benefits, insurance can provide pre-disaster security necessary to take on productive but risky investments (Höppe and Gurenko, 2007). Due to high uninsured risk exposure, households, businesses and farmers may adopt low-risk, low-return strategies, like giving relatives low-paid but secure employment, or planting low-yield but drought-resistant seeds. This reduces their ability to accumulate the assets needed to escape poverty through savings and investment. In the words of agricultural insurance experts:

... those with few assets may accurately perceive that time is not an ally in their daily struggle to climb out of poverty ... (those with assets may) suffer uninsured asset losses that suddenly cast them into poverty and possibly onto a downward spiral from which they have a difficult time re-emerging. These themes from the emerging literature on poverty traps underscore the relation between risk and persistent poverty, as well as the opportunities afforded by innovations in risk management. (Barnett *et al.*, 2008)

To summarise, insurance is not appropriate in all contexts. Insurance will generally increase, rather than decrease, the expected (average) financial cost of disasters to those in the insurance pool because of transaction and capitalisation costs, and absent outside support. In cases where individuals or organisations have lower cost alternatives for providing post-disaster liquidity, insurance may not be advisable. In many contexts, however, these alternatives are ineffective (especially for very devastating disasters affecting large regions), in which case these individuals or organisations should weigh the benefits of insurance against the costs. These benefits include, first and foremost, security against the wholesale loss of assets, livelihoods and even lives in the post-disaster period. Insurance not only provides the liquidity to smooth out disaster shocks, but by enabling productive investments has the added benefit of helping high-risk individuals escape disaster-induced poverty traps.

13.4 What is the experience with insurance instruments and programmes in developing countries?

The seriousness of the post-disaster capital gap, as well as the emergence of novel insurance instruments for pricing and transferring catastrophe risks, has motivated development institutions, NGOs and other donor organisations to consider pre-disaster financial instruments as an alternative to post-disaster humanitarian assistance (Linnerooth-Bayer *et al.*, 2005). Donor-supported pilot insurance programmes are already demonstrating their potential to pool economic losses and smooth incomes of the poor, who face weather variability and climate extremes. These schemes provide insurance to farmers, property owners and small businesses, as well as transfer the risks facing governments to the global capital markets. A few examples serve to illustrate:

(i) In *Malawi*, smallholder farmers can purchase affordable index-based drought insurance, where, unlike traditional claims-based insurance, indemnity is based on an index of rainfall measured at a local weather station. By making farmers more creditworthy, this pilot insurance scheme enables farmers to purchase hybrid seeds, and thus greatly increase their productivity (Suarez *et al.,* 2007; Hess and Syroka, 2005);

(ii) Herders participating in a pilot project in *Mongolia* can purchase an index-based insurance policy to protect them against livestock loss due to extreme winter weather or *dzuds*. Herders bear smaller losses (self-insurance), while larger losses are transferred to private insurers, and the final layer of catastrophic losses is borne by the government with backing from the World Bank (social insurance) (Skees *et al.,* 2008; Skees and Enkh-Amgalan, 2002); Micro-insurance has also been demonstrated in crop pilot projects in Inner-Mongolia (see Chapter 9, Werners *et al.*);

(iii) The World Food Programme (WFP) designed an index-based insurance product for the Ethiopian government that provides capital in the case of extreme drought. The amount is based on contractually specified catastrophic shortfalls in precipitation measured in terms of the Ethiopia drought index (EDI). In 2006, the WFP successfully obtained an insurance contract based on the EDI through a Paris-based reinsurer (Hess, 2007);

(iv) The *Mexican Government* was the first developing country to issue a catastrophe bond to partly insure its catastrophe fund and thus reduce its risk of large fiscal deficits following disasters. This bond transfers sovereign risk directly to the world's capital markets (Cardenas *et al.*, 2007);

(v) The *Caribbean island states* have recently formed the world's first multi-country catastrophe insurance pool to provide governments with immediate liquidity in the aftermath of hurricanes or earthquakes. There is a largely untapped potential for pooling uncorrelated risks of country governments that would be ill-prepared to respond to disasters with their own means (Ghesquiere *et al.*, 2006).

At an ADAM workshop on *Insurance Instruments for Adaptation to Climate Risk*[2] these risk pooling and risk transfer programmes were examined by those most familiar with them. The participants noted that experience is too short to judge if these internationally backed public–private systems are viable in the long haul, but as pioneering 'test balloons' (and some are beyond the testing phase) they might radically change the way development organisations provide disaster aid and support adaptation to climate change (see Chapter 12, Gupta *et al.*). For the most part, these programmes directly or indirectly target the most vulnerable, which is made possible by the technical and/or financial support that each has received from international development and donor organisations.

The potential for insurance is large. Over 40 per cent of farmers in the developing world face weather-related threats to their livelihood (World Bank, 2005), which will predictably increase with climate change, and yet only a small percentage

---

[2] The workshop, *Insurance Instruments for Adaptation to Climate Risks*, organised jointly by the International Institute for Applied Systems Analysis (IIASA), Munich Re, the German Agency for Technical Co-operation (GTZ) and the World Bank, was part of the activities of the EU-funded project on Adaptation and Mitigation (ADAM). It took place at IIASA in Laxenburg, Austria, on September 24 and 25, 2007.

benefit directly from micro-insurance systems. Many issues in scaling up small pilot projects remain unresolved. On a positive note, however, an index-based weather insurance programme operating in India has scaled up to serve over a half million clients (Gunaranjan, 2008).

13.5 Can weather insurance be designed in a way that contributes to adaptation instead of maladaptation?

In the context of weather variability and extremes, adaptation can be thought of as reducing risks to property, assets, livelihoods and lives. This can take many forms, including:

  (i)   physical interventions, for example, flood defences or early warning systems;
 (ii)   lifestyle changes, for example, relocating or changing livelihoods;
(iii)   training for early warning systems; and
(iv)   strategies for recovery, for example, formal and informal insurance.

Insurance is generally not viewed as a measure to prevent loss of life and property, and for this reason it is commonly regarded as an alternative to adaptation, rather than an integral part as suggested here. This view, however, overlooks the long-term preventative benefits of insurance. By enabling recovery, insurance can significantly reduce long-term indirect losses – even human losses – which do not show up in the disaster statistics.

   The view that insurance is an alternative to adaptation also overlooks the propensity of *well-designed* insurance programmes to provide incentives for physical interventions and lifestyle changes that reduce disaster risks. A few examples illustrate this: in Istanbul, apartment owners who choose to disaster-proof their properties pay a lower insurance premium, thus making investments in safety more attractive; in Mongolia, herders who insure their livestock will face increasing premiums as climate change worsens weather conditions, giving them an added incentive to change livelihoods if animal husbandry becomes unproductive; in Thailand, designers of an index-based flood insurance system anticipate that middle-class property owners will relocate out of the high-risk areas. Well-designed insurance is thus not an alternative to adaptation measures, but it is in itself an adaptation measure in the strictest sense. Although the potential for behavioural changes induced by insurance are well documented, there is little empirical evidence to support these claims, and for this reason more attention needs to be given to the role of insurance in promoting loss-reducing behaviour.

   Poorly designed insurance contracts, on the other hand, can discourage investments in loss prevention or even encourage negligent behaviour, commonly referred to as 'moral hazard', or maladaptation. Insurers guard against moral hazard by requiring deductibles or co-insurance, such that the insured incur part of their losses.

A major advantage of index-based insurance schemes is their avoidance of moral hazard. Mongolian farmers can only gain by taking measures to protect their herds against adverse winter weather, since insurance claims are based on average livestock loss in designated regions. While index-based insurance discourages moral hazard (taking risks that are not socially beneficial, for example moving one's couch to the basement in the case of a flood), paradoxically it can encourage risks that are socially beneficial. In Malawi, for example, the insurance contract enables farmers to plant riskier but higher yield crop varieties with a higher expected return. Not only does insurance smooth the incomes of Malawian farmers facing weather variability, but it actually provides them the safety net necessary for riskier and more productive activities, ultimately reducing vulnerability to weather shocks and thus contributing to adaptation. This point cannot be overemphasised. In the words of an expert on the Malawi index insurance project:

We want farmers to adopt high return technologies that allow them finally to make the leap and accumulate earnings over time. Systemic risk is *the* factor impeding this and so far banks cannot handle the risk and the high transaction costs in rural areas. The Malawi transaction shows that there is a sustainable way to take the big rocks out of the way – drought risk – and clear the path to development. (U. Hess, personal communication, 21 February 2006)

This same 'investment effect' operates at the national level. If governments can reassure outside investors that disasters will only temporarily disrupt critical infrastructure, this will create a more secure environment for attracting international capital.

In summary, well-designed insurance reduces disaster losses in two ways: by providing early liquidity, it prevents long-term loss of livelihood and lives; and by pricing risk, it provides strong incentives for pre-disaster preventive behaviour. However, ill-conceived external assistance for risk premium and inflexible insurer behaviour can result in an incentive for maladaptation. Because of the absence of moral hazard, index-based systems are particularly promising as instruments for adaptation.

13.6 What principles should guide outside support for insurance programmes?

While the benefits of catastrophe safety nets are for the most part uncontested, the role of outside assistance to make insurance affordable to the poor is controversial. Outside assistance, especially in the form of premium subsidies, can distort the price signal and in this way weaken incentives for taking preventive measures and perpetuate vulnerability by making it possible to remain in high-risk occupations or locations. Moreover, critics argue that other types of support, like providing re-insurance to small insurers, can crowd out the role of the private market. Proponents counter these charges by noting that the market often fails to provide 'correct' price signals, and that enabling insurance for the poor can create a role for the private sector by encouraging public–private partnerships (Linnerooth-Bayer and Mechler, 2007a).

We discuss the arguments for and against outside support by examining the current role of the private sector, the advantages and disadvantages of public–private partnerships, and conclude with principles for guiding outside assistance.

Private market provision of insurance

Is outside support always necessary to insure the poor, or will insurance emerge autonomously from market forces? The scattered examples of micro-insurance schemes that offer catastrophe cover without outside support appear to be viable mainly due to very low cover. For example, Proshika, a large micro-finance institution in Bangladesh, offers compulsory group-based disaster insurance to its clients. Under this programme, two per cent of the savings balance is transferred annually to a fund that will pay twice the amount of the savings deposit in the case of property damage due to disasters, while savings stay intact. The scheme operates without re-insurance or donor support. With more than two million clients in 20 000 villages and 2000 slums, this insurance pool has wide geographical diversification. Yet, the indemnity payments are only twice the amount in the savings account, which is likely to be only a small percentage of any damages arising from a disaster.

Observers point to the Malawi case for a demonstration of a micro-insurance system that operates with only minimal start-up assistance from the international community. Insurance in the Malawi pilot project protects the bank against loan default (a loan guarantee), and the premium can be paid easily by the five-fold foreseen productivity increase of the hybrid seeds made possible by the loan. However, extending cover to provide security against drought-induced food scarcity (livelihood insurance) would be largely unaffordable for Malawi's smallholder farmers. This risk cannot be covered by the private market acting alone, and is currently absorbed by post-disaster emergency food programmes on the part of the WFP, and UN Food and Agriculture Organization and other donors. Emergency food assistance, while indispensable for humanitarian reasons, disrupts local food markets and gives farmers little incentive to diversify their crops or livelihoods, two activities that would render them less vulnerable to droughts.

Outside support through internationally backed public–private partnerships

The inability of the poor to afford sufficient insurance cover and the reluctance of the private market to commit capital and expertise to the low-income market can be overcome by forming partnerships with insurers, governments and NGOs, and providing support from bilateral and multilateral development and donor organisations. The pilot programmes discussed above illustrate the diverse roles that these partners can play.

In Mongolia, a syndicate pooling arrangement, the Mongolian index-based livestock insurance programme, protects the under-developed insurance industry against extreme losses and insolvency. The government supports this syndicate by absorbing the losses from very infrequent extreme events (over 30 per cent animal mortality), and it can call upon a World Bank contingent debt arrangement to back this commitment (Skees *et al.*, 2008). The designers of this programme argue that subsidising the 'upper layer' is less price-distorting than subsidising lower layers of risk because the market may fail to provide insurance for this layer. The reasons are related to demand and supply. On the demand side, most people tend to underestimate very low probability events; on the supply side, insurers tend to charge premiums above the market price because of the large ambiguities in the risk estimates (Kunreuther, 1998).

Up to now, commercial reinsurers have been reluctant to commit significant capital and underwriting expertise to develop micro-insurance programmes, although they are absorbing the low-probability/high-consequence layers of many recent public–private programmes (for example, in Ethiopia and Mexico). Exceptionally, Swiss Re, in partnership with a non-governmental organisation (NGO) and an academic research institute, has insured about 150 000 smallholder farmers in Kenya, Mali and Ethiopia against drought through an index-based product. The insurance is purchased by the NGO with international backing, and other partners are being solicited to provide further financial support (Allen, 2007).

The re-insurance and catastrophe bonds that transfer risks from Mexico and Ethiopia to the international capital markets were made possible by outside technical support from international financial institutions (IFIs) and other types of start-up assistance. The same is true for the largely self-financing Caribbean catastrophe risk insurance facility (CCRIF), although, here too, donors have pledged significant capital to the reserve fund.

Advantages of international support for disaster insurance systems

Governments and donor organisations can reap large gains by moving away from providing post-disaster humanitarian aid towards enabling public–private insurance systems. By sharing responsibility with individuals and the state, donors leverage their limited budgets and substitute a calculable annual commitment for the unpredictable granting of post-disaster aid. With donor-supported risk transfer programmes, developing country governments will rely less on debt financing and international donations, while assurances of the timely repair of critical infrastructure will attract foreign investment.

Moreover, for many in the developing world an insurance contract is preferred to humanitarian assistance. According to a developing country participant at the ADAM meeting:

'Communities value disaster insurance not because it rewards them or makes them richer after a disaster. They value insurance because they see it as an instrument of dignity. Financial support to recover from a disaster becomes their right without sacrificing their self respect. It is far more dignified to claim your right for recovery than to find yourself dependent on the ad hoc generosity of donors.' (Krishna, 2007)

Most importantly, by making outside assistance or premium reductions contingent on taking preventive measures as part of a comprehensive risk management programme, pre-disaster assistance can ultimately reduce the human and economic toll that disasters take on the poor. This means that switching to pre-disaster donor aid, even at extra cost, can be an efficient, long-term strategy because of its potential, ultimately, to reduce the need for humanitarian assistance.

Challenges of international support for disaster insurance systems

Despite compelling arguments for internationally supported insurance providers, there are concerns that excessive public and international assistance will:

(i) distort market prices;
(ii) greatly jeopardise the incentive effects of insurance;
(iii) crowd out private initiatives; and
(iv) create unstable systems, due to the inability of donor institutions to make long-term commitments.

Critics point out that subsidised premiums in the US farm insurance programme have weakened incentives to plant more robust crop varieties, or to move away from farming in high drought or flood risk areas (Skees, 2001). In the words of a US insurance expert, participating at the ADAM meeting:

'If the intent is to improve the well-being of farmers, it may be preferable to give them direct monetary transfers than to subsidise insurance premiums. A particularly 'bad' subsidy is one that is proportional to the premium since the disincentive to change crop practices becomes greater as the risk (and premium) increases. Furthermore, given the political economy of subsidies, it is likely that any subsidy will benefit the larger farmers more than the smaller farmers.' (Skees, 2007)

Tempering this argument is the fact that even donor-supported insurance has a possibly greater incentive effect than the current practice of extending free, post-disaster aid to disaster victims, although the disincentives of post-disaster aid may be weakened by the fact it is ad hoc and uncertain. Moreover, if existing risk markets do not reflect the social cost of transferring risk, insurance premiums will not give

---

Box 13.3 Incentive compatible 'smart' subsidies

Subsidies, especially if they are constructed as a percentage of the premium, can lead to perverse behaviour, like cropping in ever more risky areas. Direct subsidies, however, are only one way of lowering premiums. Other possibilities include funding start-up costs or absorbing a layer of the risk. These indirect subsidies may be more incentive compatible since they do not increasingly reward behaviour as it becomes riskier.

Moreover, 'smart' subsidies might be those that correct rather than distort market prices. One idea is to subsidise, whilst ensuring that premiums remain above the social costs, which include the expected loss, administrative expenses and capital costs. This cost is less than actual premiums, which include what Cummins and Mahul (2008) refer to as 'frictional' and 'uncertainty' loads. Since global capital markets have less information about the insurer's exposure to catastrophic risk and the adequacy of its loss reserves than do the firm's managers, the capital market may charge a higher cost of capital to provide a margin for the informational asymmetry. Adding to this 'frictional' load, insurers may charge a higher premium to account for uncertainty and ambiguity in the risk (it can be argued that *uncertainty* is a social cost, but not *ambiguity*, which stems from imperfect information about the uncertainty). Imperfect information is a major cause of market failure. For lines of insurance covering significant, sporadic events, Cummins and Mahul (2008) claim that these extra charges can be a large component of the premium. In the long run, these loads will lead to increasing insurer profit. Arguably, thus, smart subsidies would reduce these loads.

'correct' price signals. This opens a window for providing incentive compatible or 'smart' subsidies (see Box 13.3).

A related concern is the possibility that international involvement in the provision of insurance, even with 'smart' subsidies, will impede the development of the private insurance market. Private insurers and re-insurers are constrained in their ability to offer worldwide, catastrophe risk coverage; they worry that public assistance will crowd out private operations. As expressed by a private sector participant at the ADAM meeting:

'The weather index industry is tiny and climate-related aid could be significant. Pouring large amounts of 'smart' aid in at the top will put too much pressure on the system – like a 'sausage machine'. It will be impossible to handle all the cash, and eventually the pressure to spend would render the subsidies as not smart at all. Even though my business potentially stands to gain from a significant influx of climate-related aid, the prospect terrifies me, due to the potential for it to destroy or distort the commercial market.' (Leftley, 2007.)

Private sector actors also worry about outside support unfairly advantaging some private companies over others, or crowding out competing private companies altogether. While partnerships, like the public–private Turkish Catastrophe Insurance Pool, actually create an opportunity for the private market to carry out business, ill-conceived public–private partnerships can prevent commercial companies from entering the market (Gurenko, 2004; Gurenko *et al.*, 2006). This issue underlines the importance of the following proposed principles to guide public and international bodies offering support:

(i) for those who can afford the price of insurance, internationally supported partnerships should be limited to ensuring conditions for private insurance provision through competitive markets;

(ii) for those who cannot afford adequate insurance cover, internationally supported partnerships can legitimately intervene, but care should be taken not to significantly distort prices or competition;

(iii) for those who cannot afford *any* insurance cover, intervention may cover the premium value, with the possibility of requiring in-kind payments through risk reduction activities.

13.7 What role might insurance instruments play in a climate-adaptation regime?

The case for intervention is greatly strengthened by recent evidence that greenhouse gas emissions are contributing to increased weather variability and the risks of extreme events, and disproportionately burdening vulnerable countries. In accordance with the UNFCCC principle of 'common but differentiated responsibilities and respective capabilities', industrialised countries are arguably obligated to support adaptation. Numerous proposals for insurance instruments have been put forward and even tabled in the negotiation process; yet, their precise role in an adaptation regime is still largely undetermined. Most recently, the Swiss government reinforced many earlier calls by proposing a multi-lateral adaptation fund, which would be spent on two pillars: *prevention and insurance* (UVEK, 2008). This built on a UNFCCC workshop in 2008 on Investment and Financial Flows, where the G77 and China called for establishing a risk insurance fund (Müller, 2008). The issue landed more solidly on the Copenhagen agenda with a technical paper (UNFCCC, 2008b) and an in-session workshop at COP 14. Building on this momentum, the Munich Climate Insurance Initiative (MCII) and the Alliance of Small Island States (AOSIS), an NGO and delegate group, respectively, have submitted two separate but similar proposals for consideration at COP 15. Table 13.2 compares the main features of these proposals, each of which is briefly discussed below.

13.7.1 The MCII proposal

The MCII proposal builds on the Swiss submission by suggesting a risk management module that includes two pillars, *prevention* and *insurance*, which would act

Table 13.2. *Comparison of MCII and AOSIS proposals*

MCII Risk Management Module	AOSIS Multi-window mechanism
(1) Prevention Pillar	(1) Risk management component
Facilitates risk assessment, risk reduction (data collection, investment advice, capacity building, etc.)	-same-
(2) Insurance Pillar	(2) Insurance component
Tier 1: A Climate Insurance Pool, which is a solidarity fund based on insurance principles, will compensate vulnerable countries for *high-layer risks*.	-no equivalent-
Tier 2: A Climate Insurance Assistance Facility will provide technical and financial support for risk sharing and transfer schemes covering *middle-layer risks*.	-Same in the form of two separate facilities- Technical advisory facility Financial vehicle/facility
-No equivalent-	(3) Rehabilitation/compensatory component
	To address progressive negative impacts, for example, sea level rise. Compensates victims for damages.

together to reduce the human and economic burdens of extreme and variable weather on developing countries. The pillars would be fully financed by an adaptation fund as part of the Copenhagen Agreed Outcome.

The prevention pillar

The first pillar of the MCII proposal provides for comprehensive risk reduction with a specific focus on the most vulnerable countries, communities and sectors. Disaster prevention can take many forms: reducing exposure to risks, reducing vulnerability, and creating institutions for better response. While details need to be worked out, this pillar would provide technical and financial support for such activities as risk assessments, investment projects, capacity building, institutional reform and education to reduce the human and economic toll of disasters on the most vulnerable. It is essential that risk-reducing measures build on detailed risk assessments, which can uncover otherwise unforeseen possibilities for risk reduction.

The insurance pillar

Not all risks can be cost-effectively reduced, and MCII's proposed insurance pillar would complement preventive activities by spreading and transferring remaining risks. The insurance pillar has two tiers, as shown in Figure 13.2,

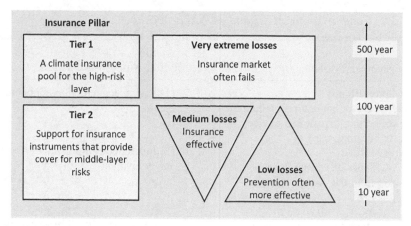

Figure 13.2. A two-tiered insurance pillar as part of an adaptation fund.

reflecting the different layers of risk that need to be addressed for effective climate adaptation.

(i) 'high-level' risk that exceeds the ability of any given country to pay in the case of an extreme event; and

(ii) 'middle-level' risk that is within the ability of any given country to cope if the proper facilitating framework were in place.

'Low-level' risks occurring, for instance, every 10 years or more frequently, are not addressed in the MCII Insurance Pillar, as these risks can often be more cost-effectively met by prevention measures (they are addressed in the Prevention Pillar).

Insurance pillar tier 1 The first tier would provide insurance cover as part of a climate insurance pool (CIP) to developing countries falling victim to infrequent and severe weather-related events (e.g. the 100- to 500-year events)[3]. Similar to a solidarity fund, the premiums would be fully financed from an adaptation fund or other financial mechanism agreed in Copenhagen. To become eligible for the CIP, it is recommended that:

(i) vulnerability assessments are carried out based on the government's risk exposure and ability to cope; and

(ii) governments fulfil basic standards of fiscal and budgetary transparency and commit themselves to risk reduction measures.

Details of this proposal are discussed (MCII, 2008); it should be noted here only that there are many options for the CIP operations, including how to define an extreme

---

[3] Tier 1 of the insurance pillar builds on earlier proposals by Bals *et al.* (2007) and Höppe (2008).

event, what losses to indemnify and how much, and how to link with prevention. One main advantage of creating an international entity, like the CIP, is that, by pooling the risks of extraordinary losses, far less reserve capital is needed than if each country created its own catastrophe fund for this same purpose.

Although post-disaster assistance, especially if the affected people or countries do not contribute to the pool, can be criticised for distorting market risk prices and crowding out commercial insurance, as discussed in Section 13.6, the market often fails in providing cover for very low-probability, high-consequence risks, especially in poorer countries.

Insurance pillar tier 2 A second tier of the MCII proposal, as pictured in Figure 13.2, would take the form of a Climate Insurance Assistance Facility (CIAF) that would provide support for the middle layer of risk not compensated by the CIP described above[4]. It would *not* directly provide insurance to households, farmers or governments, but would offer support to nascent micro-, meso- and macro-scale disaster insurance systems, like those now operating (or formerly operating) in Mongolia, Ethiopia and the Caribbean, respectively.

The core of this second tier is the provision of capacity building and technical support, which might include such activities as collecting and disseminating weather data, financing risk assessments or weather stations, or supporting delivery systems, all of which render these systems more accessible and affordable to poor communities. In addition, this tier can provide more direct support by offering or brokering pooling and reinsurance arrangements, or even, if appropriate, by subsidising premiums (see Box 13.3).

The MCII two-pillar proposal meets the challenge of providing support to promote sustainable, affordable and incentive-compatible insurance programmes with minimal crowding out of private sector involvement. While the first tier, the CIP, arguably distorts prices by offering premium-free insurance for an upper layer of risk, it can be justified by market failure for this risk layer. By enabling insurance for the poor, the CIP opens opportunities for capitalisation through risk transfer programmes involving the private market. Since the market currently fails for this risk layer, this proposal actually creates market opportunities for the private sector in the form of reinsurance for the global pool. The second tier, the CIAF, imposes affordable prices on heretofore un-priced risks, thus replacing the negative incentives and moral hazard created by post-disaster aid, and creates ample opportunities for the private sector in insuring and reinsuring these programmes.

---

[4] This tier is based on a proposal by Linnerooth-Bayer and Mechler (2007b), and is similar to a recent proposal set out by the International Fund for Agricultural Development (IFAD) and the WFP for a IFAD-WFP Weather Risk Management Facility, funded by the Bill and Melinda Gates Foundation.

13.7.2 The AOSIS proposal

The AOSIS proposal suggests a multi-window mechanism to be included in an adaptation regime. As shown in Table 13.2, this mechanism has strong similarities and also important differences with the MCII proposed risk management module. The two pillars (prevention and insurance) of the MCII proposal correspond with two components (risk management and insurance) of the AOSIS proposal. Importantly, however, the AOSIS insurance component does not have an equivalent CIP (first tier of the MCII proposal), which compensates for high-level risks. Otherwise, AOSIS proposes that adaptation funding includes a prevention component and assistance to micro- and other insurance schemes operating in developing countries, both of which are suggested by MCII. According to AOSIS, a technical advisory facility, as its name suggests, would provide technical assistance and capacity building; a financial vehicle/facility would provide financial support in the form of capitalisation or subsidies.

The AOSIS proposal differs from the MCII proposal in another important aspect. The scope of the AOSIS multi-window mechanism goes far beyond the MCII risk management module by including compensation for foreseeable and thus uninsurable damages, such as sea level rise, ocean acidification and desertification. The former is of particular concern to small island states, and AOSIS proposes a rehabilitation/compensatory component for this and other foreseeable climate damages.

13.7.3 Other financing mechanisms

It should be emphasised that insurance mechanisms are not the sole candidate for support from an adaptation fund. In Section 13.3.1 we discussed alternatives to insurance, for example, post-disaster emergency lending, all of which could be deserving of support alongside insurance. Although insurance has become topical in the climate discussions, it might be more appropriate to replace the focus on insurance, with a focus on risk financing instruments. This would include, besides risk sharing and transfer mechanisms, such financing practices as remittances, micro-lending and savings. A more complete discussion can be found in Linnerooth-Bayer (2008).

13.8 Conclusion

The messages to the climate-adaptation policy community and Party negotiators are twofold:

(i) insurance mechanisms have a promising and legitimate role in an adaptation regime; and
(ii) practical options exist for including insurance mechanisms in the post-Kyoto adaptation strategy.

Insurance mechanisms have a promising and legitimate role in an adaptation regime

There is now broad scientific consensus that climate change is contributing to worsening climate variability and extremes, which are imposing disproportionately large human and economic burdens on developing countries. These losses can be amplified by the inability of households, SMEs and governments to raise sufficient post-disaster capital for the recovery process. They should weigh the benefits and costs of insurance strategies, recognising that insurance to meet this liquidity gap will not be appropriate in all contexts, especially if reliable informal arrangements are in place.

There are large, potential benefits for insurance in the developing world: providing security against the wholesale loss of assets, livelihoods and even lives in the post-disaster period; changing the way development organisations provide disaster assistance whilst engaging the private sector in vast markets; ensuring reliable and dignified post-disaster relief; setting powerful incentives for prevention; and not least, spurring economic development. There are also many challenges: assuring sustainability and affordability in light of covariant risks; defining an appropriate role for donors in light of the inefficiencies of subsidies; and assuring that systems avoid moral hazard and contribute to 'good' investments.

While the benefits and challenges of catastrophe safety nets are uncontested, the role of outside assistance for insurance instruments is highly controversial. Opponents rightly argue that support in the form of subsidies can distort the price signal and encourage maladaptation; support in the form of reinsurance can crowd out the role of the private market. Yet, most experts agree that even subsidised insurance systems are in this regard preferable to post-disaster aid, and the reinsurance market is not yet prepared to commit sufficient and affordable capital to markets serving the poor. Experts also agree that outside support should be closely coupled with a risk management programme including a vulnerability assessment. Pilot programmes are offering a testing ground for the efficacy of international assistance (a subject dealt with in depth in Chapter 12, Gupta *et al.*), and these programmes should be carefully monitored and built upon by governments, international development organisations, NGOs, private insurers, and the climate adaptation community.

The case for intervention as part of an adaptation regime is legitimised by the failure of the market, and greatly strengthened by recent evidence that greenhouse gas emissions are contributing to increased weather variability and risks of extreme events. According to the Climate Convention's principle of 'common but differentiated responsibilities and respective capabilities', industrialised countries are obliged to absorb a portion of this burden.

*Practical options exist for including insurance mechanisms in the
post-Kyoto adaptation strategy*

As a practical way forward, this discussion has laid out two proposals recently put
forward by MCII and AOSIS. MCII proposes a two-pillar risk management module
as part of an adaptation regime, and partly identical to this is the three-component
multi-window AOSIS mechanism proposal. Both include provisions for supporting
preventive measures. They also enable implementation of micro- and national
insurance systems in vulnerable developing countries by providing technical assistance, capacity building and possibly absorbing a portion of the insurance costs.
Both proposals, however, also have important elements not shared by the other;
MCII suggests a climate insurance pool that indemnifies victims of extreme cata-
strophes in vulnerable countries by a percentage of their losses, where premiums are
paid fully by an adaptation funding mechanism; AOSIS suggests a rehabilitation/
compensatory component that would compensate victims for sea level risk and
other uninsurable damages.

By clarifying the opportunities and challenges of insurance as an instrument for
adaptation, and outlining a practical way forward, it is hoped that this discussion can
contribute to the opportunities and challenges facing negotiators at COP 15 and
beyond, in adopting a comprehensive adaptation strategy that enables risk management and insurance to be implemented for the most vulnerable in the developing
world.

References

Allen, M. (2007) African farmers offered drought insurance. *Swissinfo,* Sept. 29, Geneva.
    www.swissinfo.org/eng/front/detail/African_farmers_offered_drought_insurance.
    html
Alliance of Small Island States (AOSIS) (1999) *Proposal to Intergovernmental Negotiating
    Committee for the Creation of an International Insurance Pool.* A/AC.237/15.
Alliance of Small Island States (AOSIS) (2008) Proposal to the ad hoc working group on
    long-term cooperative action under the Convention (AWG-LCA). *Multi-window
    Mechanism to Address Loss and Damage from Climate Change Impacts.* Submission
    to the UNFCCC on 6 December 2008. A/AC.237/15. http://unfccc.int/files/
    kyoto_protocol/application/pdf/aosisinsurance061208.pdf.
Bals, C., Butzengeiger, S. and Werner, K. (2007) Insuring the uninsurable: design options
    for a climate change funding mechanism. Special Issue on insurance and climate
    change, ed., Gurenko, E., *Climate Policy,* **6**, 637–47.
Barnett, B. J., Barrett, C. B. and Skees, J. R. (2008) Poverty traps and index-based risk
    transfer products. *World Development,* **36**, 1766–85.
Carter, T., Jones, R., Lu, X. *et al.* (2007) New assessment methods and the characterisation
    of future conditions. In *Climate Change 2007: Impacts, Adaptation and Vulnerability.
    Contribution of Working Group II to the Fourth Assessment Report of the
    Intergovernmental Panel on Climate Change,* ed. Parry, M., Canziani, O., Palutikof, J.,
    van der Linden, P. and Hanson, C. Cambridge, UK: Cambridge University Press,
    pp. 133–71.

Cardenas, V., Hochrainer, S., Mechler, R., Pflug, G. and Linnerooth-Bayer, J. (2007) Sovereign financial disaster risk management: the case of Mexico. *Environmental Hazards*, **7**, 40–53.

Cohen, M. and Sebstad, J. (2003) *Reducing Vulnerability: The Demand for Microinsurance* MicroSave-Africa http://www.microinsurancecentre.org.

Cummins, D. and Mahul, O. (2008) *Catastrophe Risk Financing in Developing Countries: Principles for Public Intervention*. Washington DC: World Bank Publications.

Ghesquiere, F., Mahul, O., Forni, M. and Gartley, R. (2006) *Caribbean Catastrophe Risk Insurance Facility: A Solution to the Short-term Liquidity Needs of Small Island States in the Aftermath of Natural Disasters*. IAT03–13/3 www.aidandtrade.org.

Gunaranjan, U. (2008) *Micro insurance for protecting livelihoods*, BASIX paper for Microfinance Summit Nepal, February. www.microfinancegateway.org/files/53200_file_23.pdf

Gurenko, E. (2004) *Catastrophe Risk and Reinsurance: A Country Risk Management Perspective*. London: Risk Books.

Gurenko, E., Lester, R. and Mahul, O. (2006) *Earthquake Insurance in Turkey: History of the Turkish Catastrophe Insurance pool*. Washington, DC: World Bank Publications.

Harmeling, S. and Bals, C. (2008) *Adaptation to Climate Change – Where Do We Go from Bali? An Analysis of the COP 13 and the Key Issues on the Road to a New Climate Change Treaty*. Bonn: Germanwatch. www.germanwatch.org/klima/adapt08e.htm.

Hess, U. (2007) *Risk Management Framework – the big LEAP in Ethiopia*, presentation, Insurance in catastrophe risk management, November 14, 2007, Bonn: Gustav-Stresemann Institut.

Hess, U. and Syroka, J. (2005) *Weather-based insurance in southern Africa: the case of Malawi*. Agriculture and Rural Development Discussion Paper 13, Washington DC: The World Bank.

Höppe, P. and Gurenko, E. (2007) Scientific and economic rationales for innovative climate insurance solutions, Special issue on insurance and climate change, ed. Gurenko, E. *Climate Policy*, **6**: 579–99.

Höppe, P. (2008) *Climate Risk Insurance: suggestions for compensation-based climate risk insurance*, presentation at the MCII side event Climate Risk Insurance at SB 28, 11 June 2008, Bonn.

Klein, R. J. T., Eriksen, S. E. H, Naess, L. O. *et al.* (in press) Portfolio screening to support the mainstreaming of adaptation to climate change into development assistance. *Climatic Change*

Krishna, H. (2007) *Presentation Discussion, Expert Workshop on Insurance Instruments for Adaptation to Climate Risks*. International Institute for Applied Systems Analysis (IIASA) Laxenburg, Austria, Sept. 24–5.

Kunreuther, H. (1998) Introduction. In *Paying the Price: The Status and Role of Insurance Against Natural Disasters in the United States*, ed. Kunreuther, H. and Roth, S. Washington, DC: Joseph Henry Press.

Leftley, R. (2007) *Presentation discussion, Expert workshop on Insurance Instruments for Adaptation to Climate Risks*. International Institute for Applied Systems Analysis (IIASA), Laxenburg, Austria, Sept. 24–5.

Linnerooth-Bayer, J. (2008) *Non-insurance Mechanisms for Managing Climate-related Risks*, Mechanisms to manage financial risks from direct impacts of climate change in developing countries. Technical paper. FCCC/TP/2008/9, Geneva: United Nations Office at Geneva. http://unfccc.int/4159.php

Linnerooth-Bayer, J. and Mechler, R. (2007a) Disaster safety nets for developing
    countries: beyond public-private partnerships. Special issue on financial
    vulnerability, Linnerooth-Bayer, J. and Amendola, A. *Environmental Hazards*, **7**(1),
    54–61.
Linnerooth-Bayer, J. and R. Mechler (2007b) Insurance for assisting adaptation to climate
    change in developing countries: a proposed strategy, Gurenko, E. (ed.) Special issue on
    insurance and climate change. *Climate Policy*, **6**, 621–36.
Linnerooth-Bayer, J., Mechler, R. and Pflug, G. (2005) Refocusing disaster aid. *Science*,
    309, 1044–6.
MCII (2008) Munich Climate Insurance Initiative. *Insurance Instruments for Adapting to
    Climate Risks: a proposal for the Bali Action Plan*. MCII submission to Accra climate
    change talks, Ghana, http://www.climate-insurance.org/upload/pdf/MCII_submission_
    Poznan.pdf
Mechler, R. (2004) *Natural Disaster Risk Management and Financing Disaster Losses in
    Developing Countries*. Karlsruhe: Verlag für Versicherungswissenschaft.
Mechler, R., Linnerooth-Bayer, J. and Peppiatt, D. (2006) *Microinsurance for Natural
    Disasters in Developing Countries: Benefits, Limitations and Viability*. Geneva:
    ProVention Consortium. www.proventionconsortium.org/themes/default/pdfs/
    Microinsurance_study_July06.pdf.
Moench, M., Mechler, R. and Stapleton, S. (2007) *Guidance Note on the Costs and Benefits
    of Disaster Risk Reduction*. Geneva: ISDR High level Platform on Disaster Risk
    Reduction.
Müller, B. (2008) *International Adaptation Finance: The Need for An Innovative and
    Strategic Approach*. Oxford, UK: Oxford Institute for Energy Studies EV 42.
Munich Reinsurance Company (2005) *NatCatSERVICE, Natural Disasters According to
    Country Income groups 1980–2004*. Munich: Munich Reinsurance Group.
Munich Reinsurance Company (2006) *Report of the Workshop on Climate Change and
    Disaster Losses: Understanding and Attributing Trends and Projections*, 25–26 May
    2006, Hohenkammer. http://www.sciencepolicy.colorado.edu/sparc/research/projects/
    extreme_events/munich_workshop/executive_summary.pdf
Munich Reinsurance Company (2007) Topics: natural disasters 2006. *Annual Review of
    Natural Disasters*. Munich: Munich Reinsurance Group.
Skees, J. R. (2001) The bad harvest: more crop insurance reform: a good idea gone awry.
    The CATO Review of Business and Government, **24**, 16–21.
Skees, J. R. (2007) *Presentation discussion, Expert workshop on Insurance Instruments for
    Adaptation to Climate Risks*. International Institute for Applied Systems Analysis
    (IIASA). Laxenburg, Austria, Sept. 24–5.
Skees, J. R., Barnett, B. J. and Murphy, A. G. (2008) Creating insurance markets for natural
    disaster risk in lower income countries: the potential role for securitisation.
    Agricultural Finance Review, **68**, 151–7.
Skees, J. R. and Enkh-Amgalan, A. (2002) *Examining the Feasibility of Livestock Insurance
    in Mongolia*. World Bank Working Paper 2886, September 17, Washington DC: The
    World Bank.
Solomon, S., Qin, D., Manning, M. *et al.* (2007) Technical Summary. In *Climate Change
    2007: The Physical Science Basis. Contribution of Working Group I to the Fourth
    Assessment Report of the Intergovernmental Panel on Climate Change*, ed. Solomon, S.,
    Qin, D., Manning, M. *et al.* Cambridge, UK: Cambridge University Press.
Suarez, P., Linnerooth-Bayer, J., and Mechler, R. (2007) *The Feasibility of Risk Financing
    Schemes for Climate Adaptation: The Case of Malawi*. DEC-Research Group,
    Infrastructure and Environment Unit, Washington DC: The World Bank.

UNFCCC (2008a) United Nations Framework Convention on Climate Change, *Bali Action Plan*, Decision 1/CP.13, FCCC/CP/2007/6/Add.1, 14 March 2008.

UNFCCC (2008b) United Nations Framework Convention on Climate Change, *Mechanisms to Manage Financial Risks from Direct Impacts of Climate Change in Developing Countries*. Technical paper. FCCC/TP/2008/9, Geneva: United Nations Office at Geneva, unfccc.int/4159.php.

UNISDR (2007) United Nations International Strategy for Disaster Reduction Disaster statistics 1991–2005 www.unisdr.org/disaster-statistics/impact-economic.htm.

UVEK (2008) Swiss Federal Department of the Environment, Transport, Energy and Communications Funding Scheme for Bali Action Plan, *A Swiss Proposal for Global Solidarity in Financing Adaptation*, 'Bali Paper' updated for SB28 Bonn, Berne: Federal Office for the Environment. www.environment-switzerland.ch/climate.

World Bank (2005). *Managing Agricultural Production Risk*. Washington, DC: World Bank.

Part IV

Synthesis

14

What can social science tell us about meeting the challenge of climate change? Five insights from five years that might make a difference

Lead authors:

ANTHONY PATT[1], DIANA RECKIEN, RICHARD J. T. KLEIN,
DETLEF P. VAN VUUREN, MARKUS WROBEL, NICO BAUER,
GUNNAR S. ESKELAND, THOMAS E. DOWNING

[1]Co-ordinating lead author

Summary

In this chapter, we describe five themes that emerge from the work within ADAM on mitigation of and adaptation to the climate change problem. Firstly, the relationship between adaptation and mitigation is complex, and indeed the extent to which the optimum level of each depends on the other cannot be determined at this time. Secondly, the challenge of mitigation is primarily that of stimulating investments in low-carbon technologies that will ultimately allow society to eliminate greenhouse gas emissions. While initially this may be a costly strategy, the faster that investment patterns change, the faster the costs of new technologies may fall to below those of fossil fuels. Thirdly, it is important to balance commitment to action, with keeping options open. For mitigation, policies need to send clear signals to actors that carbon intensive activities will become rapidly less attractive. For both mitigation and adaptation, there is a tremendous amount of learning yet to take place that could influence which technologies and behavioural changes people view as appropriate. Fourthly, giving the private sector flexibility in the form of mitigation actions they can take has its uses, but it can also stand in the way of inducing necessary technological change. Many of the least cost ways of reducing carbon dioxide equivalent emissions are not those that will contribute to the elimination of emissions altogether, which we now understand needs to be the objective for this century. Fifthly, some of the entrenched principles of key actors in international climate policy, such as the polluter pays principle, may in fact stand in the way of forging rapid consensus around necessary policies. We describe each of these five themes with the use of a metaphor or analogy, in order to create a clear image of how we ought to view the climate change problem.

Making Climate Change Work for Us: European Perspectives on Adaptation and Mitigation Strategies,
ed. Mike Hulme and Henry Neufeldt. Published by Cambridge University Press © Cambridge University Press 2010.

Introduction

The preceding 13 chapters have presented a number of quantitative and qualitative conclusions about the best strategies for European climate policy, covering adaptation, mitigation, and in some cases the interaction between the two. They are based on 3 years of research within the ADAM integrated project, funded by the European Commission's Sixth Framework Programme. It was in 2004 that the scientists who designed that project began discussing their research objectives. These objectives were shaped by how they and others at the time viewed climate change as a set of social and policy challenges. In the intervening 5 years, many of these views have evolved, partly as a result of ADAM research. In this chapter, we describe how.

Where was the climate change problem in 2004? Russia had finally ratified the Kyoto Protocol, allowing it to enter into force. The European Union was finalising the design of the emissions trading scheme (ETS), which would go into effect that same year. A vigorous academic debate was under way concerning the cost and feasibility of reducing future carbon dioxide emissions. A paper in *Science* suggested that emissions could actually be frozen at current rates using a portfolio of existing technologies (Pacala and Socolow, 2004). This challenged the wisdom from some integrated assessment models: that only small reductions – in the order of 7–10 per cent from 'business as usual' growth scenarios – could be achieved at modest cost (Nordhaus and Boyer, 2000). Meanwhile, participants in a symposium to be held in February 2005, 'Avoiding dangerous climate change,' were about to argue that policy would have to achieve much more than simply a stabilisation in emissions if the effects of climate change were to become manageable. The issue of adaptation was starting to rise to prominence. Numerous international and non-governmental organisations, such as the Red Cross /Red Crescent Climate Center, started to investigate how to address growing climate risks at the programmatic level. There was a profound lack of clarity as to whether the challenges of mitigation (reducing carbon dioxide emissions to minimise the climate change problem) and adaptation (responding intelligently to the effects of rising carbon dioxide concentrations in the atmosphere) were linked at the policy level, and if so how.

Participants in the ADAM project pushed forward on each of these issues, examining adaptation and mitigation both separately and together. In the following sections, we present five insights gleaned from this work, and the work of our peers over the same 5-year period. We present each of these insights in the form of a metaphor or image. The way that we visualise a problem has a profound influence on how we communicate, and how we try to solve it (Gigerenzer and Selten, 2001; Morgan *et al.*, 2001), and our hope is that, by offering a set of new metaphors and images for the climate problem, we can stimulate more effective policy responses than those that have taken place so far.

14.1 Turn off the tap with one hand, and start mopping with the other

If a tap is on, and running into a hopelessly clogged sink, water will spill over the edge. Turning off the tap has first priority, but finding a mop to clean up the mess, and minimise the damage, is also important. If many hands are available, the two can start at once. But the activities are essentially separate. Do not necessarily expect to find synergies between them.

This is a core insight about the relationship between adaptation and mitigation; it is impossible to pinpoint an optimal mix of adaptation and mitigation effort, even as decisive action is needed on both. Indeed, it captures what the authors of the Fourth Assessment Report (AR4) of the Intergovernmental Panel on Climate Change (IPCC) said when they described this relationship (Klein *et al.*, 2007b: 747). Firstly, '[e]ffective climate policy aimed at reducing the risks of climate change ... involves a portfolio of diverse adaptation and mitigation actions ...' Even were climate change mitigation efforts to begin immediately at full capacity, there would still be effects of climate change which would demand adaptation in response (see Chapter 3, van Vuuren *et al.*). Secondly, '[d]ecisions on adaptation and mitigation are taken at different governance levels ... Effective mitigation requires the participation of the major greenhouse-gas emitters globally, whereas most adaptation takes place from local to national levels.' In the case of climate change, there are many hands available, but the arms of government that are involved in mitigation are quite different from those that are working on adaptation. Third, 'synergies between adaptation and mitigation can increase the cost-effectiveness of actions and make them more attractive to stakeholders, [but] such synergies provide no guarantee that resources are used in the most effective manner when seeking to reduce the risks to climate change.' This statement can be a bit confusing. On the one hand, many isolated actions can be improved by looking for synergies. For example, the cost-effectiveness of a new windmill or hydroelectric dam (primarily mitigation) will be better if their designs take into account future climate conditions. On the other hand, the most cost-effective actions for either adaptation or mitigation may be the ones for which the synergies are not important. In other words, make sure that the person who is mopping up the floor doesn't get in the way of the person turning off the tap, but don't expect the two of them to work together closely to get their respective jobs done. Fourthly, '[i]t is not possible to answer the question as to whether or not investment in adaptation would buy time for mitigation ... In particular, the notion of an 'optimal mix' of adaptation and mitigation is difficult to make operational, because it requires the reconciliation of welfare impacts on people living in different places and at different points in time into a global aggregate measure of well-being.' Would buying a bigger mop mean that you can wait longer before turning off the tap? Maybe, but it is hard to tell, because it could be that the water is already leaking through the roof of your downstairs neighbour. Fifthly, '[p]eople's capacities to adapt and mitigate are driven by similar sets of factors.' Both involve taking deliberate action, and wealthier, better-organised societies will probably be able to do them more easily than poorer or more chaotic ones.

Within the ADAM project, a group of researchers have looked more closely at how to model the potential trade-offs and synergies between adaptation and mitigation (Patt *et al.*, 2009), providing further insight into the fourth of the Klein *et al.*'s (2007b) five main points. The authors of this paper looked at how adaptation has been represented so far in integrated assessment models (IAMs) – particularly those that analyse climate change in a cost–benefit framework – and what additional insights could be learned from improving that representation. The authors found that, in many ways, existing IAMs assume some measure of optimal adaptation; smart farmers will change the crops they grow as rainfall regimes change; people will stop buying snow tyres, and instead buy air conditioners, as temperatures rise. These adaptations reduce the net costs of climate impacts, and hence the marginal benefits of mitigation and the 'optimal' level of mitigation. What the models fail to take into account is the fact that collective or government action is often necessary to facilitate such private adaptive efforts. Such actions include educating farmers about the seed varieties that are now appropriate, or relaxing the requirement for car owners to have snow tyres. Given that these government actions can be just as difficult to implement as mitigation actions, adaptation can be expected to be far less than optimal. The authors then point to two IAMs that have attempted to explore the sensitivity of the optimal mitigation target to different levels of adaptation, and note that the target moves by a few percentage points (Hope, 2006; de Bruin *et al.*, 2007). If, in all likelihood, adaptation will not achieve optimal amounts (there will be too much of it, or too little, in terms of particular indicators of change), the actual sensitivity of the optimal mitigation target to changes of adaptation within the feasible range is likely to be very small. The authors then note two reasons why more explicit attention to adaptation within such models can be useful. The first is simply to add greater specificity to the adaptation options explored, which will make the costs of climate change impacts more concrete and more salient to actors who would have to incur those costs. This could have the paradoxical result of making mitigation, which also has salient costs, more attractive relative to adaptation. The second is more complicated: if the estimates of the total costs of climate impacts increase, then a more accurate appraisal of adaptation can suggest how much additional mitigation may be necessary. Figure 14.1 illustrates this. If the estimates of climate change impacts rise, as some suggest is happening (Yohe *et al.*, 2007), then the effect of more accurately representing adaptation could be important, and imply a much larger shift in the optimal mitigation target. Figure 14.1 also illustrates that the optimal mitigation target does, in fact, depend on the level of adaptation that is expected to occur. It is important to note, however, that this depends

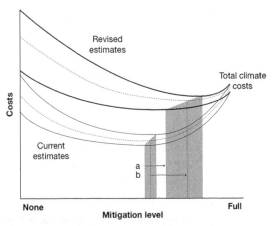

Figure 14.1. Sensitivity of mitigation target to adaptation and to different estimates of the magnitude of the entire climate problem. The bottom set of curves represent current cost estimates of climate change, as a function of mitigation level, where the lower solid curve is the case with optimal adaptation, and the upper solid curve is the case with no adaptation. In between the two curves is a dashed line, which is the case with a realistic amount of adaptation, given how difficult adaptation will be to optimise. The upper curves represent revised estimates, in which total damage costs from climate change are higher. In this case, the benefits from adaptation – and the distance between the two solid curves – are also likely to be greater. Under reasonable assumptions about the shape of the two curves, the difference in optimal mitigation level implied by a move from optimal to sub-optimal adaptation is also greater. The net change in the optimal mitigation target, represented by arrows a and b, depends in turn on whether one assumes optimal or realistic amounts of adaptation. In this case, it becomes more important to estimate how much adaptation is actually possible, as this could imply a difference in the shift of the optimal mitigation target. (Source: Patt *et al.*, 2008.)

on the result that mitigation becomes costlier, the more that is done, and that those costs can be estimated in advance. As the subsequent lessons in this chapter describe, this may not be the case.

14.2 Move your pawns forward

The number of ways to play, and indeed to win, at the game of chess is virtually infinite, but some strategies are better than others. Gary Kasparov, one of the best chess players ever, explained his strategy: early in the game he tries to move the game into a situation that allows him to move his pawns forward. A pawn that makes it to the other side becomes a queen, the most powerful player on the board. Nobody can predict which pawn will reach the other side, or what the board will look like when it gets there. The art of playing this aggressive style is to develop new opportunities and keep options open. If one pawn does make it to the other side, it is usually the decisive factor for winning a match.

Applying this thinking to climate change sees the challenge of mitigation policy at this stage in the game, primarily as one of stimulating the development of new technologies for de-carbonising, or the development of lower-cost ways of implementing existing technologies. This kind of development happens when firms are forced to try to reduce their emissions, set their engineers or their suppliers with tasks to accomplish, learn by doing, and achieve what is known as induced technological change.

While the idea of learning by doing is not new, the last 5 years have seen it playing a much stronger role in the economic analysis of mitigation, with some surprising results. Until recently, economic models (Manne and Richels, 1997; Nordhaus and Boyer, 2000) examined the costs and benefits of fairly modest changes to the energy system. To assess the costs of these changes, the modellers typically assumed an exogenous rate of technological change, meaning that, over time, technologies would become available or less expensive independent of the level of mitigation undertaken by society. This may not have been a poor assumption, as long as the range of mitigation scenarios considered was small. One of the interesting results of these models, however, was that it made sense to take a wait-and-see approach to climate change mitigation, for two reasons. Firstly, in the intervening years more may be learned about the sensitivity of the climate system, including the possibility that climate change is not a problem. Secondly, technological advances in the intervening years would make the cost of mitigation, once that mitigation started in earnest, lower. Why not wait for the price of alternative technologies, like solar and wind power, to come down, before implementing them at full scale?

The change of modelling approach came about as natural scientists, engineers and economists began to see the task of mitigation policy not as maximising economic welfare, but as reducing the likelihood of unacceptable risks, or 'dangerous climate change' (Barker, 2008). Policy makers converged on a scientifically justified threshold, 2 °C total average warming, that could avoid dangerous climate change and serve as the target for mitigation policy (Welp *et al.*, 2008). Achieving the 2 °C target would mean substantially greater mitigation effort than that ever identified as optimal by the earlier models, and the assumption of exogenous technological change became less valid (Manne and Richels, 2004). Models such as MIND (Edenhofer *et al.*, 2005) began to pay explicit attention to learning by doing, and their results were startling, in two respects. Firstly, the differences in costs between modest and very large emissions reductions might be smaller. Secondly, the estimated costs of achieving very large emissions reductions in the models that incorporated learning by doing were *lower* than the costs of very modest emissions reductions in the models that treated technological change as exogenous. The story behind this is quite simple: aggressive policies that reduce emissions dramatically also promote faster learning by doing, and this drives down the costs of new

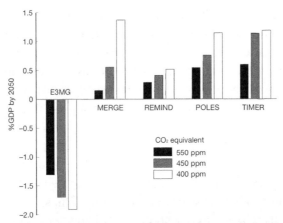

Figure 14.2. – Cost estimates associated with achieving different stabilisation scenarios, from five different models, compared to a business as usual scenario. The models E3MG, MERGE, and REMIND express costs in terms of a net reduction in global GDP brought about by mitigation, while POLES and TIMER express costs in terms of the fraction of global GDP required for abatement activities in the energy sector. All models make use of a 3% annual discount rate, to express costs in net present value. Negative cost values, projected by the E3MG model, imply that mitigation would generate net additional economic growth. A 550 ppm scenario would limit climate change to 2 °C with about 15% probability, a 450 ppm scenario with about 50% probability, and a 400 ppm scenario with about 75% probability. (Source: Chapter 11, Knopf *et al.*)

technologies faster. It can even drive them down to below the costs of conventional fossil fuels, meaning that the sooner this threshold is passed, the greater the overall cost savings. Reviewing the complete literature, both the Stern Review (Stern, 2006) and the Working Group III report of the AR4 (Metz *et al.*, 2007) concluded that the costs of reducing emissions by 50 to 80 per cent over the next several decades will likely be modest, less than two per cent of economic output. According to the Stern Review, this would be far less than the costs of inaction. Moreover, it is important to start now, rather than wait.

ADAM researchers have continued to push the state of the art of this kind of modeling, through the analysis of low stabilisation scenarios, comparing results from different models (Knopf *et al.*, 2008; and see Chapter 11, Knopf *et al.*). Figure 14.2 shows the estimates of the total costs on global economic product by 2050 associated with three different stabilisation scenarios. There are several interesting results. All of the estimated costs are close to or below one per cent of global GDP. This is small, perhaps even trivial. One way of thinking of it is that, in *one* of the next few years, economic growth would be one per cent lower than it otherwise would have been, meaning that in all subsequent years, economic output would remain one per cent lower. One of the models, E3MG, even suggests a negative cost associated with aggressive climate policies. Like the other models, E3MG incorporates learning by doing, but it differs in that it does not rest on a foundation of a general economic equilibrium. Rather, it assumes that the energy system is path dependent, meaning that its current state may be quite far from a lowest cost equilibrium, and that policy intervention could push the energy system towards a state that on the one hand has no carbon dioxide emissions, and on the other hand costs less than the current system. While such a conclusion is controversial, several energy system studies show that this is possible (e.g. Czisch, 2005).

A final result, not indicated by the figure, is that there is no clear prediction about the future energy mix. The different models meet the stabilisation targets through very different future energy systems, but the costs of those pathways are all quite low. This is like the chess player who does not know which pawn will reach the other side, but that one of them will do so, and help to win the game. There is a great deal of uncertainty about how fast learning by doing will take place, and for which technologies it will take place the fastest, although across all models renewable energy and carbon capture and storage (CCS) outperform nuclear power. In each of the models, however, learning by doing drives down the costs of the dominant technology the fastest, making the costs of substantial emissions reductions much smaller than we used to believe. The ultimate objective of mitigation policy is to reduce or eliminate carbon dioxide emissions: turning off the tap. But the strategy for doing so, at this stage, needs to be one of stimulating investment to build up at least one alternative energy source to the point where it becomes less expensive than fossil fuels, rather than simply reducing the use of fossil fuels by small incremental amounts.

14.3 Burn the right bridges

There are two different bridge burning metaphors, and we think both are important for climate change. The first version of the metaphor holds that you should keep on good terms with people or places you are leaving, because you may decide to return to them in the future, even if you cannot foresee doing so now. The second version of the metaphor has the opposite meaning: if you are a military general advancing on the enemy, you may choose to burn the bridges behind you, in order to force your army to achieve its goals. You know that, when the battle gets difficult, your soldiers will want to retreat, and preventing that option will eliminate that distraction, and force them to fight harder. Both of these bridge-burning metaphors provide intuition about what to do in the face of uncertainty, and both are relevant for climate change policy.

The first metaphor, not to burn bridges, suggests that there may be benefits from avoiding irreversible decisions while valuable information is yet to be received, and is formalised in real options theory (Dixit and Pindyck, 1994). Early analysis suggested that not mitigating could have irreversible consequences on the climate system, while mitigating could have irreversible consequences on the economy (O'Neill, 2008). At a time when the economic consequences of relatively small mitigation efforts were expected to be quite large, a great deal of effort went into analysing the economic irreversibility of mitigation. One result was to suggest that it makes sense to engage in 'no regrets' mitigation – modest measures that make sense even if climate change is not a problem – until we learn more about how sensitive the climate is, and hence how much mitigation is optimal (Manne and Richels, 1992). But this argument has become less valid over the last few years. Firstly, climate scientists have made very little, if any, progress towards reducing the uncertainty about climate sensitivity (Oppenheimer *et al.*, 2008). Although modest learning through continued observation could cause us to revise our estimate of what level of mitigation is optimal (Webster *et al.*, 2008), it is also quite possible that the complexity of the climate system means that the sensitivity of climate to changes in carbon dioxide concentration might only be possible to measure *ex post*, and never predicted with very much certainty *ex ante* (Allen and Frame, 2007; O'Neill and Melnikov, 2008). Secondly, and more importantly, the results from the recent economic models suggest that marginal mitigation costs across the full range of uncertainty about climate sensitivity are so low as to be below marginal damage costs (Stern, 2006). This means that there is no value in waiting for that uncertainty to grow smaller.

A very different kind of uncertainty lies in beliefs about how other people, or even oneself, will react to a particular problem in a particular social setting. For this it is useful to consider the insights from game theory (Schelling, 1978; von Neumann and Morgenstern, 1944), showing that the deliberate elimination of options can move players from one equilibrium to a superior one, and from behavioural decision theory (Loewenstein and Elster, 1992), showing that certain commitments to action can improve the committer's welfare by eliminating counter-productive behavioural patterns, like procrastination. Both of these relate to induced technological change, which we discussed in the last section, because they suggest the kinds of policies that are needed to force economic actors to engage in behaviours which promote learning. As discussed in the last section, government policies can stimulate investment in one or more new technologies, creating future options, but they must do so by giving actors no choice but to try to find new ways of using less energy, or producing energy with less carbon. In climate policy in general, ADAM researchers have shown that when government commitments are ambiguous, they generate far less than the expected behavioural change (see Chapter 6, Berkhout *et al.*; Chapter 12, Gupta *et al.*). Governments have to make a clear commitment to upholding these policies into the future, so that nobody will try to wait until high carbon dioxide emissions are again possible. Policy makers need to burn the bridges behind them, to avoid leading back to high carbon dioxide emissions.

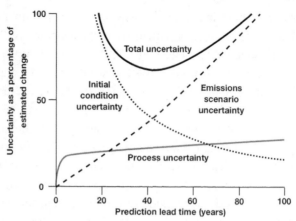

Figure 14.3. Schematic of the sources of future climate uncertainty. Three factors influence the uncertainty of climate projections. Uncertainty in the initial conditions does not increase over time, and so its size relative to the magnitude of predicted change decreases. The other two factors – climate feedback process uncertainty, and emissions scenario uncertainty – play a small role in near term projections, but grow as the time scale of the projections increases. Total uncertainty about the magnitude of change, as a share of the median estimate of that change, is highest for near- and long-term projections. (Source: Cox and Stephenson, 2007.)

Adaptation, by contrast, remains a task of keeping as many bridges as possible passable, both literally and figuratively. It is essential to see adaptation as a learning process, for two reasons. The first stems from the combination of adaptation being primarily a local process and the very high range of uncertainty concerning the direction and magnitude of local level impacts (Aaheim *et al.*, 2007, see also Chapter 4, Aaheim *et al.*). Figure 14.3 shows the relative influence of different sources of climate uncertainty, and how they change over the time period for which projections or scenarios are being made. In the short term, the greatest source of uncertainty derives from uncertainty over initial conditions (Cox and Stephenson, 2007). The ADAM project has delivered some of the first estimates of the need to climate-proof infrastructure within Europe (Genovese *et al.*, 2007; Wreford *et al.*, 2007), and has also examined the efforts at climate proofing that have been under way in developing countries. As Klein *et al.* (2007a) describe, numerous organisations engaged in development assistance have started adaptation programmes. These are mainly aimed at identifying those development projects that are vulnerable to the physical impacts of climate change, and estimating the costs associated with reducing that vulnerability. The authors find this practice of portfolio screening, to have been initiated at several organisations or agencies, including the World Bank, the German Technical Co-operation Agency (GTZ), the Organisation for Economic Co-operation and Development

(OECD), the Swiss Agency for Co-operation and Development, and the UK Department for International Development. But the result of portfolio screening has been to show how wide the range of uncertainty, and thus the potential for misdirecting investments, actually is. A background paper prepared by the United Nations Framework Convention on Climate Change secretariat estimated the need for annual global public expenditure of US$ 46–182 billion, but even to make the range this narrow they had to reconcile estimates made by the World Bank, the OECD Development Assistance Committee (DAC), and other organisations (UNFCCC, 2007). Importantly, the uncertainty over future climate, and the climate proofing required, is often highest in developing countries, where the relative scarcity of historical weather records makes it extremely difficult to construct and validate downscaled climate models (Washington *et al.*, 2006). This, in turn, is an essential step in making predictions at the same spatial scale at which adaptations occur. It suggests that it may often make good sense to take a 'wait and see' approach, as long as one makes sure that action can be taken, and taken quickly, after 'seeing'.

The second reason that adaptation is about keeping options open is that the social process of adaptation is so poorly understood. Many adaptation options are not about climate proofing, but about helping people to change behaviour in response to evolving climate risks (see Chapter 8, Mechler *et al.*). There are competing models and theories within economics (Mendelssohn *et al.*, 2000), psychology (Grothmann and Patt, 2005), geography (Adger *et al.*, 2003; Pelling and High, 2005), and other social science disciplines for the factors that cause people to change their behaviour. They also consider whether public policy ought to try to accelerate this process (Mendelssohn, 2006), and if so, what policies are likely to be most effective (Klein *et al.*, 1998; Kahn, 2003). ADAM researchers have pushed this body of knowledge by examining alternative theories of social change with expert and key stakeholder interviews in Europe (McEvoy *et al.*, 2007, see also Chapter 2, Russel *et al.*). Their preliminary findings indicate a link between adaptation and perceptions of social justice, the need for more information about climate change, and the link between cultures of organisational learning and levels of adaptation (see Chapter 9, Werners *et al.*). In short, adaptation is a process that moves forward or not according to complicated features of social interaction, and not simply because a single analyst sees it as a good idea (although having a champion can be necessary; see Chapter 5, Hinkel *et al.*). Systematic efforts to monitor the process of climate adaptation, and to influence it through policy, are both new; there are few data available on which to identify successful approaches (see Chapter 6, Berkhout *et al.*). In 30 years' time, we may have a better understanding of what works. For now, it is largely a process of trial and error, which means that we should remain open to learning from, and responding to, our mistakes.

14.4 Build the ladder

If you need just enough apples or cherries to make a pie, then you should look for some low-hanging fruit. The metaphor of the low-hanging fruit has dominated climate change discourse for well over a decade, and could guide policy makers through 2020, by which time Europe will have met its pledged emissions reductions of 20 to 30 per cent at low cost. The goals from 2020 on, however, are substantially more challenging. To avoid dangerous climate change, the world needs to reduce carbon dioxide emissions globally by about 100 per cent over the next century, give or take a few percent or a few decades, and Europe needs to reduce emissions by 80 per cent by 2050 (Metz *et al.*, 2007; Alfsen and Eskeland, 2007). Work since then, some of it within the ADAM project, has suggested that even more rapid cuts may be necessary and possible (van Vuuren *et al.*, 2007). Thus, from the perspective of the challenge of the century, we will have to pick the tree clean, and we do not benefit much by looking for low-hanging fruit. Rather, we need to build a ladder to reach the top of the tree, and pick all of the fruit that are there.

Applying this thinking in practice means re-evaluating the role of flexible policy instruments. It was possible to achieve Kyoto targets, and will likely be possible to achieve 2020 targets, without any substantial restructuring of the world's energy system. Within Europe, the least expensive emissions reductions could come about through improvements in efficiency, both end-use and in power-plant design; the European Emissions Trading Scheme (ETS) provides the flexibility to focus on these activities. As shown in the ADAM project's research (Eskeland *et al.*, 2008a and 2008b), 2020 goals in Europe can be met at low cost by combining energy efficiency with substitution towards gas and renewables, and some downscaling of energy intensive manufacturing. Indeed, those goals can be met without making much progress in terms of far-reaching technological change, and thus without building a bridge towards emission reductions in subsequent decades. Globally, Annex 1 countries will be able to meet their Kyoto targets by purchasing excess emissions credits from the countries of the former Soviet Union – supplemented by project-based credits from Joint Implementation (JI) and the Clean Development Mechanism (CDM). JI and CDM projects, similarly, have largely achieved their reductions from baseline emissions through improvements in energy efficiency, as well as through the cutting of greenhouse gases other than carbon dioxide, such as methane and chloro-fluoro carbons (Michaelowa and Jotzo, 2005). With the exception of some European countries, very few of the Annex 1 emissions reductions have taken place through investment in carbon-free energy infrastructure, such as renewables or carbon capture and storage. The problem is that efficiency improvements, as well as reductions in non-carbon dioxide gases, can never take us to the level of emissions reductions needed by 2050 in Europe, and globally by 2100 (Hasselmann *et al.*, 2003). The changes that flexible policy instruments have so far promoted, in Europe and the rest of the world, are not the kind that will help us to achieve these longer-term goals (see Chapter 7, Eskeland *et al.*).

Flexible policy instruments – ones that change behaviour by placing a price tag on carbon emissions – need to be viewed in terms of both their short-term and long-term benefits. Flexibility is a good thing if it lessens the disruptions, to the economy as a whole, associated with the transformation of the energy system from carbon intensive to carbon free. It can do this in two ways. Firstly, it can stop people stranding productive capital, both physical and human. The first tradable permit system ever deployed was as part of the complete phase-out of leaded gasoline in the United States, and was designed precisely to soften a costly transition (Portney and Stavins, 2000). In a carbon trading system today, a power company that owns a coal plant can continue to operate it for another 10 or 15 years, buying the permits to do so, until it is fully depreciated. In turn, today's coal miners can continue working until they retire. Secondly, flexibility can promote innovation, by making it possible for firms that find radically new ways of transforming the energy system at low cost, ways never envisioned by regulators, to make a profit from their discoveries. But flexibility is a bad thing if it allows us to continue investing in the old energy system (Franco *et al.*, 2008).

It may be clear what the goals of flexibility need to be – to smooth and accelerate a transition, rather than to avoid it – but it is still far from clear what policy instruments can best achieve this. Economic theory suggests quite convincingly that policies that put a price on carbon, either in the form of a tax or a tradable emissions permit, will create incentives for all economic actors to explore least cost ways of reducing emissions. Most economic models, including those used in the ADAM project, assume the imposition of a carbon price to be the primary policy instrument, and they deliver the result that expanding the carbon market – from national to regional to global – increases its benefits and reduces the cost of major emissions reductions (Knopf *et al.*, 2008). At the same time, however, there remain doubts that a carbon price alone can stimulate long-term investment in renewable energy, or stop new investment in conventional energy technologies (Hanemann, 2007; Alfsen and Eskeland, 2007; Alfsen *et al.*, 2008). The empirical literature on the factors that drive the direction of new investment in the energy sector is extremely thin, but there is reason to believe that uncertainties associated with new technologies in developing and developed countries, established business practices associated with old technologies, and the fluctuating risk aversion of financial markets may play at least as large a role as the price mechanism (Gillissen *et al.*, 1995).

What may be needed is a portfolio of policy instruments that address the issue of long-term energy investment from several different angles. Given current carbon prices, renewable support schemes such as quotas and feed-in tariffs are clearly necessary to stimulate investment in particular technologies (Menz and Vachon, 2006; Mendonça, 2007; Rickerson and Grace, 2007). The EU has recently promulgated a Directive for the support of renewables, establishing a continent-wide quota that leaves room for a diversity of national-level policies. It is also developing a Directive to support the development of CCS, which would establish the legal framework for site development and national emissions credits. Both could be extremely important complements to the ETS, building the ladder to reach zero carbon dioxide emissions in the future. It now seems both likely and necessary to expand far-reaching R&D on carbon-hostile technology with multiple instruments, including public expenditures. As Europe and the United States consider this – building the technologies ladder – there is also increased scope for international co-operation, including being specific about technology in international treaties (Alfsen *et al.*, 2008).

14.5 It takes at least two to tango

To mitigate to the full extent necessary, Europe needs to transform its energy system completely within the next two to four decades. Within a region committed to mitigation, like Europe, there is still the challenge of agreeing on how to do it, sharing burdens and benefits across countries and stakeholders. Looking further, globally and over more decades, meeting the climate challenge requires co-operation on a grand scale.

The primary challenge is to start moving, and keep moving, quickly enough to complete the job in the time that we have. Successful scenarios to reach low carbon dioxide concentrations rely on a complete redirection of new energy system investment starting by 2013, only 4 years away (van Vuuren *et al.*, 2007). Yet few European countries have achieved the level of political commitment that is necessary for this to happen. At the time of writing, for example, the most recent government to form was in Austria, and one result of the negotiations forming the governing coalition was the abandonment of new plans to reduce carbon dioxide emissions. Austria is not alone. As the economic outlook has worsened during the third and fourth quarters of 2008, several countries have called on the European Commission to soften European emission reduction and renewable energy targets. What lies behind the political hurdles to rapid action? It is probably not the cost to citizens, in their role as energy consumers, of redirected investment, since estimates of these costs indicate they will be rather trivial (Barker, 2008; Knopf *et al.*, 2008). Instead, it may be that effective policies will create a mix of winners and losers, and some sectors that may lose are large employers and those who have high political influence.

Those industries and sectors most reliant on fossil fuels are likely to be those losers, but the impacts can vary substantially, depending on the policy measure adopted. In a case study of the electricity sector, ADAM researchers examined the effect of different policy instruments on the profitability of current power-plant owners (Eskeland *et al.*, 2008a). The demand curve for electricity slopes downward with higher electricity prices, just like in an economics textbook. The supply curve, however, is anything but textbook, because of regulatory history of the sector, differences in marginal costs between different suppliers, and the need to meet demand at all moments. One result of the supply curve's unique shape is that a modest tax on carbon (equivalent to auctioned permits) will not change at all the consumer price of electricity, the quantity consumed, or the mix of supply (e.g. coal, oil, gas, renewables). Rather, it will have the sole effect of shifting profits from fossil fuel power producers to the government and to the producers of less carbon-intensive power. Only when the tax is substantial enough to completely eliminate all profits of fossil fuel power producers – making their assets and their value to shareholders worthless – will there be an effect on consumer price and quantity, and a shift towards the supply of renewable power. By contrast, other policy instruments (such as renewable energy quotas) generate greater immediate shifts in supply to renewable energy sources, and the same disincentive to build new coal-fire power plants, while preserving the profits from existing infrastructure.

A carbon tax, or the complete auctioning of emissions permits within the ETS (Hepburn *et al.*, 2006), is easily justified on both moral and economic grounds. Morally, it is consistent with the polluter pays principle, which states that those who cause environmental damage should bear the costs of avoiding it, or compensating for it (OECD, 1975; European Commission, 1999). A recognised implication of the polluter pays principle is that 'public financing of environmental policy is in most cases to be avoided, as it should be financed by the polluters themselves as far as they can be identified' (European Commission, 1999: 5). Economically, in a legal framework where polluters have to bear the costs that their activity creates and the cost of halting that activity, individuals and firms will have an efficient and equitable (i.e. independent of their wealth) incentive not to engage in pollution in the first place (Portney and Stavins, 2000). At least, this is the case in theory. In specific practice, as the ADAM researchers argue, the outcome may be different:

[T]he full polluter pays principle ... combines relatively large income transfers with a small responsiveness, at least in the short to intermediate term, of emissions. In fact, the transitions that could be observed in the short term, such as erection of windmills, can be facilitated with a smaller cost to users, and a smaller 'insult' to fossil plant owner, by policy combinations that involve less of a windfall profit for non-fossil (or low carbon) asset owners.

(Eskeland *et al.*, 2008a, p. 54).

Whether one views the 'insult' to fossil plants owners as important in part depends on how one views the climate change problem. One view is that it is an externalities problem; certain industries did not have to bear the full costs of their carbon dioxide emissions, and so had an economic incentive to pollute, and the prioritisation of profits over the good of the planet is seen as immoral. An alternative view sees the energy sector as having evolved at a time when few imagined the dangers that carbon dioxide emissions would cause. Large parts of that energy sector were either publicly owned or regulated as public utilities, and their success has been the engine of economic growth and rising standards of living. The climate problem is about re-designing that sector to serve a modified public purpose, powering the planet while saving it. In this latter perspective it can be viewed as unfair, and indeed unneces-sary, to place the cost of transformation on the shoulders of a small set of firms and industries.

ADAM research has been relatively silent with respect to the moral arguments for or against applying the polluter pays principle in the climate case. On the economic side, the research has supported its application in the form of auctioned permits on efficiency grounds (Harstad and Eskeland, 2008). At the same time it has pointed out that free permits often serve the purpose of facilitating the transition. The electricity case study demonstrated how particular principles (e.g. the polluter pays) and theories (e.g. that auctioning of permits is more efficient than grand-fathering them) may appear to be sound in the abstract, but the complexity of the climate change problem can make their application counter-productive in practice (Eskeland *et al.*, 2008b). What is clear from ADAM research is the need to begin achieving results quickly (van Vuuren *et al.*, 2007), and for this it may be expedient to seek policy solutions that will diffuse political opposition, rather than polarise it, building alliances between today's carbon intensive industries and climate protec-tion advocates. The two can dance together.

Conclusion

This is a critical time for European climate policy, and the issues are complex. ADAM research has, in some ways, added to that complexity. The premise for this chapter is that complexity should not prevent us from attempting to distill fairly simple insights. There is no single lesson that can guide our thinking to cover all of the important questions of climate change and policy, and we have presented a set of different mental models for different aspects of the climate problem. Prioritising between mitigation and adaptation may be like turning off a tap and finding a mop. But the image of turning off a tap is unhelpful, and indeed impedes people's thinking for designing a strategy to reduce emissions throughout the century. For that, one needs to think like a chess player, moving ahead a number of pawns without knowing which of them will present to us the most important opportunities. Flexibility can be helpful, but not if it allows economic actors to put off the kinds of investment that will eventually be necessary. Rigour should be sought in the commitment to reducing emissions and rewarding those who bring solutions forward, while flexibility should be sought in cultivating all potential solutions. All the same, policy makers need to be careful about how they apply seemingly straightforward principles, like the polluter pays principle. Our emphasis on building a ladder while reducing emissions highlights the challenges of emission reductions throughout the century, and the fact that technology investments may be important, even if not paying off in a decade or two, in terms of emission reductions.

Taken together, the five highlighted insights do suggest a way of thinking about the climate change problem that is different from that dominating as little as 5 years ago. Five years ago, many still believed that the greatest obstacle to solving the climate change problem was the cost of doing so. We suggest that this is not the case: the cost of confronting climate change is not really a problem. Instead, the challenge lies in understanding how to engineer co-operative arrangements and large-scale social change in a complex world.

Human society is inexperienced at trying to steer itself, deliberately and quickly, in fundamentally new directions. The social sciences – economics, political science, and others – can guide us, but can also lead us down blind alleys. But the new way of thinking can also be a source of optimism, because it leaves open the possibility that society may have already achieved the necessary consensus to take action. It may be that it is misguided thinking that is preventing meaningful action about how to effectively deal with the particular complexity of social relationships that climate change presents (Verweij and Thompson, 2006); when we change this, we will start to see meaningful progress. Perhaps in the last five years we have learned the lessons that society can use to meet the climate change challenge. These lessons can find resonance with established insights in human societies, and the connections to these insights will be helpful in meeting the evolving challenges.

References

Aaheim, A., Berkhout, F., Kundzewicz, Z. *et al.* (2007) *Why We will Need Adaptation and How It can be Implemented*. Norwich UK and Brussels: ADAM Project and Centre for European Policy Studies.

Adger, W. N., Huq, S., Brown, K. and Hulme, M. (2003) Adaptation to climate change in the developing world. *Progress in Development Studies*, **3**(3), 179–95.

Alfsen, K. H. and Eskeland, G. S. (2007) *A Broader Palette: The Role of Technology in Climate Policy*. Report to the Expert Group for Environmental Studies 2007:1, Sweden: Ministry of Finance.

Alfsen, K. H. Dovland, H. and Eskeland, G. S. (2008). *Elements for an Agreement on Climate and Energy Technology Development (ACT)*. *Policy Note* 2008:1. Oslo: Cicero.

Allen, M. R. and Frame, D. J. (2007) Atmosphere: call off the quest. *Science*, **318**(5850), 582–3.

Barker, T. (2008) The economics of avoiding dangerous climate change. An editorial essay on The Stern Review. *Climatic Change*, **89**(3), 173.

Cox, P. and Stephenson, D. (2007) A changing climate for prediction. *Science*, **317**, 207–8.

Czisch, G. (2005). *Szenarien zur zukünftigen Stromversorgung: kostenoptimierte Variationen zur Versorgung Europas und seiner Nachbarn mit Strom aus erneuerbaren Energien*. Kassel, Germany: Universität Kassel.

de Bruin, K., Dellink, R. and Tol, R. S. J. (2007) *AD-DICE: An Implementation of Adaptation in the DICE Model* (Nota di lavoro 51.2007). Milan: Fondazione Eni Enrico Mattei.

Dixit, A. and Pindyck, R. (1994) *Investment under Uncertainty*. Princeton NJ: Princeton University Press.

Edenhofer, O., Bauer, N. and Kriegler, E. (2005) The impact of technological change on climate protection and welfare: insights from the model MIND. *Ecological Economics*, **54**(2–3), 277–92.

Eskeland, G., CICERO, FhISI, DIW, UEA, PIK and IVM. (2008) *Policy Appraisal for the Electricity Sector* (D-P3c.2). Oslo: Cicero.

Eskeland, G. S. and Linnerud, K. (2008). *An Appraisal of the EU Climate Policy Mix Affecting the Electricity Sector*. Mimeographed, http://www.Cicero.uio.no

Eskeland, G. S. Mideksa, T. and Rive, N. (2008a) *European Emissions, Carbon Leakage, a Case for Border Tax Adjustments?* Mimeographed, http://www.Cicero.uio.no.

Eskeland, G. S. Mideksa, T. and Rive, N. (2008b) *European Climate Goals for 2020 and the Role of the Electricity Sector*. Mimeographed, http://www.Cicero.uio.no.

European Commission (1999) *Application of the Polluter Pays Principle: differentiating the rates of Community assistance for structural funds, cohesion funds and ISPA*. Brussels.

Franco, G., Cayan, D., Luers, A., Hanemann, M. and Croes, B. (2008) Linking climate change science with policy in California. *Climatic Change*, **87**(0), 7.

Genovese, E., Lugeri, N., Lavalle, C., Barredo, J., Bindi, M. and Moriondo, M. (2007) *An Assessment of Weather-related Risks in Europe* (DA2.1). Norwich UK: ADAM (Adaptation and Mitigation Strategies) Project Consortium.

Gigerenzer, G. and Selten, R. (eds.). (2001) *Bounded Rationality: The Adaptive Toolbox*. Cambridge, MA: MIT Press.

Gillissen, M., Opshoor, J. C. M., Farma, K. and Blok, K. (1995) Energy conservation and investment behavior of firms. Amsterdam: Vrije Universiteit.

Grothmann, T. and Patt, A. G. (2005) Adaptive capacity and human cognition: the process of individual adaptation to climate change. *Global Environmental Change*, **15**(3), 199–213.

Hanemann, M. (2007) Innovative climate change mitigation policy: the case of California. *The Future Climatic Window: Local Impacts of Climate Change*. Austria: Seggau Castle, Leibnitz.

Harstad, B. and Eskeland, G. S. (2008) *Trading for the Future: Signalling in Permit Markets*. Working paper. http://www.kellogg.northwestern.edu/faculty/harstad/htm/trading.pdf

Hasselmann, K., Latif, M., Hooss, G. *et al.* (2003) The challenge of long-term climate change. *Science*, **302**(5652), 1923–5.

Hepburn, C., Grubb, M., Neuhoff, K., Matthes, F. and Tse, M. (2006) Auctioning of EU ETS phase II allowances: how and why? *Climate Policy*, **6**, 137–60.

Hope, C. (2006) The marginal impact of CO_2 from PAGE2002: an integrated assessment model incorporating the IPCC's five reasons for concern. *The Integrated Assessment Journal*, **6**(1), 19–56.

Kahn, M. (2003) Two measures of progress in adapting to climate change. *Global Environmental Change*, **13**, 307–12.

Klein, R. J. T., Eriksen, S., Naess, L. O. *et al.* (2007a) Portfolio screening to support the mainstreaming of adaptation to climate change into development assistance. *Climatic Change*, **84**(1), 23–44.

Klein, R. J. T., Huq, S., Downing, T., Richels, R., Robinson, J. and Toth, F. (2007b) Inter-relationships between adaptation and mitigation. In *Climate change 2007: Impacts, Adaptation and Vulnerability. Contribution of Working Group II to the Fourth Assessment Report of the Intergovernmental Panel on Climate Change*. ed. Parry, M. L., Canziani, O. F. Palutikof, J. van der Linden, P. and Hanson, C. Cambridge, UK: Cambridge University Press, pp. 745–77.

Klein, R. J. T., Smit, M., Goosen, H. and Hulsbergen, C. H. (1998) Resilience or vulnerability: coastal dynamics or Dutch dikes? *The Geographical Journal*, **164**(3), 259–268.

Knopf, B., Edenhofer, O., Turton, H. *et al.* (2008) *Report on the First Assessment of Low Stabilisation Scenarios* (D-M2.6). Potsdam, Germany: Potsdam Institute for Climate Impact Research.

Loewenstein, G. and Elster, J. (1992) *Choice Over Time*. New York: Russell Sage Foundation.

Manne, A. S. and Richels, R. (1992) *Buying Greenhouse Insurance: The Economic Costs of GHG Limitations*. Cambridge MA: MIT Press.

Manne, A. S. and Richels, R. (1997) On stabilising CO_2 concentrations: cost effective emissions reduction strategies. *Environmental Modeling and Assessment*, **2**, 251–265.

Manne, A. S. and Richels, R. (2004) The impact of learning-by-doing on the timing and costs of CO2 abatement. *Energy Economics*, **26**, 603–19.

McEvoy, D., Lonsdale, K., Takama, T. *et al.* (2007) *A Draft Synthesis Report of the Actor-based Analysis of Adaptation to Climate Change*. Norwich UK: ADAM Project Consortium: Adaptation and mitigation strategies: supporting European Climate Policy.

Mendelsohn, R. (2006) The role of markets and governments in helping society adapt to a changing climate. *Climatic Change*, **78**, 203–15.

Mendelsohn, R., Morrison, W., Schlesinger, M. E. and Andronova, N. G. (2000) Country-specific market impacts of climate change. *Climatic Change*, **45**(3–4), 553–69.

Mendonça, M. (2007) *Feed-in Tariffs: Accelerating the Deployment of Renewable Energy*. London: Earthscan.

Menz, F. C. and Vachon, S. (2006). The effectiveness of different policy regimes for promoting wind power: Experiences from the states. *Energy Policy*, **34**(14), 1786.

Metz, B., Davidson, O., Bosch, P., Dave, R. and Meyer, L. (eds.). (2007) *Climate Change 2007: Mitigation. Contribution of Working Group III to the Fourth Assessment Report of the Intergovernmental Panel on Climate Change*. Cambridge, UK: Cambridge University Press.

Michaelowa, A. and Jotzo, F. (2005) Transaction costs, institutional rigidities and the size of the clean development mechanism. *Energy Policy*, **33**(4), 511.

Morgan, M. G., Fischhoff, B., Bostrom, A. and Atman, C. J. (2001) *Risk Communication: a Mental Models Approach*. Cambridge, UK: Cambridge University Press.

Nordhaus, W. and Boyer, J. (2000) *Warming the World: Economic Modeling of Global Warming*. Cambridge, MA: MIT Press.

OECD (1975) *The Polluters Pays Principle: Definition, Analysis, Implementation*. Paris.

Oppenheimer, M., O'Neill, B. and Webster, M. (2008) Negative learning. *Climatic Change*, **89**(1), 155.

O'Neill, B. (2008) Learning and climate change: an introduction and overview. *Climatic Change*, **89**(1), 1.

O'Neill, B. and Melnikov, N. (2008) Learning about parameter and structural uncertainty in carbon cycle models. *Climatic Change*, **89**(1), 23.

Pacala, S. and Socolow, R. (2004) Stabilisation wedges: solving the climate problem for the next 50 years with current technologies. *Science*, **305**(5686), 968.

Patt, A. G., van Vuuren, D. P., Berkhout, F., *et al.* (2009) Adaptation in integrated assessment modeling: where do we stand? *Climatic Change, (in press).*

Pelling, M. and High, C. (2005) Understanding adaptation: what can social capital offer assessments of adaptive capacity. *Global Environmental Change*, **15**, 308–19.

Portney, P. and Stavins, R. (eds.). (2000) *Public Policies for Environmental Protection* (2nd edn.). Washington: Resources for the Future.

Rickerson, W. and Grace, R. (2007) *The Debate over Fixed Price Incentives for Renewable Electricity in Europe and the United States: Fallout and Future Directions.* Washington DC: Heinrich Böll Foundation.

Sanden, B. and Azar, C. (2005) Near term technology policies for long-term climate targets–economy wide versus technology specific approaches. *Energy Policy*, **33**(12):1557–76.

Schelling, T. (1978) *Micromotives and Macrobehavior.* New York: Norton.

Stern, N. (2006) *The Economics of Climate Change.* Cambridge, UK: Cambridge University Press.

UNFCCC (2007) *Analysis of Existing and Planned Investment and Financial Flows Relevant to the Development of Effective and Appropriate International Response to Climate Change.* Bonn: United Nations Framework Convention on Climate Change Secretariat.

van Vuuren, D. P., Criqui, P., Barker, T., Isaac, M., Kitous, A. and Scrieciu, S. (2007) *Preliminary ADAM Scenarios* (D-S.1). Bilthoven: Netherlands Environmental Assessment Agency.

Verweij, M. and Thompson, M. (eds.). (2006) *Clumsy Solutions for a Complex World: Governance, Politics, and Plural Perceptions.* New York: Palgrave Macmillan.

von Neumann, J. and Morgenstern, O. (1944) *Theory of Games and Economic Behavior.* Princeton: Princeton University Press.

Washington, R., Harrison, M., Conway, D. *et al.* (2006) African climate change: taking the shorter route. *Bulletin of the American Meteorological Society*, **87**(10), 1355–66.

Webster, M., Jakobovits, L. and Norton, J. (2008) Learning about climate change and implications for near-term policy. *Climatic Change*, **89**(1), 67.

Welp, M., Battaglini, A. and Jaeger, C. C. (2008) Defining dangerous climate change: the Beijing exercise. In Patt, A. G., Schröter, D. Klein, R. J. T. and de la Vega-Leinert, A. C. (eds.). *Assessing Vulnerability to Global Environmental Change: Making Research Useful for Adaptation Decision Making and Policy.* London: Earthscan, pp. 215–30.

Wreford, A., Hulme, M. and Adger, W. N. (2007) *Strategic Assessment of the Impacts, Damage Costs, and Adaptation Costs of Climate Change in Europe* (D-A2.7). Norwich, UK: Tyndall Centre for Climate Change.

Yohe, G., Lasco, R., Ahmad, Q. *et al.* (2007) Perspectives on climate change and sustainability. In Parry, M. L., Canziani, O. F. Palutikof, J., van der Linden, P. and Hanson, C. (eds.). *Climate Change 2007: Impacts, Adaptation and Vulnerability. Contribution of Working Group II to the Fourth Assessment Report of the Intergovernmental Panel on Climate Change.* Cambridge UK: Cambridge University Press, pp. 811–41.

Appendix: Description of models

This Appendix provides short descriptions of 22 models which have been used in the ADAM project, or developed by the ADAM project, and results from which are reported in various chapters in this book. These models address various aspects of global change: for example, energy-economics, technology efficiency, adaptation, malaria risk, coastal management. For each entry, a short model description is provided, together with some of the main limitations. Published sources offering more detailed descriptions of the models, and their applications, are listed where relevant.

AD-DICE (Chapters 3 and 4)

Adaptation in DICE is an integrated assessment model based on the DICE model (Dynamic Integrated model of Climate and the Economy) (Nordhaus, 2000; de Bruin et al., 2009). The DICE model is a global model and includes economic growth functions as well as geophysical functions. The use of adaptation is assumed to be optimal and is already included in the damage function. AD-DICE, on the other hand, includes adaptation as a decision variable. Estimates from empirical literature on the costs and benefits of adaptation are used to calibrate the model and derive the adaptation cost curve that is implicit in the DICE model. So AD-DICE unravels the damage function given in DICE into residual damages and adaptation costs. Adaptation reduces the residual damages, making decisions at the levels of adaptation and mitigation separable. Adaptation costs and residual damages both depend on the level of adaptation, but the costs are independent of each other.

Limitations

AD-DICE assumes that the level of adaptation is chosen every time period (10 years). This means that the same problem is faced each decade, and the same trade-off between adaptation and mitigation holds. It also means that both the costs and benefits of adaptation are 'instantaneous', i.e. they fall within the same time period. The important implication of this assumption is that as long as adaptation is applied optimally, the benefits of adaptation will always outweigh the costs and hence the adaptation decision will never draw away funds from mitigation policy. Although this way of modelling adaptation benefits and costs has its limitations, many adaptation measures have this characteristic. However, some adaptation measures, especially in the category of anticipatory adaptation, have a time-lag in costs and benefits. An analysis which adds an adaptation capital stock to represent anticipatory adaptation would be interesting and is deferred to future work.

De Bruin, K. C., Dellink, R. B. and Tol, R. S. J. (2009) AD-DICE: an implementation of adaptation in the DICE model. *Climatic Change*, **95**, 63–81.
Nordhaus, W. D. and Boyer, J. (2000) *Warming the World: Economic Models of Global Warming*. Cambridge, MA: MIT Press.

ASTRA (Chapter 7)

ASTRA is a strategic integrated assessment model that includes a core macro-economic model, a trade model, a population model, a transport model, vehicle fleet models and transport energy and emissions models. The model builds on recursive simulations following the system dynamics concept and enables scenarios to be run until 2050. The economic models apply different theoretical concepts; for example, endogenous growth by linking total factor productivity to investments; neo-classical production functions; and Keynesian demand driven investment functions. ASTRA incorporates its own bottom-up models for the European transport system and has been extended to link with other bottom-up models for the household sector, industry and services sector, energy and renewables sector. The ASTRA transport model is implemented as classical four-stage model, but with endogenous reactions on all stages and a limited assignment stage. Changes in the economic system immediately feed into changes of the transport behaviour and alter origins, destinations and volumes of European transport flows. The ASTRA model has been applied in various transport policy studies (e.g. pricing, infrastructure, integrated programmes), employment studies of new technologies and renewables, climate policy analysis in general and in the transport sector.

Krail, M., Schade, W., Fiorello, D. *et al.* (2007) Outlook for Global Transport and Energy Demand. Deliverable 3 of TRIAS (Sustainability Impact Assessment of Strategies Integrating Transport, Technology and Energy Scenarios). *Funded by European Commission 6th RTD Programme*. Karlsruhe, Germany.

Schade, W. (2005) *Strategic Sustainability Analysis: Concept and Application for the Assessment of European Transport Policy.* Baden-Baden: NOMOS-Verlag. ISBN 3–8329–1248–7.

Schade, W., Fiorello, D., Beckmann, R. *et al.* (2008) High oil prices: quantification of direct and indirect impacts for the EU. Deliverable 3 of HOP! (Macro-economic impact of high oil price in Europe). *Funded by European Commission 6th RTD Programme.* Karlsruhe, Germany.

CATSIM (Chapter 8)

The catastrophe simulation model (CATSIM) is a risk-based economic framework for studying the probabilistic economic impacts of natural disasters, as well as the benefits of risk management strategies to be taken by government entities. By systematically assessing disaster risks, economic vulnerability and adaptive capacity of governments to extreme disaster events, it can be used to illustrate the trade-offs and choices governments must make in managing the economic risks due to natural disasters and assists policy makers in developing equitable and efficient risk management strategies. CATSIM introduces stochastic disasters shocks (in terms of capital stock losses) into a simple economic growth framework in order to study the follow-on indirect economic consequences, such as adverse fiscal or macroeconomic effects. Equipped with a graphical user interface, it has been used as a capacity building tool in various workshops with policy makers in Europe, Latin America, Africa and Asia for helping to examine pre-disaster risk financing strategies (see Hochrainer, 2006; Mechler *et al.*, 2006).

Hochrainer, S. (2006) *Macroeconomic Risk Management Against Natural Disasters.* Wiesbaden: German University Press (DUV).

Mechler, R., Linnerooth-Bayer, J., Hochrainer, S. and Pflug, G. (2006) Assessing financial vulnerability and coping capacity: the IIASA CATSIM Model In Birkmann, J. (ed.) *Measuring Vulnerability and Coping Capacity to Hazards of Natural Origin. Concepts and Methods.* Tokyo: United Nations University Press, pp. 380–398.

Cropsyst (Chapter 8)

CropSyst (Stöckle *et al.*, 2003) is a multi-year, multi-crop, daily time-step crop growth simulation model. It simulates soil–water budget, soil–plant nitrogen budget, crop canopy and root growth, phenology, dry matter accumulation and partitioning, yield, residual production and decomposition and erosion. The user can input management parameters such as sowing date, cultivar genetic coefficients (photoperiodic sensitivity, duration of grain filling, maximum LAI, etc.), soil profile properties (soil texture, thickness), fertilizer and irrigation management, tillage, atmospheric carbon dioxide concentration, etc. The core of the model is the determination of the biomass potential growth under optimal conditions (without water–nitrogen stress) based both on crop potential transpiration and crop intercepted photosynthetic active radiation. The potential growth is then corrected by water and nitrogen limitations, if any, and the actual daily biomass gain is thus determined. The simulation of crop development is mainly temperature dependent and it is based on the thermal time required to reach specific development stages.

Stöckle, C., Donatelli, M. and Nelson, R. (2003) CropSyst, a cropping systems simulation model. *European Journal of Agronomy*, **18**, 289–307.

DIVA (Chapters 2 and 5)

Dynamic and interactive vulnerability assessment (DIVA) is an integrated model of coastal systems that was developed, together with its coastal database, within the EU-funded project DINAS-COAST[1] (DINAS-COAST Consortium, 2006; Hinkel and Klein, 2007). DIVA produces quantitative information on a range of ecological, social and economic coastal vulnerability indicators from sub-national to global scales, covering all coastal nations. The model consists of a number of modules developed by experts from various engineering, natural and social science disciplines. Based on climatic and socio-economic scenarios, the model assesses coastal erosion (both direct and indirect), coastal flooding (including rivers), wetland change and salinity intrusion into deltas and estuaries. DIVA also considers coastal adaptation in terms of raising dikes and nourishing beaches and includes several predefined adaption strategies such as no protection, full protection or optimal protection.

Limitations

DIVA excludes the following processes that are likely to affect coastal impacts, but currently can not be modelled with confidence: changes in storm frequency and intensity, local distribution of gross domestic product and population growth due to rapid coastal development and urbanisation, and salinity intrusion into coastal aquifers. Further important uncertainties arise due to the coarse resolution and the inaccuracy of elevation data

DINAS-COAST Consortium (2006). *DIVA 1.5.5*. Potsdam, Germany: Potsdam Institute for Climate Impact Research.
Hinkel, J., and Klein, R. J. T. (2007) Integrating knowledge for assessing coastal vulnerability. In Fadden, L. M., Nicholls, R. J. and Penning-Rowsell, E. (eds.) *Managing Coastal Vulnerability*. London: Earthscan.

---

[1] Dynamic and interactive assessment of national, regional and global vulnerability of coastal zones to sea-level rise; http://www.pik-potsdam.de/dinas-coast/

E3MG (Chapter 7)

Energy–environment–economy modelling at the global level (E3MG) is a macro-econometric non-equilibrium hybrid simulation model of the global E3 system, estimated on annual data 1971–2002 and projecting annually to 2020 and every 10 years to 2100 (Barker *et al.*, 2006, Barker *et al.*, 2008). It is designed to address the issues of energy security and climate stabilisation, both in the medium and long terms, with particular emphasis on dynamics, uncertainty and the design and use of economic instruments. E3MG represents a novel approach to the modelling of technological change in the literature on the costs of climate stabilisation. It is based upon a 'new economics' view of the long-run drawing on post-Keynesianism, adopting a 'history' approach of cumulative causation and demand-led growth, and incorporating technological progress through gross investment enhanced by R&D expenditures. In addition to formal co-integration econometric techniques that identify long-run macro-trends of the global energy–economy system, a bottom-up energy-technology simulation module has been incorporated allowing for the explicit modelling of 28 energy technologies. Hence, a treatment of sub-stitution between fossil and non-fossil fuel technologies is included, accounting for non-linearities resulting from investment in new technology, learning-by-doing, and innovation. The modelling of a two-way feedback between the economy, energy demand/supply and environmental emissions is an undoubted advantage over other models, which may either ignore the interaction completely or only assume a one-way causation.

Limitations

One of the model's limitations may be that the parameters based on 33-year historical data may not be appropriate for solutions covering a highly uncertain distant future of 100 years. However, the E3MG modelling approach assumes that understanding the future is best done by first understanding the past; hence the econometric basis of the model. A more detailed specification of future technologies may be nevertheless required to improve long-term forecasts.

Barker, T., Haoran, P., Köhler, J., Warren, R. and Winne, S. (2006) Decarbonising the global economy with induced technological change: scenarios to 2100 using E3MG. *Energy Journal*, **27**, 143–60.
Barker, T., Scrieciu, S. S. and Foxon, T. (2008) Achieving the G8 50% target: modelling induced and accelerated technological change using the macro-econometric model E3MG. *Climate Policy, special issue*, **8**, S30–S45.

E3ME (Chapters 11 and 14)

E3ME is an econometric, annual, dynamic, simulation, structural, post-Keynesian energy–environment–economy (E3) model of Europe based on a social accounting matrix and ESA95.

The model is intended to meet an expressed need of researchers and policy makers for a framework for analysing the long-term implications of E3 policies, especially those concerning R&D and environmental taxation and regulation. The model is also capable of addressing the short-term and medium-term economic policies as well as, more broadly, the long-term effects of such policies. The current version of E3ME is capable of forecasting annual macroeconomic effects, energy use and emissions in the period up to 2030.

Exogenous inputs include the world oil price, regional gas and coal prices, energy supplies, population, participation rates, exchange and interest rates and other fiscal and monetary policies. The model includes the economic instruments of carbon dioxide emission allowances (auctioned or grandfathered), energy and carbon taxes, employment taxes, and other direct and indirect taxes.

The main outputs of the model are annual to 2030 for 27 European countries. They include energy demands by fuel users and fuels, greenhouse gas emissions by fuel users and fuels (CO_2, NO_2 and methane) and sources (all six GHGs), other polluting gases by fuel users and fuels (SO_2, NOx, PM10), gross output, GDP, the structure of GDP in terms of industrial outputs and expenditure components by 41 products/industries in constant (year 2000) and current prices, employment in the same detail as gross output, external and internal trade in the same detail as gross output, macroeconomic aggregates, prices, wage rates, tax revenues, and income flows consistent with the gross output detail.

Cambridge Econometrics (2006) *E3ME: An energy–environment–economy model for Europe*. E3ME Version 4.1: A Non-Technical Description, Cambridge Econometrics, Cambridge, July 2006.

EMELIE (Chapter 7)

The model EMELIE assesses the European electricity market with 25 regions and 58 electricity producers, in a computational partial equilibrium framework with strategic investment and supply of dominant companies. Producers are represented by 12 existing production technologies and have the possibility to invest in six fossil fuel and seven renewable energy technologies. Regions are imperfectly linked by international transmission capacities. Furthermore, the model includes an endogenous emissions market simulation with market clearing of the European

Emissions Trading Scheme (ETS). EMELIE is applied to the analysis of effects of policy combinations–like emissions trading plus renewable energy support (Traber and Kemfert, 2009); to the projection of technology diffusion under different emissions cap regimes (Kemfert and Traber, 2007); and to assess the effects of market power on these relationships. Model results are electricity and emissions prices, investments in fossil fired and renewable energy technologies, international trade in electricity and profits of the main players on the market until the year 2050.

Kemfert, C. and Traber, T. (2007) *Impacts of the German Support for Renewable Energy on Electricity Prices, Emissions and Profits: An Analysis Based on a European Electricity Market Model*. Berlin: DIW, (Diskussionspapiere / Deutsches Institut für Wirtschaftsforschung 712.)
Traber, T. and Kemfert, C. (2009) Impacts of the German support for renewable energy on electricity prices, emissions, and firms, *The Energy Journal*, volume 30, No. 3, 155–78.

EuroMM (Chapter 7)

The European Multi-regional MARKAL (EuroMM) energy-conversion model is a bottom-up, perfect-foresight optimization model. EuroMM is part of the MARKAL (MARKet ALlocation) family of models that is typically used to determine the least-cost energy system configuration over a given time horizon under a set of assumptions about technologies, resource potentials and demands (Fishbone *et al.*, 1983; Loulou *et al.*, 2004). EuroMM provides a detailed representation of technologies in the electricity and heat production and fuel conversion sectors in Europe, including carbon capture and storage (CCS) and thermal power plant cooling system technologies, along with trade networks for energy carriers. The model represents 18 distinct regions covering the 27 EU Member States plus Norway and Switzerland, and is calibrated to 2005 statistics with a time horizon up to 2050 (see also Jochem *et al.*, 2007).

The EuroMM energy-conversion model is used to compute impacts of climate change on the energy conversion sector as well as to analyse policy instruments, such as carbon taxes or emissions targets and their related effects.

Limitations

EuroMM relies on uncertain assumptions regarding future potential technology deployment rates and exogenous technological learning. The model uses the perfect foresight optimisation approach assuming perfect information and limited transaction costs. Furthermore, EuroMM only partly identifies the impacts from climate change and the necessary adaptation needs since it lacks a representation of detailed spatial impacts and extreme events.

Fishbone, L. G., Giesen, G., Goldstein, G. A. *et al.* (1983) *User's guide for MARKAL A Multi-period, linear programming model for energy systems analysis (BNL/KFA Version 2.0)*. BNL 51701, Brookhaven, USA: Brookhaven National Laboratory and Kernforschungsanlage Jülich.

Jochem E., Barker T., Scrieciu S. *et al.* (2007) *EU-Project ADAM: Adaptation and Mitigation Strategies: Supporting European Climate Policy–Deliverable M1.1: Report of the Base Case Scenario for Europe and full description of the model system*. Fraunhofer ISI: Karlsruhe.

Loulou, R., Goldstein, G. and Noble, K. (2004). *Documentation for the MARKAL Family of Models*. Energy Systems Technology Analysis Programme (ETSAP). International Energy Agency (IEA). <http://www.etsap.org/MrklDoc-I_StdMARKAL.pdf>

FAIR (Chapters 3, 10 and 11)

The climate policy model FAIR (den Elzen *et al.*, 2008) is used in conjunction with the IMAGE model to determine the reduction rates across different emission sources. Global climate calculations make use of the simple climate model, MAGICC 4.1 (Wigley and Raper, 2001; Wigley, 2003). Required global emission reductions are derived by taking the difference between the baseline and a global emissions pathway. The FAIR cost model distributes these between the regions following a least-cost approach using regional marginal abatement costs curves for the different emissions sources. Recently, the FAIR model has been extended with damage and adaptation costs curves (based on the AD-DICE model (De Bruin *et al.*, 2007)) and the ability to estimate macro-economic impacts on growth of Gross Domestic Product (Hof *et al.*, 2008). This allows the model to explore the economic impacts of combined mitigation and adaptation strategies.

Limitations

In its aim to be flexible, the FAIR model does not include a sectoral macro-economic model or an energy model. The model thus works from a partial equilibrium approach – and more underlying consequences of climate policy can only be studied by forwarding the FAIR results to other (linked) models.

De Bruin, K. C., Dellink, R. and Tol, R. S. J. (2007) *AD-DICE: An implementation of adaptation in the DICE Mode*. FEEM Working Paper No. 51.2007.

den Elzen, M. G. J., Lucas, P. L., and Van Vuuren, D. P. (2008) Regional abatement action and costs under allocation schemes for emission allowances for achieving low carbon dioxide equivalent concentrations. *Climatic Change*, **90**(3), 243–68.

Hof, A. F., den Elzen, M. G. J. and van Vuuren, D. P. (2008) Analysing the costs and benefits of climate policy: value judgements and scientific uncertainties. *Global Environmental Change*, **18**(3), 412–24.

Wigley, T. M. L. (2003) *MAGICC/SCENGEN 4.1: Technical Manual*. Boulder, CO: UCAR–Climate and Global Dynamics Division.

Wigley, T. M. L. and Raper, S. C. B. (2001) Interpretation of high projections for global-mean warming. *Science*, **293**, 451–454.

GRACE (Chapters 4, 7 and 8)

The model for global responses to anthropogenic change in the environment (GRACE) is a multi-sector, multi-region, recursively dynamic global computable general equilibrium model (CGE), which is mainly based on the Solow–Swan neoclassical growth model (Solow, 1956). The model has been previously applied to integrated air quality and climate policy analysis (Rypdal *et al.*, 2007), and analysis of climate change impacts on the forestry sector (Rive *et al.*, 2005). GRACE is calibrated around the global trade analysis project (GTAP) v6 database which represents the global economy in 2001 using the input output table of 87 regions for 57 sectors. As with previous database versions, this database contains bilateral trade and transport along protection data representing the linkages among the 87 regions of the global economy. The structure of production and consumption of GRACE is based on a number of other models. The quantity and price flows within the economy are based around the GTAP6 database. The structure of production and consumption (i.e. the demand trees) is based on the MIT EPPA model. The disaggregated electricity sector is based on work by Wing (2008). The dynamics and treatment of investments were adapted from the GTAP-Dyn model (Ianchovinchina *et al.*, 2007).

As a top-down macroeconomic model, GRACE is useful for analysing the economy-wide costs and impacts of environmental policy. This includes the cost of emission reduction, consequential price effects and feedbacks, trade effects, and welfare costs. However, the trade-off is that economic activities are treated in an aggregated fashion. As such, GRACE lacks the technological, sectoral, and spatial detail that can be offered by bottom-up technology-oriented models.

GRACE-EL, developed for the ADAM project, (i) establishes a vintage structure in long lived assets; (ii) provides technology detail in particular for electricity generation; (iii) provides other developments to improve analysis and interpretations for energy-intensive production and consumption.

Ianchovinchina, E. and McDougall, R. (2007) *Theoretical Structure of Dynamic GTAP*. GTAP Technical Paper 17. Purdue University: Center for Global Trade Analysis.

Rive, N., Aaheim, H. A. and Hauge, K. (2005) *Adaptation and world market effects of climate change on forestry and forestry products*. Presented at annual GTAP Conference, Lübeck.

Rypdal, K., Rive, N., Åström, S. *et al.* (2007) Nordic air quality co-benefits from European post-2012 climate polices. *Energy Policy*, **35**(12), 6309–22.

Solow, R. M. (1956) A contribution to the theory of economic growth. *Quarterly Journal of Economics*, **70**(1), 65–94.

Sue Wing, I. (2008) The synthesis of bottom-up and top-down approaches to climate policy modelling: electric power technology detail in a social accounting framework. *Energy Economics*, **28**, 539–62.

IMAGE 2.4 (Chapters 3 and 11)

The IMAGE 2.4 Integrated Assessment model (Bouwman *et al.*, 2006) consists of a set of linked and integrated models that together describe important elements of the long-term dynamics of global environmental change, such as air pollution, climate change, and land-use change. The global energy model used in IMAGE is TIMER (see separate entry), which describes the long-term dynamics of demand and production of primary and secondary energy and the related emissions of greenhouse gases and regional air pollutants. The agricultural model of IMAGE models the productivity of seven crop groups and five animal categories (Leemans and Born, 1994). The regional production of agricultural goods is distributed spatially (at 0.5×0.5 degree) on the basis of a set of allocation rules (Alcamo *et al.*, 1998). IMAGE computes land use changes based on regional production of food, animal feed, fodder, grass and timber, with consideration of local climatic and terrain properties, and changes in natural vegetation due to climate change. Consequently, emissions from land use changes, natural ecosystems and agricultural production systems, and the exchange of carbon dioxide between terrestrial ecosystems and the atmosphere are calculated. Both the land use change maps and the agricultural activity data are used to model emissions from land use (change).

The emissions of greenhouse gases (from the energy system and from land use) are then used by the MAGICC (simple) climate model to calculate global-mean temperature change (Wigley and Raper, 2001). Patterns of temperature change are obtained by making a link to climate change patterns generated by a general circulation models (GCM). The IMAGE model is particularly strong in the detailed description of energy technologies and the geographically explicit land use. The integration of land and energy use in one model is also noteworthy.

Limitations

IMAGE is provides a physically orientated description of human activities (use of tonnes of oil; production of tonnes of cereals, etc.). A fuller macro-economic description only emerges from co-operation with other models. The broad coverage of IMAGE as an integrated assessment model implies that many critical uncertainties influence the model outcomes. In this context, use of a single base-line (as in the ADAM project) does not do full justice to the fundamental uncertainties involved.

Alcamo, J., Kreileman, E., Krol, M. *et al.* (1998) Global modelling of environmental change: an overview of IMAGE 2.1. In Alcamo, J., Leemans, R. and Kreileman, E. (eds.) *Global Change Scenarios of the 21st Century.* Results from the IMAGE 2.1 model. Oxford, UK: Elsevier Science Ltd.

Bouwman, A. F., Kram, T. and Klein Goldewijk, K. (Eds.). Integrated modelling of global environmental change. An overview of IMAGE 2.4. *Netherlands Environmental Assessment Agency (MNP)*, the Netherlands: Bilthoven.

Leemans, R. and Born, G. J. v. d. (1994) Determining the potential global distribution of natural vegetation, crops and agricultural productivity. *Water, Air and Soil Pollution*, **76**, 133–61.

Wigley, T. M. L. and Raper, S. C. B. (2001) Interpretation of high projections for global-mean warming. *Science*, **293**, 451–4.

ISIndustry (Chapter 7)

The model ISIndustry belongs to the class of energy system or bottom-up models, which compute, based on technological information about distinct conservation options and industrial processes. A distinction is made between process-specific technologies and cross-cutting technologies. Blast furnaces in steel making are one example of the former, being sector and even process specific. For process-specific technologies, the main driver is the projection of physical production (e.g. tonnes of crude steel from blast furnaces). The 40 most energy-intensive and greenhouse gas-intensive processes were considered separately in the model. For each of these processes, the specific energy consumption/GHG emissions and the physical production output per country are model parameters.

In contrast, cross-cutting technologies are widespread over very different industrial sectors. Examples are electric motors or lighting equipment, which are applied throughout all industrial sectors. Although cross-cutting technologies are usually smaller, there are huge numbers involved due to their widespread application and so they are responsible for a huge share of industrial electricity consumption. Electric motor systems and lighting account for more than 70 per cent of industrial electricity consumption. These are implemented in the model as a share of the total sector's electricity consumption and their main driver is the projected development of value added per industrial sector. The technological detail of the model allows simulation of the long-term industrial energy demand based on distinct technological energy efficiency options whilst considering main economic trends.

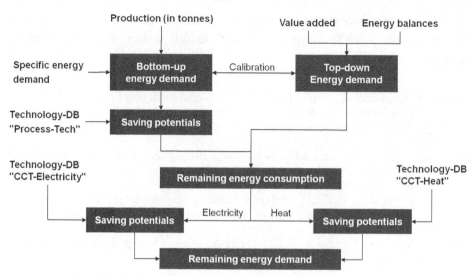

Figure A-1. Simplified structure of the Model ISIndustry.

MARA / ARMA (Chapter 3)

Malaria vectors, the mosquitoes spreading the malarial infection, can only survive in suitable climates with high average temperatures, no frost and sufficient precipitation. The MARA/ARMA malaria suitability model (Craig *et al.*, 1999) incorporates these climatic factors to determine climatic suitable areas. The climatic levels required for the maximum suitability of one, and for the minimum suitability of zero, are shown in Table 1. For indicators with levels between those required for zero or one suitability, a level is calculated using a simple function (Craig *et al.*, 1999). All these factors are calculated at half by half degree grid level, making use of the output from the IMAGE-model (Bouwman *et al.*, 2006). Total climatic malaria suitability for each grid cell is determined by the lowest of these three indices.

Limitations

The MARA/ARMA model describes suitability for malaria vectors. It does not provide a process description of the spread of mosquitoes, nor does it explicitly describe how people may react to increased risk levels.

Bouwman, L., Kram, T. and Klein-Goldewijk, K. (2006) *Integrated Modelling of Global Environmental Change. An Overview of IMAGE 2.4*. Bilthoven: Netherlands Environmental Assessment Agency

Craig, M. H., Snow, R. W. and le Sueur, D. (1999) A climate-based distribution model of malaria transmission in Africa. *Parasitology Today*, **15**(3), 105–11.

Table App. 1. *Malaria suitability indices for climatic determinants*

	Suitability = 0	Suitability = 1
Monthly temperature (degrees Celsius)	<18	>22
	>40	<32
Annual minimum monthly temperature (degrees Celsius)	<0	>4
Precipitation (mm/month)	0	>80

MERGE-ETL (Chapters 3 and 11)

The model for evaluating regional and global effects (MERGE) is an integrated assessment model that provides a framework for assessing climate-change policy. MERGE-ETL represents a modified version of MERGE5 described by Kypreos and Bahn (2003) and Manne and Richels (2004a,b). Key features include: a nine-region global disaggregation; a combined 'top-down' Ramsey-type economic and 'bottom-up' engineering modeling approach; a simple climate model with damage function; and international trade. Regional technological learning with global spillovers, climate-change impacts and damages further enhance the regional links and interactions (Magne *et al.*, in press).

Technologies for electricity generation (including options for carbon capture and storage–CCS), and secondary fuel production (synthetic fuels from coal and biomass; hydrogen from a range of sources; including options for CCS) are explicitly included in MERGE-ETL. Technological learning in MERGE-ETL (see Barreto and Kypreos, 2004 and Kypreos, 2005a,b) is represented by two-factor learning curves for technology investment costs. The paradigm of technology clusters described in Seebregts *et al.* (2000) is applied, and thus we assume that development and adoption of technologies occurs as a collective evolutionary process.

Limitations

MERGE relies on assumptions about perfect competition and information, production/utility function continuity, representative agents and so on. The level of technology detail also enables only a generic representation of end-use energy efficiency (i.e. explicit end-use technologies are not represented). Further, there is substantial uncertainty concerning many of the parameters used in the model related to resources, economic development, and demands.

Barreto, L., Kypreos, S. (2004) Endogenizing R&D and market experience in the 'Bottom-up' energy-systems ERIS model. *Technovation*, **24**(8), 615–29.

Kypreos, S. (2005a) Modeling experience curves in MERGE. *Energy,* **30**(14), 2721–37.

Kypreos, S. (2005b) *Impacts of RD&D on carbon mitigation cost.* PSI mimeography.

Kypreos, S. and Bahn, O. (2003) A MERGE model with endogenous technological progress. *Environmental Modeling and Assessment,* **8**, 249–59.

Magne, B., Kypreos, S. and Turton, H. (in press). Technology options for low stabilization pathways with MERGE. *The Energy Journal* (Special Issue on Low Stabilization Pathways).

Manne, A. and Richels, R. (2004a) *MERGE: An integrated assessment model for global climate change.* http://www.stanford.edu/group/MERGE/biblio.htm

Manne, A. and Richels, R. (2004b) The impact of learning-by-doing on the timing and costs of carbon dioxide abatement. *Energy Economics,* **26**(4), 603–19.

Seebregts, A., Bos S., Kram T. and Schaeffer, G. (2000) Endogenous learning and technology clustering: analysis with MARKAL model of the Western European energy system. *International Journal of Global Energy Issues,* **14**(1/2/3/4), 289–319.

POLES (Chapters 11 and 14)

The POLES model (European Commission, 1996) is a global sectoral model of the world energy system. It has been developed in the framework of a hierarchical structure of interconnected sub-models at the international, regional and national level. The POLES model works in a year-by-year recursive simulation and partial equilibrium framework, with endogenous international energy prices and lagged adjustments of supply and demand by world region. The model provides comprehensive energy balances for 47 countries and regions, among which are the OECD countries and key developing countries. The model produces detailed long-term (2050 and beyond) energy outlooks with demand, supply and price projections by main region; greenhouse gas emissions, marginal abatement cost curves by region, and emission trading systems analyses, under different market configurations and trading rules; technology improvement scenarios – with exogenous or endogenous technological change – and analyses of technological progress in the context of greenhouse gas abatement policies. The key issues addressed include the long-term simulation of world energy scenarios on the regional and nation level and international energy markets analysis and the impacts of energy prices and tax policies. Target users of the model are international organisations and policy makers, energy R&D experts, energy analysts and strategy departments in the field of global energy markets and related environmental issues.

Limitations

POLES uses currently known or well described prospective technologies. However, by 2100, future development in fundamental science may trigger the development of completely new technological concepts. In addition, even within the scope of current knowledge, some technologies could develop or be adapted to new uses, leading to substantial modifications of the way energy systems evolve.

European Commission (1996) *POLES 2.2*. European Commission DG XII, EUR 17358 EN.

PowerACE-ResInvest (Chapter 7)

The agent-based sector model PowerACE-ResInvest simulates the future development of energy conversion technologies based on the use of renewable energy (RET) in the electricity sector. Capacity expansion decisions of RET are modelled from an investor's perspective. The corresponding investment decisions are mainly driven by the heterogeneous techno-economic characteristics of RET, on the one hand, and on available financial support for RET on the other hand. In turn, techno-economic characteristics are represented by cost resource curves, describing a combination of the available resource potential and the corresponding electricity generation costs. To cite an example, detailed cost resource curves have been derived combining land availability and wind regimes in a geographical information system for wind onshore energy. Technology options are integrated dynamically into the model taking into account future cost developments of RET in terms of experience curves.

Limitations

Aiming at a detailed representation of RET, PowerACE-ResInvest does not integrate a representation of the conventional electricity sector. Interactions with the conventional power system are rather considered in terms of softlinks to the optimising energy system model EuroMM. This fact, and the long-term horizon of consideration, imply that operational aspects of an increased use of RET characterised by a fluctuating power output are only accounted for in a simplified way.

RESIDENT, RESAPPLIANCES and SERVE (Chapter 7)

The three models – RESIDENT, RESAPPLIANCES and SERVE – are bottom-up simulation models to calculate energy demand and additional investments due to energy policies. These models are developed at CEPE, ETH Zurich (Aebischer and Catenazzi, 2007; Jochem *et al.*, 2007). The models are designed to be integrated into a model system, thus adapting the inputs from new macroeconomic data, and giving an economic estimation of technical policies. Each model, in the European version, calculates energy demand and additional investment, for every year from 2004 to 2050, for 25 European regions covering the 29 European countries (EU-27, Norway and Switzerland). RESIDENT simulates the non-appliances demand in the residential sector (heating, hot water, cooking and lighting), modelling two household types (single and multi-family houses) each divided into four building cohort (ages). It evaluates the floor area demand from population and gross domestic product dependent specific needs, and the specific energy demand per floor area. RESAPPLIANCES simulates the appliances in the residential sector (seven types plus ventilation/air conditioning), using appliance diffusion and specific demand. SERVE calculates the energy demand and investment for heating, hot water, air conditioning and other electricity uses, in seven sectors (six in the tertiary sector plus one the primary sector). It simulates the floor area demand from employment, and from grade of technology in the different sectors.

Limitations

The lack of uniform data (and sometime lack of data) restricts the result details on some countries, and the precision of policies implementation. The models simulate 'samples' (e.g. average buildings), thus missing some energy demands.

Aebischer, B. and Catenazzi, G. (2007) Der Energieverbrauch der Dienstleistungen und der Landwirtschaft, 1990–2035. Ergebnisse der Szenarien I bis IV und der zugehörigen Sensitivitäten BIP hoch, Preise hoch und Klima wärmer, Bern, Report for the Swiss Federal Office of Energy.
Jochem E., Catenazzi, G., Jakob, M. *et al.* (2007) *Adaptation and Mitigation Strategies: Supporting European Climate Policy – Deliverable M1.1: Report of the Base Case Scenario for Europe and full description of the model system.* Karlsruhe: Fraunhofer ISI.

REMIND-R (Chapters 10, 11 and 14)

The global multi-region model REMIND-R (Leimbach *et al.*, 2008) represents an intertemporal optimizing energy–economy–environment model which maximises global welfare subject to equilibrium conditions on different markets. REMIND-R is a hybrid model which couples an economic growth model with a detailed energy system model and a simple climate model. The hard-link between the energy system and the macroeconomic system follows the method by Bauer *et al.* (2008). A technical description of REMIND-R is given on our website[2]. The main advantage of REMIND-R is a high technological resolution of the energy system with more

---

[2] On http://www.pik-potsdam.de/research/research-domains/sustainable-solutions/remind-code-1 the technical description of REMIND-R is available.

than 50 conversion technologies and inter-temporal trade relations between regions. Both features expand the range of mitigation options. The latter allows for investment-intensive mitigation options to be financed by credit from abroad. The present version of REMIND-R distinguishes 11 world regions. Each region is modelled as a representative household with a utility function that depends upon per capita consumption. The individual regions are linked by trade relations. Trade is modelled in the goods: coal, gas, oil, uranium, composite good and emission permits.

Macro-economic output is determined by a nested 'constant elasticity of substitution' function of the production factors labour, capital and several end use energy types. The switch between energy technologies is a crucial element of endogenous technological change in REMIND-R. This is supplemented by learning curve effects that impact the investment costs of wind and solar technologies.

Limitations

While providing a first best solution based on the perfect foresight assumption, REMIND-R widely ignores market imperfections and is restricted to an exogenous representation of technological change in the macroeconomic sector.

Bauer, N., Edenhofer, O. and Kypreos, S. (2008), Linking energy system and macroeconomic growth models. *Journal of Computational Management Science*, **5**, 95–117.

Leimbach, M., Bauer, N., Baumstark, L. and Edenhofer, O. (2009) Mitigation costs in a globalized world: climate policy analysis with REMIND-R, *Environmental Modeling and Assessment*, in press.

TIMER (Chapter 11)

Within IMAGE 2.4 (see separate entry), the global energy system model TIMER (van Vuuren *et al.*, 2006; 2007) describes the investment in, and the use of, different types of energy options influenced by technology development (learning-by-doing) and resource depletion. It describes the long-term dynamics of demand and production of primary and secondary energy and the related emissions of greenhouse gases and regional air pollutants. The model behaviour is mainly determined by substitution processes of various technologies on the basis of long-term prices and fuel-preferences. The output of TIMER provides regional energy consumption, energy efficiency improvements, fuel substitution, supply and trade of fossil fuels and renewable energy technologies. On the basis of energy use and industrial production TIMER computes emissions of greenhouse gases, ozone precursors and acidifying compounds.

Limitations

A drawback of TIMER is that economic development is treated as an exogenous driver, and there is thus no feedback from changes in energy and land use to changes in development of gross domestic product. It is also worth mentioning that this is a energy simulation model, not an optimisation model, and the resulting scenarios may thus not be optimal when viewed overall.

van Vuuren, D. P., van Ruijven, B., Hoogwijk, M., Isaac, M. and De Vries, B. (2006) TIMER 2: Model description and application. In Bouwman, L., Kram, T., and Klein-Goldewijk, K. (eds.) *Integrated Modelling of Global Environmental Change. An Overview of IMAGE 2.4*. Bilthoven, the Netherlands: MNP–Netherlands Environmental Assessment Agency.
van Vuuren, D. P., den Elzen, M. G. J., Lucas, P. L. *et al.* (2007) Stabilizing greenhouse gas concentrations at low levels: an assessment of reduction strategies and costs. *Climatic Change*, **81**(2), 119–59.

TIMER-COOLING/HEATING ENERGY DEMAND
(Chapters 3, 11 and 14)

The TIMER cooling/heating energy demand model (Isaac and van Vuuren, 2009) describes the energy use for cooling and heating as a function of several factors, including population levels, changing income levels and climate. For both heating and cooling, empirical data are used to calibrate a set of system-dynamic demand functions. Climate (cooling and heating degree days) plays an important role. The model is able to account for the impacts of climate change.

Limitations

The empirical basis on which the model is calibrated is relatively poor for developing countries. The model does not contain a description of different ways cooling and heating demand can be supplied and the costs involved in substituting one technology for the other.

Isaac, M., and van Vuuren, D. P. (2009) Modeling global residential sector energy demand for heating and air conditioning in the context of climate change. *Energy Policy*, **37**(2), 507–21.

WATER RESOURCES IMPACT MODEL (Chapter 3)

The water resources impact model (Arnell, 2003, 2004) has two components. The first simulates river runoff across the entire global land surface (at 0.5 degrees × 0.5 degrees) using the macro-scale hydrological model Mac-PDM, and the second determines indicators of water resource stress at the watershed level by calculating per capita water resource availability. A watershed is assumed to be exposed to water resource stress if it has an annual average runoff equivalent to less than $1000\,m^3$/capita per year, a semi-arbitrary threshold widely used to identify water-stressed regions. Climate change leads to an increase in exposure to water resource stress if it causes runoff in a water-stressed watershed to decrease significantly, or causes the watershed to fall below the threshold. Climate change leads to an apparent reduction in exposure to the opposite trends. These changes cannot be directly compared; whilst a reduction in runoff (and an increase in exposure) is highly likely to be adverse, an increase in runoff (and apparent decrease in exposure) may not be beneficial if the additional water cannot be stored or if it occurs during high flow seasons as increased flooding. The number of people living in watersheds exposed to an increase in water resource stress can be used as an indicator of exposure to climate change. The actual impacts (in terms of real water shortages) will depend on water management structures in place.

Limitations

The hydrological model does not simulate perfectly the volume of river runoff, and in particular tends to overestimate runoff in semi-arid regions. The water resources indicator is a measure of exposure to impact, not actual impact; it can be seen as a surrogate for the demand for adaptation.

Arnell, N. (2003) Effects of IPCC SRES emissions scenarios on river runoff: a global perspective. *Hydrology and Earth System Sciences*, **7**(5), 619–41.

Arnell, N. (2004) Climate change and global water resources: SRES emissions and socio-economic scenarios. *Global Environmental Change*, **14**(1), 31–52.

Index

Italics refer to figures.

ADAM, i, xix–xx, 349, 354, 355, 383
  scenarios, 61–69
adaptation
  autonomous, 14, 25, 77, 99, 122
  awareness, 120
  barriers, 25, 89, 98
  benefits, 127
  constraints, 239, 244, 245, 249, 251, 253
  costs, 8, *8*, *56*, *76*, *80*, 90–95, *93*, *95*, *279*, 280
  dimensions, 253–255
  governance, 266, 278–283
  IPCC perspective, 5, 6
  learning process, 378
  national, 88, 116–119, 124
  opportunities, 239, 244, 245, 249, 251, 253
  over time, 21, *21*, 57
  perceptions, 246
  perspectives, reflexive, 114, 115
  planned, 25, 117
  policy, *12*, *23*, 89, 109, 116–119, 201, 204, 266, 268
  practices, 23, 124, 231, 232, 236, 237, 242, 252, 256
  regional, 124, 125
  research, 113
  response capacity, 7
  responsibility, 121
  spatial scale, 56, 88
  stimulation, 129
  see also adaptation/mitigation linkages; mainstreaming
  see also governance
adaptation/mitigation linkages, 4, 5
  balance, 376–379
  cost-effectiveness, 371
  damages, 92
  European, 12, 20–24
  IPCC perspective, 6
  mix, 59, 371–373
  policy, 15, 89
  sectors, 22, *22*
  sensitivity, *373*
  synergies, 24, 371
  trade-offs, 5, 24, 372

Adaptation and Mitigation Strategies: Supporting
  European Climate Policy *see* ADAM
agriculture, 73, 99, *102*, 124, 211, 215, 216, 235, *235*,
  237, 240, 246
air conditioning, *77*, 97, 174, 175, 176
Alxa region, 232, 240–246, *241*, *242*, 245
Association of Small Island States (AOSIS) multi
  window mechanism, 357, 360
Austria, 105, 218, *219*, *221*
see ADAM

Bali Action Plan, 341
Bangladesh, 352
barriers
  economic, 292
  market, 89, 98
  policy appraisal, 48–49
  political, 292
  technological, 292
  to synergies, 17
  see also adaptation; mitigation
behaviour change, 127, 139, 377, 379
  incremental, 153
biomass energy, 179, 184, 213, 291, 292, 300, 306
building
  design, 17, 122
  efficiency, 173, 174
  regulations, 17, 122
  stock, retrofit, 175, 176
  technology, 98

carbon
  intensity, 15, 302, *302*
  leakage, 193
  price, 189, 305, *305*
  tax, 149, 154, 196, 383
carbon capture and storage (CCS), 67, 186, 301, 306, 308
carbon dioxide concentrations, *296*, 382
carbon dioxide emissions, 14, *65*, 66, *66*, 144, 145,
  170, 185, 187, 188, 196, 294, *296*, 376, 377,
  380, 383

energy sector, *171*, *188*, *315*
  reductions, 171, *171*, *187*, 192, *302*
    long term, 197
    regional, 171
  see also greenhouse gas emissions
carbon dioxide equivalent, *see* greenhouse gas
carbon reductions, *see* carbon dioxide emission
    reductions
Caribbean, 349
cities, 15–17, 119–122
  climate-proofing, 16
  decarbonisation, 15
  density, 16
  heat impact, 16, 119
  planning, 16
Clean Development Mechanism (CDM), 189, 276, 323
climate change
  as a wicked issue, 114
  as market failure, 147, 148
  as state failure, 147
  communication, 129
  costs and benefits, 142, 154–156
  damages, 90, *91*
  developing countries, 282, 283, 320, 343, 379
  induced migration, 280
  insurance, 341
  position in 2004, 370
  projections, 205
  responses, 32, 231
  risks, 57, 78, 124
  scientific evidence, 118, 121, 129
  uncertainty, 55, 57, 67, 81, 90, 225, 377, *378*
  see also global warming; mainstreaming
  see also risk
climate change impacts, 69–81, 99–101, *102*
  economic, *104*
  regional, 100, 103
  relationships, 102
  sectoral, 100, 103
  variability, 101–104
climate governance, *see* climate policy; governance;
    policy
climate policy, 32
  effectiveness, 159
  European, 17–20, 32, 140, 143–144, 158
  evolution, 370
  harmonisation, 158
  insights, 384
  integration, 237
  international, 63
  meta-analysis, 146
  models, 396
  objectives, *12*
  rationale, 144
climate refugees, 280
climate-proofing, 322, 324, 378, *see also* cities
Commission for the Convention Development and
    Application (CDAC), 238
community engagement, 127
computable general equilibrium (CGE) models, *see*
    models

conservation
  rangeland, 127
  wetlands, 127
cooling, 14, 77, 96, 99, 175, 178, 180, 181, 196, 406
costs as proportion of GDP, 77, 80, *80*, 93, 94, *95*,
    303, 375
crops
  drought-resistant, 243
  yield, *73*, 212, 216, 236
  yield risk, *214*

decarbonization, 297–302, 309
decision-making levels, 18, 88
desertification, 241, 242
disaster
  funding, 216–224, *219*, 341, 345–347, 348
  knowledge, 204
  loss prevention, 345–347
  security, *see* security, 345
disaster risk
  losses, 105, *217*, 219, 343, *343*
  mapping, 206, 209
  methodological approach, 206
  see also risk
Driving forces Pressures State Impacts Response
    (DPSIR) framework, *57*, 59
drought, 79, 205, 213, 241
  financial risk, *213*

ecological migration, 243, 244
economics of climate change, 79–80, 103, *103*, 130, 189
  activities, 245, 246
  cost–benefit analysis, 79
  developing countries, 328–329
  electricity generation, 187–196
  European Emission Trading Scheme (EU ETS), *see*
    EU Emission Trading Scheme (EU-ETS)
  extreme weather events, 104
  feed-in tariffs, 196
  instruments, 381, 383
  loss, 215
  macroeconomic models, 89, 172
  models, 168, 391, 393, 394, 404
  polluter pays principle, 383
  public good, 88, 89
  Stern Review, (The Economics of Climate Change),
    82, 119, 375
  welfare, 189
  see also barriers; carbon tax
electricity
  demand, 185
  efficiency, 176
  Europe, 180–187
  markets, 190
  mix, 184
  models, 394
  price, 189, 190
  supply, 97, 167
electricity generation, 178, *183*, *184*, 185–187, *186*, *193*
  carbon capture and storage (CCS)
  coal-based power plants, 182

electricity generation (cont.)
  disincentives to technological change, 197
  economic analysis, 187–196
  technology, 191, 192
  see carbon capture and storage (CCS)
energy
  consumption, 167
  efficiency, 15, 66, 166, 170, 192
  intensity, 302, *302*
  mix, 64–67, *65*, 179, *299*, 300, 309, *310*, 376
  modelling, 167–169, *168*
  policy, 197
  production, *66*, 167
  regulation, 197
  structural change, 192
  substitution, 166
  supply, 13
  technological change, 194
  see also energy demand; energy sector; energy
    system; security
energy demand, 14, 77, *77*, 167, *173*
  district heat, 179, *180*
  per capita, 314, *314*
  reduced, 170
  sector, 96, 311, *312*
energy sector, 169, 171
  apportionment of abatement, 191–192, *192*
  Europe, 13–15, *170*, 309–314, *310*
  industry, 178, 311
  primary energy conversion, 179
  residential, 172, *173*, 311
  service, 175, 311
  transport, 176, *177*, 311
energy system, 166
  European, 95–98, 169, *170*, 172
  investment, 382
  models, 395, 399, 402, 403, 404, 405, 406
  research and development, 89, 97, 178
  transformation, 382–384
environmental degradation, 127
EU 20% by 2020 GHG emissions reduction target, 146,
  166, 188
  costs of achievement, 189, 191, 194
EU 2008 Climate and Energy Package, 146
EU 2°C target. *See* global warming
EU Adaptation Green Paper (GP), 204
EU Burden Sharing Agreement (BSA), 145
EU Climate Change Programme (ECCP), 143
EU Common and Coordinated Policies and Measures
  (CCPMs), 143, 146, 150
EU Emission Trading Scheme, 18, 145, 151, 156, 179,
  189, 194
EU Global Climate Change Alliance, 325
EU Solidarity Fund, 106, 222
EU White Paper on Adaptation, 89
Europe
  central, 176, 212
  electricity, 180–187
  northern, 14, 26, 96, 99, 173, 175, 180, 204,
    205, 213
  renewable electricity, 183–185

  southern, 13, 14, 26, 69, 96, 97, 99, 173, 175, 180,
    182, 204, 205, 212, 213, 234
  western, 90, *91*, 92, *93*
European Climate Change Programme (ECCP II), 11, 18
European Commissions Sustainable Development
    Strategy, 37
European Union
  INTERREG programme, 123
  see also climate policy; energy sector; energy
    system; policy appraisal
extreme weather events, 78, 81, 89, 120
  as shocks, 118, 129
  assessment, 203–206
  attribution to climate change, 344
  economic modelling, 104
  information, 201, 204
  losses, 201, 343
  probabilistic analysis, 208
  response, 231
  risk, 104–109, 206
  spatial scales, 208
  see also risk; disaster

financial aspects
  export credits, 323
  fiscal space, 106, 220, *221*
  insurance mechanisms, 360
  resources, *108*
  see carbon tax; carbon price; disaster; drought;
    extreme weather events; flood; risk; vulnerability
Finland, 153
fisheries, 99
flood, 107, *210*
  disasters, 78
  insurance, 108
  losses, *106*, *212*, *219*, *223*
flood risks, *see* risk
floodplain revitalization, 248
forests, 99
fossil fuels, 65, 167, 178, 179, 185
fuel substitution, 191, 192
fuels
  coal, 191
  fossil fuels
  natural gas, 182, 192
  non-fossil, 65
  non-fossil fuels, 67
  nuclear energy
  prices, *296*
  renewables
  see fossil fuels, nuclear energy, renewable energy

Global Environment Facility (GEF), 327
global warming, 281
  2 °C above pre-industrial levels, 66, 68, 78, 93, 143,
    147, 166, 212, 292, 374
  4 °C above pre-industrial levels, 78
  see also climate change
governance, 275
  agency, 265, 268, 274–278
  architecture, 265, 268, 272

climate institutions, 272
  dilemmas, 139, 140–143
  evolution, 144
  frameworks, 10, 140
  institutional fragmentation, 269–274, *270*
  long-term, 264
  modes, 142, 152–153
  partnerships, 276
  problem framing, 141, 147
  scales, 148–150
  transboundary, 238
  UN climate regime, 273–274
  see also climate policy; mainstreaming; policy
greenhouse gas concentrations, 67, *68*, 292
greenhouse gas emissions, 13, 15, 64–67, *66*, 138, *139*,
      144, 149, 185, 186, 278, 307, 314, 331
  targets, 138, 146
  see also carbon dioxide emissions
Guadiana river basin, 123, 215, 232, 233–240, *234*,
      235, 239

heating, 14, 77, *77*, 96, 99, 175, 178, 196, 197, 406
heatwaves, 13, 71, 120, 121, 129, 131, 181, 205, *213*,
      228, 241, 344
human health, 71
  heat stress, 119–122
  malaria, 72, *72*
  temperature related mortality, 71, 120
Hungary, 107, 232, 246, *247*

Inner Mongolia, 232, 240
institutions
  changes, 128
  complexities, 250
  incentives, 130
  international, 128
  national, 128
  scenarios, 272
insurance
  against risk, 344–348
  appropriateness, 344
  contribution to adaptation, 350–351, 356
  developing countries, 348–350, 354
  mechanisms, 347–348
  outside support, 351–356
  pre-disaster finance, 353
  private market provision, 352
  public–private partnerships, 352
  subsidies, 351, 355
  see also Association of Small Island States; climate
      change; financial aspects; flood; Munich
      Climate Insurance Initiative
integrated assessment (IAM) models. *See* models
Intergovernmental Panel on Climate Change (IPCC)
  Fourth Assessment Report, 4, 56, 69, 88, 201, 204,
      231, 251, 281, 293, 304, 371, 375
  Special Report on Emission Scenarios (SRES), 55
  Third Assessment Report, 21, 231

Kyoto Protocol to the United Nations Framework
      Convention on Climate Change, 138, *139*, 264

land use, *67*, 232, 235, *235*, *242*, 252
learning
  adaptation, 378
  by doing, 373–376
  social, 4, 9–10, *10*, 34, 35, 115, 125, 130
  stakeholder, 35
Lesotho, 126–128
Living Tisza Alliance' project, 248

mainstreaming
  adaptation, 231, 238, 249, 321, 324, 325
  climate change, 124, 283, 320
  concept, 321–323
  conditions for success, 331–332, 333, 334
  current practice, 324–326
  development economics, 328–329
  evolution, 323–324
  in ODA, 320, *322*, *323*, 327, 329
  mitigation, 321, 323, 325
  political aspects, 322, 326–328
  sustainability, 329–331
  technical aspects, 322
  see also governance; policy integration
Malawi, 348, 351, 352
Meteorological (Met.) Office Hadley Centre, 118
Mexico, 349
mitigation
  barriers, 25
  costs, *56*, *80*, *279*, 302–305, *303*, *308*, 309, *311*, *375*
  IPCC perspective, 5, 6
  measures, 23
  over time, 21, *21*, 57
  policy, *23*, 309–314, 373–376
  response capacity, 7
  spatial scale, 56, 88
  welfare costs, 169
  see also adaptation/mitigation linkages; climate
      policy; economics of climate change
models
  baselines, 297–300
  climate policy, 268, 396
  coastal system, 392
  computable general equilibrium (CGE), 98, 397
  crop simulation, 391
  decarbonisation, 300–302
  economic, 304, 391, 393, 394
  electricity market, 394
  energy-economy, 268, 404
  energy system, 395, 399, 402, 403, 405, 406
  framework for analysis, 267
  improvements, 316
  integrated assessment (IAM), 4, 7–9, 90, 92, 389,
      390, 398, 401
  low stabilisation, 293–309
  malaria vectors, 400
  uncertainty, 99
  water resources, 406
  see also Appendix for model descriptions
Mongolia, 349, 350, 351
Munich Climate Insurance Initiative (MCII),
      356–359, *358*

National Adaptation Plan of Action (NAPA),
    Lesotho, 127
national heatwave action plan, UK, 121
networks, 130
nuclear energy, 301, 306, 308

ODA, *see* Official Development Assistance
Official Development Assistance (ODA), 319
    funding, 328–329
Orange-Senqu river basin, 126–128
Organization of Economic Cooperation and
    Development (OECD) 2006 declaration, 324

policy
    analysis, 10, 266, 268
    capacity building, 11
    change, 36
    choices, 140
    development, 20
    early-warning mechanisms, 157
    effectiveness, 138, 149, 152, 153
    environmental governance, 11
    framework, 11, 31, 140–143
    implementation, 156–158
    integration, 19, 41, 43, 45, 46, 250
    level, 141, 148
    local, 240
    loopholes, 156
    mix, 195
    monitoring, 157, 158
    national, 150
    options, 264, 284, 285
    sanctions, 142, 156–158
    temporality, 150–152
    timing, 141, 142, 150–152
    see also governance; mainstreaming; policy
        appraisal; policy instruments; stakeholders
policy appraisal, 19, 32, *155*, 195
    barriers, 48–49
    by state institutions, 48
    definition, 32
    European, 33, 40–41, 146–158
    evaluation, 33–38
    instrumental, 33, 34–38, *39*
    integrated, 37
    learning, 34–36, 41, 42, 44, 46, 47, 51
    participatory approaches, 267, 268
    reflexive, 33, 34–38, *39*, 50
    Sweden, 42–43
    trade-offs, 41, 43, 44, 46
    United Kingdom, 43–46
policy instruments, 152, 194, 381, 383
    cap and trade, 195
    flexible, 380–382
    innovation incentives, 194
    neutrality principle, 194
    option values, 316
    polluter pays principle (FPPP), 194
    voluntary/negotiated agreements, 152
Portugal, 124, 232, 233
precipitation change, 68, *69*, 182, 205, *234, 241, 247*

rangeland, 240, 244
regions
    Alxa, Inner Mongolia, 232, 233, 240, *241, 242,* 245
    Austria, 105, 218, *219, 221*
    Bangladesh, 352
    Caribbean, 349
    Finland, 153
    Guadiana river basin, Iberia, 123, 215, 232, 233–240,
        234, 235, 239
    Malawi, 348, 351, 352
    Mexico, 349
    Mongolia, 349, 350, 351
    Orange-Senqu river basin, Lesotho, 126–128
    Portugal, 124
    Spain, 124, 213
    Sweden, 42–43
    Tisza river basin, Hungary, 107, 232, 233, 246–251,
        247, 249, 251
    United Kingdom, 43–46, 117
renewable energy, 151, 179, 182, *184,* 306
    biomass. *See* biomass energy
    certificate schemes, 151
    feed-in-tariffs, 151
    hydropower, 99, 181, 184
    promotion mechanisms, 151
    targets, 149
    technologies (RET), 183
    wind energy, 184
residual damages, 8, *8, 56, 76, 80,* 92, *93,*
    279, 280
resources
    financial, *108*
    management, 125–128
    pastures, 126, 246
    wetlands, 126, 246
risk
    assessment, 209–216
    awareness, 125
    definition, 203
    direct, 207, *207*
    disaster, *106,* 202, 203
    drought, 211
    dynamic analysis, 208
    financial consequences, 207, 216, 346
    flood, 78, 205, 209, *210, 212, 223*
    government liability, 217, 218
    heat, 211
    indirect, 213
    management, 107, 119–122, 202, 203, 204
    planning, 104–109
    pooling, 349
    sharing, 218, 222, 342
    transfer, 342, 349
    see climate change; crops; disaster risk; drought;
        extreme weather events; insurance

scenario
    ADAM, 61–*69*
    adaptation, 55, 69–81, 166, 169
        definition, 58
        analysis, 55

baseline, 55, *63*, *65*, 293–309, *295*, *296*
  business-as-usual, 58
  definition, 58
  IPCC SRES B2, 90, *91*
  development, 58–61
    approaches, 60–61
    integration, 59–60
  economic growth, 62, 64
  low-stabilisation, 58
  mitigation, 55, 56, 166, 169
    definition, 58
  population projections, 62
  stabilisation, 292
sea level rise, 76
security
  disaster, 345, 347, 352
  economic, 341
  energy, 192, 197
  environmental, 248
  food, 281, 284, 352
Spain, 124, 213, 232, 233
stabilisation
  lower limit, 293
  targets, 295–297
  see also scenario; stabilisation levels
stabilisation levels, *82*
  400 ppm, 292, *299*, 300, 306, 311
  440 ppm, *305*
  450 ppm, 292
  550 ppm, 292, *299*, 300, *305*, 306
  models, 307
stakeholder
  communication, 34
  engagement, 12, 35, 40, 42, 44, 45, 47,
    122, 226
  representation, 250
sustainable rural development, 248
Sweden, 42–43

technology
  induced change (ITC), 304, 373–376
  lock-in, 97
  options, *187*, 306–309
  see also barriers; building; electricity generation;
    energy; renewable energy
temperature change, 68, *68*, *69*, *82*, 101
  electricity transmission, 182
  impact on crops, 73

Tisza river basin, 107, 232, 246–251, *247*, *249*, 251
tourism, 101
transport, 101
triptych approach, 278
  see also energy sector

UK Climate Impacts Programme (UKCIP), 118
uncertainty
  behaviour, 377
  extreme weather event occurrence, 208
  financial, 91, 95, 154
  initial conditions, 378
  knowledge base, 201
  learning speed, 376
  mitigation targets, 32, 61
  models, 99
  policy effectiveness, 159
  regulatory, 271
  see also climate change
United Kingdom, 43–46, 117
United Nations Development Assistance Framework
  (UNDAF), 325
United Nations Framework Convention on
  Climate Change (UNFCCC), 92,
  144, 320
  Adaptation Fund, 95
urban areas *see* cities
urban design, 16
urban heat island *see* cities: heat impact

vulnerability
  coastal zones, 76
  definition, 203
  energy system, 25
  financial, 107, 207, 219, 220, 358
  food production, 246
  heat stress, 120
  physical, 207

water
  irrigation, 74
  management, 232, 243, 248, 252
  models, 406
  research and development, 97
  resources, 74–76, *75*, 97, 123–125
  snow/ice melt, 182
  system, 97, 246
World Food Programme (WFP), 329, 349